Lecture Notes in Computer Science 7728

Commenced Publication in 1973
Founding and Former Series Editors:
Gerhard Goos, Juris Hartmanis, and Jan van Leeuwen

Jong-Il Park   Junmo Kim (Eds.)

Computer Vision – ACCV 2012 Workshops

ACCV 2012 International Workshops
Daejeon, Korea, November 5-6, 2012
Revised Selected Papers, Part I

 Springer

Volume Editors

Jong-Il Park
Hanyang University
Computer Science and Engineering
222 Wangshimni-ro, Seongdong-gu
Seoul 133-791, Korea
E-mail: jipark@hanyang.ac.kr

Junmo Kim
KAIST
Department of Electrical Engineering
291 Daehak-ro, Yuseong-gu
Daejeon 305-701, Korea
E-mail: junmo@ee.kaist.ac.kr

ISSN 0302-9743   e-ISSN 1611-3349
ISBN 978-3-642-37409-8   e-ISBN 978-3-642-37410-4
DOI 10.1007/978-3-642-37410-4
Springer Heidelberg Dordrecht London New York

Library of Congress Control Number: 2013934227

CR Subject Classification (1998): I.4, I.5, I.2.10, I.2.6, I.3.5, F.2.2, H.5, J.5

LNCS Sublibrary: SL 6 – Image Processing, Computer Vision, Pattern Recognition,
and Graphics

Typesetting: Camera-ready by author, data conversion by Scientific Publishing Services, Chennai, India

Printed on acid-free paper

Springer is part of Springer Science+Business Media (www.springer.com)

Preface

The 11th Asian Conference on Computer Vision (ACCV), held in Daejeon, South Korea, during November 5–9, 2012, was accompanied by a series of nine high-quality workshops covering the full range of state-of-the-art research topics in computer vision.

The workshops consisted of six full-day workshops and three half-day workshops. Their topics diversely ranged from traditional issues to novel current trends. On November 5, four workshops took place: the Workshop on Computer Vision with Local Binary Pattern Variants and the Workshop on Computational Photography and Low-Level Vision (both full-day workshops), and the Workshop on Developer-Centered Computer Vision and the Workshop on Background Models Challenge (both half-day workshops). The remaining five workshops were held on November 6: the Workshop on e-Heritage (Electronic Cultural Heritage), the Workshop on Color Depth Fusion in Computer Vision, the Workshop on Face Analysis: The Intersection of Computer Vision and Human Perception, and the Workshop on Detection and Tracking in Challenging Environments (all full-day workshops), and the Workshop on Intelligent Mobile Vision (a half-day workshop).

This year, the workshops received 310 paper submissions, and 78 presentations were selected by the individual workshop committees, yielding an overall acceptance rate of 25%. All contributions to each workshop are published in the two-volume ACCV workshop proceedings. We thank everyone involved in the remarkable programs, committees, reviewers, and authors, for their contributions.

We hope that you enjoy reading these proceedings, which may inspire you to further research.

November 2012

Jong-Il Park
Junmo Kim

International Workshop on Computer Vision with Local Binary Pattern Variants

Local Binary Pattern (LBP) is a simple and efficient texture operator, unifying statistical and structural approaches in texture analysis. It is a powerful gray-scale invariant measure, derived from a general definition of texture in a local neighborhood. Due to its discriminative power and computational simplicity, the LBP operator has become a highly popular approach in various computer vision applications, including facial image analysis, visual inspection, image retrieval, remote sensing, biomedical image analysis, biometrics, motion analysis, environment modelling, and outdoor scene analysis. Especially the use of LBP in biomedical applications and biometric recognition systems has grown rapidly in recent years. LBP has been highly successful in numerous applications around the world and has inspired plenty of new research on related methods. Since the introduction of the basic LBP operator, several variants have been proposed to improve the discriminative power and robustness of the operator. The recent emergence of LBP has also led to significant progress in applying texture methods to various computer vision problems and applications.

This workshop provided a clear summary of the state of the art and discussed the most recent developments on the use of Local Binary Patterns and their variants in different computer vision applications.

The workshop received 45 submissions (16 through direct submission and 29 via dual submission with the ACCV 2012 main conference). Based on the thorough reviews by the program committee, 13 papers were finally selected. Besides the 13 interesting oral presentations, the workshop also included a keynote speech from a pioneer of LBP (Prof. Matti Pietikäinen from the University of Oulu) and a best paper award sponsored by KeyLemon – a leading face recognition software company.

The workshop organizers would like to thank all the participants of this workshop. Many thanks go also to the Program Committee for their efforts during the reviewing process and to the ACCV 2012 workshop chairs and publication chairs who dealt with the organizational aspects of this workshop.

November 2012

Abdenour Hadid
Sebastien Marcel
Jean-Luc Dugelay
Matti Pietikäinen
Mohammad Ghahramani
Stan Z. Li

Program Committee

Janne Heikkilä	University of Oulu, Finland
Norman Poh	University of Surrey, UK
Shengcai Liao	Michigan State University, USA
Hazim Kemal Ekenel	Karlsruhe Institute of Technology, Germany
Jie Chen	University of Oulu, Finland
Caifeng Shan	Philips Research, The Netherlands
Loris Nanni	University of Padua (Padova), Italy
Bill Triggs	Laboratoire Jean Kuntzmann, Grenoble, France
Liming Chen	École Centrale de Lyon, France
Karl Ricanek	University of North Carolina Wilmington, USA
Lior Wolf	Tel Aviv University, Israel
Yunhong Wang	BUAA University, China
André Anjos	Idiap Research Institute, Switzerland
Guoying Zhao	University of Oulu, Finland
Chi-Ho Chan	University of Surrey, UK
Messaoud Bengharabi	Centre de Développement des Technologies Avancées, Algeria
Mohammad Ghahramani	University of Oulu, Finland
Mark Nixon	University of Southampton, UK
John A. Ruiz-Hernandez	University of Oulu, Finland
Zhen Lei	Chinese Academy of Sciences, China
Vili Kellokumpu	University of Oulu, Finland
Fabio Roli	University of Cagliari, Italy
Timo Ahonen	Nokia Research Center, Palo Alto, California
Xilin Chen	Chinese Academy of Sciences, China
Shiguang Shan	Chinese Academy of Sciences, China
Rui Min	Eurecom, France
Yimo Guo	University of Oulu, Finland
Miguel Bordallo	University of Oulu, Finland
Juha Ylioinas	University of Oulu, Finland

Workshop on Computational Photography and Low-Level Vision

Computational Photography is an exciting new field at the intersection of computer vision, computer graphics, and photography. The goal of computational photography is to enhance or extend the capabilities of digital photography to produce new photographs that could not have been taken by a traditional camera. Fundamental low-level computer vision techniques will be particularly useful to this end.

The goal of this workshop is to provide a platform for researchers in computational photography and computer vision to meet and share their ideas on recent trends and research in the two areas.

We received 17 papers and selected 8 papers for publication based on the reviews by the Program Committee. All submissions were reviewed in a double-blind fashion by at least three experts in the area. We thank all the authors who submitted their work. Topics of accepted papers spanned a wide range of areas from camera calibration to vehicle localization. Other topics represented were segmentation and colorimetric correction. We were pleased to have Michael S. Brown (National University of Singapore) as the keynote speaker and also the advisory chair at the workshop. We would also like to express our appreciation to the members of the Program Committee for their remarkable efforts and the quality of the reviews.

November 2012

Jinwei Gu
Yu-Wing Tai
Ping Tan
Sai-Kit Yeung

Program Committee

Workshop on Developer-Centered Computer Vision

The majority of research in computer vision is focused on technology and systems that advance the state of the art. However, there is very little focus on how we can make the state of the art useable by the majority of people. Recently there has been an increased interest in "Vision for HCI" and how we use computer vision to interact with the world. We proposed a parallel theme of "HCI for Vision" for this workshop, looking at how to provide accessible computer vision targeted towards mainstream software developers. We aimed to explore ideas that take existing vision methods and present them in a manner that enables users with varying degrees of vision knowledge to use them.

There has been a relatively recent surge in the number of developer interfaces to computer vision becoming available: OpenCV has become much more popular, Mathworks have released a Matlab Computer Vision Toolbox, visual interfaces such as Vision-on-Tap are available online and specific targets such as tracking (OpenTL) and GPU (Cuda, OpenVIDIA) have working implementations. Additionally, last year, Khronos (the not-for-profit industry consortium that creates and maintains open standards) formed a working group to discuss the creation of a computer vision hardware abstraction layer (tentatively titled CV HAL).

Developing methods to make computer vision accessible poses many interesting questions and will require novel approaches to the problems. DCCV is a half-day workshop aiming to bring together researchers from academia and industry in the fields of vision and HCI to discuss the direction of research into developer-centred computer vision. The workshop included an introductory talk by the organisers followed by presentations of the five accepted papers (24% acceptance), covering mainstream-developer targeted topics as well as more advanced concepts such as algorithm efficiency.

The DCCV organisers would like to thank the ACCV Workshop and Publication Chairs, in particular Junmo Kim and In Kyu Park, for their help and support throughout the workshop organisation process.

November 2012

Gregor Miller
Sidney Fels

Program Committee

Workshop on Background Models Challenge

The detection of moving objects in video sequence is an important task in many video-surveillance systems. As a matter of fact, the output of this very first stage, named *background modeling or background subtraction*, determines the quality of the rest of the pipelines developed for the detection, identification, or tracking of persons, objects, etc. Background modeling is sometimes considered either as a trivial operation, carried out by computing a simple difference between the current frame and a single background image (or with the previous frame, etc.), or a mastered technique that does not need any improvement or development nowadays. In the latter case, very famous methods, such as the Gaussian Mixture Models introduced by Stauffer and Grimson in 1999, are cited, and this is considered sufficient. Unfortunately, these kinds of algorithms are limited in outdoor environments, when used in long-term surveillance applications, because of various unpredictable circumstances such as global variation of luminance, shadows of objects, bad weather, camera tilts, etc.

Since this is a key-point of video-surveillance applications, background subtraction has become a popular topic, and many techniques have been proposed since the 1990s. For the BMC (*Background Models Challenge*), we proposed a new benchmark composed of almost 30 synthetic and real video sequences. Thanks to these data-sets, we were able to propose very complex situations, in various surveillance contexts (human activities or traffic, for example). We also developed a free software (BMC Wizard) to compute relevant criteria for the evaluation of statistical, signal, and structural information from a background subtraction algorithm.

For the BMC, six papers were accepted for publication, and an invited speaker, Thierry Bouwmans, gave a talk on the state of the art of the domain and recent advances in his personal research. We hope that our benchmark will be used as a reference for further research in background modeling. The data-sets and the BMC Wizard will remain available on the BMC website *http://bmc.univ-bpclermont.fr*. Finally, we would like to thank the Program Committee of ACCV for their support in organizing this event.

November 2012

Antoine Vacavant
Laure Tougne
Lionel Robinault
Thierry Chateau

Program Committee

Workshop on e-Heritage

Digitally archived world heritage sites are broadening their value for preservation and access. Many valuable objects have been decayed by time due to weathering, natural disasters, even man-made disasters such as the Taliban destruction of the great Buddhas in Afghanistan, or the recent destruction by fire of the 600-year-old Great South Gate in Seoul. Cultural heritage also includes music, language, dance, and customs that are fast becoming extinct as the world moves toward a global village. Furthermore, most of the sites still face a problem of accessibility. Digital access projects are necessary to overcome those problems.

Computer vision research and practices have played, and will continue to play, a central role in such cultural heritage preservation efforts. The Workshop on e-Heritage and Digital Art Preservation aimed to bring together computer vision researchers, as well as interdisciplinary researchers, working in areas related to computer vision, in particular computer graphics, image and audio research, image and haptic (touch) research, as well as presentation of visual content over the Web, and education.

In this workshop, eight contributions to the field of e-heritage were presented, covering the areas of automatic character recognition, classification based on shape and image analysis, image enhancement, virtual-reality applications, three-dimensional modeling, and reconstruction. All submissions were double-blind reviewed by at least two experts. We thank all the authors who submitted their work. It was a special honor to have Martial Hebert (Carnegie Mellon University, USA), and Jean Ponce (Ecole Normale Supérieure, France) as the invited speakers at the workshop. We are especially grateful to the members of the Program Committee for their remarkable efforts and the quality of the reviews.

November 2012

Hongbin Zha
Takeshi Oishi
Rei Kawakami
Yunsu Bok
Katsushi Ikeuchi

Program Committee

Olga Bellon	Universidade Federal do Paraná, Brazil
Asanobu Kitamoto	National Institute of Informatics, Japan
Yasuyuki Matsushita	Microsoft Research Asia, China
Shohei Nobuhara	Kyoto University, Japan
Tomokazu Sato	NAIST, Japan
Luciano Silva	Universidade Federal do Paraná, Brazil
Jun Takamatsu	NAIST, Japan
Robby T. Tan	University of Utrecht, Netherlands
Yingqing Xu	Tsinghua University, China
Toshihiko Yamasaki	University of Tokyo, Japan

Workshop on Color Depth Fusion in Computer Vision

The ambition of this workshop was to provide an opportunity to disseminate recent theories, methods, and practical algorithms that explicitly exploit the enormous potential of combining low-resolution depth cameras with high-resolution color cameras for a wide variety of computer vision tasks. The workshop brought together researchers and practitioners from various fields of study: computer vision, robotics, computer graphics, image processing, and sensor architecture.

We received 44 submissions and 12 papers were accepted for single-track oral presentation. We also had an invited demonstration on single-sensor color and depth capturing sensor and applications.

November 2012

Seungkyu Lee
Hyunjung Shim
Ouk Choi
Seung-Won Jung
Radu B. Rusu

Program Committee

Workshop on Face Analysis: The Intersection of Computer Vision and Human Perception

The analysis of faces is a very active research area within both the computer vision and the human perception communities, and there is a large array of potential applications and research topics. The two communities have traditionally worked separately, but there are clear benefits to closer collaboration. For instance, humans develop extensive experience in the processing of face identity, age, gender, and non-verbal communication signals such as facial expressions. Thus, tapping into existing knowledge from human facial perception research can enable the targeted design of computer vision facial analysis and synthesis systems with more realistic behavioural facial models and performance. Moreover, computer vision systems can provide many useful tools for research into the human perception of faces such as the generation of photo-realistic and controllable stimuli for perceptual experiments, which enables more subtle manipulation of facial appearance and dynamics than would be possible using natural video capture.

The current state of the art in facial analysis within both disciplines is now quite evolved. To go to the next level, the disciplines need to strengthen their collaboration even further. This workshop provided the forum to enable this step. Its goal was to examine existing work that straddles the border of these communities, and to map out future steps for integrative research.

We received 37 full-paper submissions which underwent a double-blind review. The multi-disciplinary nature of the workshop meant that making decisions regarding papers was more difficult than usual, and so up to 7 reviewers per paper were used (with a minimum of 3 reviewers/paper) to ensure that a balanced assessment was made. A total of 9 papers were selected for the workshop, and are collected in these proceedings.

We were fortunate to have three invited speakers at the workshop, who have all worked extensively at the intersection of computer vision and human perception: Heinrich H. Bülthoff (Max Planck Institute for Biological Cybernetics), Alan Johnston (University College of London) and Darren Cosker (University of Bath). We would like to thank the invited speakers as well as all the members of the Programme Committee for their help in organising and running this event.

November 2012

Paul L. Rosin
David Marshall
Christian Wallraven
Douglas W. Cunningham

Program Committee

Workshop on Detection and Tracking in Challenging Environments (DTCE)

Recent progress in computer vision has opened new possibilities in robust visual tracking and in human and object detection. Although these have a wide range of practical applications, there are still many challenges when applying such algorithms to real-world data. These include: complex crowded environments with many activities, challenging lighting, and frequent occlusions; large variations of pose, motion, and appearance; limited computational resources; and the need for training from large datasets. DTCE 2012 brought together researchers working on these challenging real-world problems to present their recent achievements and provided a place to share their experiences and visions with others.

We received 89 submissions jointly with ACCV and 9 independent submissions; and we selected 15 papers for publication. The review process was double-blind. The independently submitted papers were reviewed by two to three members of the workshop Program Committee, while most of the joint submissions received one independent review from this committee in addition to their three reviews and area chair summary from ACCV. We would like to thank the Program Committee members for their effort in reviewing the papers.

The workshop featured a keynote address by Ming-hsuan Yang of the University of California, Merced, as well as oral presentations of 8 of the accepted papers, and poster presentations of all 15 of the accepted papers.

November 2012

Bohyung Han
Jongwoo Lim
Bill Triggs
Ahmed Elgammal
Jason Corso

Program Committee

Serge Belongie	University of California, San Diego, USA
Terrence Chen	Siemens Corporate Research, USA
Naresh Cuntoor	Kitware, USA
Larry Davis	University of Maryland, USA
Jan Feyereisl	POSTECH, South Korea
Kikuo Fujimura	Honda Research Institute, USA
Abhinav Gupta	Carnegie Mellon University, USA
Iasonas Kokkinos	Ecole Centrale Paris, France
Tony X. Han	University of Missouri, USA
Ser-Nam Lim	GE Global Research, USA
Haibin Ling	Temple University, USA
Sangmin Oh	Kitware, USA
David Ross	Google inc., USA
Yoichi Sato	University of Tokyo, Japan
Zhuowen Tu	MSRA / University of California, Los Angeles, USA
Jianxin Wu	Nanyang Technological Univeristy, Singapore
Ming-hsuan Yang	University of California, Merced, USA
Kuk-Jin Yoon	GIST, Korea
Lei Zhang	Hong Kong Polytechnic University, Hong Kong

International Workshop on Intelligent Mobile Vision (IMV)

With the fast growth of hand-held computing platforms such as smart phones and tablet PCs, computer vision on mobile computing devices has become an important research area. There is tremendous potential for developing computer vision techniques and applications on mobile camera computing devices. In particular, more and more mobile vision applications rely on object and scene recognition or understanding techniques. It is observable that advances in mobile visual-information analyses are closely related to cutting-edge applications in robotics, human-computer interaction, smart sensors, and ubiquitous computing.

The main goal of this workshop was to identify state-of-the-art mobile vision algorithms, systems, and frameworks that are particularly suitable for intelligent visual information processing based on mobile camera computing platforms. In addition to visual information, the integration of vision with additional sensors, such as GPS, accelerometer or gyroscope, or information retrieval transmitted through communication networks, helps to develop an even more intelligent mobile vision application. The associated methodologies and applications are expected to be able to demonstrate the advantages of advanced computer vision techniques based on mobile camera computing devices.

November 2012

Shang-Hong Lai
Chu-Song Chen

Program Committee

Table of Contents – Part I

International Workshop on Computer Vision with Local Binary Pattern Variants

Workshop on Computational Photography and Low-Level Vision

Workshop on Developer-Centred Computer Vision

Workshop on Background Models Challenge (BMC)

Table of Contents – Part II

Workshop on e-Heritage

ACCV Workshop on Color Depth Fusion in Computer Vision

Workshop on Face Analysis: The Intersection of Computer Vision and Human Perception

ACCV Workshop on Detection and Tracking in Challenging Environments (DTCE)

International Workshop on Intelligent Mobile Vision (IMV)

Noise Resistant Gradient Calculation and Edge Detection Using Local Binary Patterns

Michael Teutsch and Jürgen Beyerer

Fraunhofer IOSB
Fraunhoferstr. 1, 76131 Karlsruhe, Germany
{michael.teutsch,juergen.beyerer}@iosb.fraunhofer.de

Abstract. Gradient calculation and edge detection are well-known problems in image processing and the fundament for many approaches for line detection, segmentation, contour extraction, or model fitting. A large variety of algorithms for edge detection already exists but strong image noise is still a challenge. Especially in automatic surveillance and reconnaissance applications with visual-optical, infrared, or SAR imagery, high distance to objects and weak signal-to-noise-ratio are difficult tasks to handle. In this paper, a new approach using Local Binary Patterns (LBPs) is presented, which is a crossover between texture analysis and edge detection. It shows similar results as the Canny edge detector under normal conditions but performs better in presence of noise. This characteristic is evaluated quantitatively with different artificially generated types and levels of noise in synthetic and natural images.

1 Introduction

Gradient calculation and edge detection are topics still worth to discuss as many image processing applications have to deal with input images affected by noise or weak signal-to-noise-ratio (SNR). In automatic surveillance and reconnaissance, difficult environmental conditions, high object distance, moving sensors, and sensor-specific noise lead to images, which are challenging to process. The quality of such image data can vary strongly even for the same sensor over time. Being an important step for applications such as object and image segmentation, line detection, texture analysis, contour extraction, or model fitting, edge detection has to be precise but at the same time robust against noise. The Canny algorithm [1] is a good choice to handle this problem, but it also reaches its limits when the noise level is getting high or alternates. We want to show, that in such cases, our proposed approach can outperform Canny and other tested methods. The original Canny processing chain consists of Gaussian smoothing, gradient calculation, directional non-maximum-suppression for gradient magnitudes and hysteresis thresholding to determine edge pixels. In this work, this chain is taken as a template and its modules are modified using Local Binary Patterns (LBPs). The main innovation is the filtering and gradient calculation strategy. The decrease of edge detection performance with increasing level of noise is slowed down compared to original Canny, while still providing a similar

J.-I. Park and J. Kim (Eds.): ACCV 2012 Workshops, Part I, LNCS 7728, pp. 1–14, 2013.
© Springer-Verlag Berlin Heidelberg 2013

performance in case of low noise level. Experiments with standard evaluation methods on synthetic and natural image data support this observation.

The presentation and discussion of related work is focused on other approaches towards noise resistant edge detection. Many authors altered the original Canny processing chain modifying mainly the smoothing or gradient calculation strategy or both, while adopting the idea of non-maximum-suppression and subsequent thresholding. Korn [2] combines smoothing and gradient calculation in only one convolution matrix which is an approximation of the first normalized Gaussian derivatives. Different scales are applied and the best one chosen automatically. Evaluation is performed visually on natural image data. Kitanovski et al. [3] use multi-scale undecimated Haar wavelet transform to emphasize edges. Their approach tracks for edges existing at several scales favoring edges at larger scale. Thus, robustness against noise is achieved but detailed edge structure may get lost. Agaian and Almuntashri [4] aim to segment MRI brain images. The typical Canny filter matrices for smoothing and gradient calculation are replaced to better deal with the impulsive noise of MRI images. This seems to work at least as well as normal Canny but no quantitative evaluation is given. Sun and Sun [5] consider two windows around each image pixel. In these windows, grayvalue mean and variance are calculated and a difference measure determines the pixel's edge strength. The edge direction is derived from the four different possible window arrangements which describe rotations in steps of 45°. Only a visual evaluation is given. Panetta et al. [6] introduce an adaptive switching function choosing the appropriate smoothing filter (Gaussian, Median, etc.) based on some performance evaluations. Then, a shape-dependent convolution is proposed using kernels of different size and shape (circle, ellipse, hexagon, diamond, etc.) for joint gradient magnitude calculation. Standard Canny and Sobel are outperformed on synthetic images with respect to Abdou and Pratt's figure of merit [7]. For natural images slightly better performance than Canny is visible in presence of noise.

Some authors developed a new edge detection processing chain independent of the Canny algorithm. Chen and Das [8] aim to detect edges and corners in noisy images. A pattern classification algorithm automatically identifies the noise type and chooses the right image restoration technique. After gradient calculation, fuzzy k-means based adaptive thresholding determines the edge or corner pixels. No quantitative evaluation is given but the results on natural data look convincing. Hou and Wei [9] use discrete singular convolution (DSC) to generate different filters for multi-scale edge detection in noisy images. With various levels of Gaussian noise the figure of merit [7] is taken as evaluation on a synthetic image. The performance is very similar to Canny. Chang [10] proposes contextual Hopfield neural networks (CHNNs) for edge detection. For segmentation of MRI images, CHNNs are able to perform better than various other approaches such as Canny or Wavelets in presence of strong salt-and-pepper noise.

In this paper, the focus lies on processing data as it appears in surveillance applications. Strong noise of unclear type, and small objects, which may

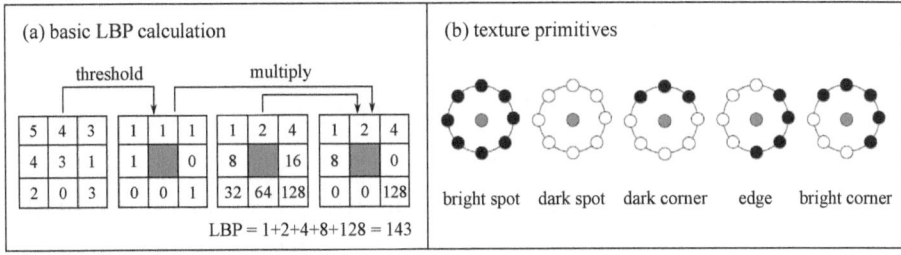

Fig. 1. Calculation and interpretation of Local Binary Patterns (LBPs) [17]

disappear after smoothing characterize the images and, thus, an approach is presented dealing with noise but keeping sophisticated edge structure. The remainder of the paper is organized as follows: the theory of LBPs and the edge detection approach using LBPs are introduced in Section 2. A description of the experimental setup as well as a demonstration of the results and some examples for processing natural image data are given in Section 3. Conclusions are presented in Section 4.

2   The Proposed Approach

The application of LBPs is widely spread in image processing research. Some examples are texture classification [11,12], face detection [13], background modeling [14], structure emphasizing filter [15], or setting up a SIFT descriptor [16]. After a short introduction to the theory of LBP, the modifications and their applicability for edge detection are presented.

2.1   Theory of LBP

LBPs describe a unique encoding for local pixel neighborhood. They are easy to implement, fast to compute, and characterized to be high-performance and robust features in the abovementioned approaches. In the following, we refer to the work of Mäenpää [17] and Ojala et al. [11] for the theory of LBP. In Fig. 1 (a), the typical way of LBP computation is shown. The gray-value of the central pixel is compared to each of the eight neighbors. In case of a higher or equal gray-value, its position will be highlighted with a 1 and, thus, considered for the LBP computation. LBP encoding is calculated by multiplying all highlighted positions with their related weights and summing them up afterwards. The result is a value between 0 and 255 describing a specific neighbor constellation. There are two basic design parameters: number of neighbors P and radius R, since neighbors are ordered circularly around the central position c. This leads to the equation:

$$LBP_{P,R} = \sum_{p=0}^{P-1} s(g_p - g_c)2^p, \text{ where } s(x) = \begin{cases} 1, \ x \geq 0 \\ 0, \ x < 0. \end{cases} \tag{1}$$

A specialization of LBPs, which will be important for edge detection, is the set of rotation-invariant, uniform LBPs:

$$LBP_{P,R}^{riu2} = \begin{cases} \sum_{p=0}^{P-1} s(g_p - g_c), & \text{if } U(LBP_{P,R}) \leq 2 \\ P + 1, & \text{else.} \end{cases} \quad (2)$$

$LBP_{P,R}^{riu2}$ can be interpreted as texture primitives [17] as seen in Fig. 1 (b). Description $riu2$ stands for rotation-invariance and uniformity measure U of 2 or less. U returns the number of bitwise 0/1 and 1/0 transitions in a LBP [11]. With $P = 8$ and $U \leq 2$, only the 58 texture primitives among the 256 LBPs are considered. Rotation-invariance is achieved by assigning all potential rotations of an uniform LBP to the same equivalence class, for example *edge*, *bright corner*, or *dark corner*. As seen in Fig. 2 (top left), there are nine equivalence classes and eight LBPs in each class, each LBP corresponding to a rotation in steps of $45°$. An exception is given by the classes *bright spot* and *dark spot* with only one representative each.

LBPs are gray-scale invariant [11] as only the sign of the gray-value difference is considered. However, further information is available in the neighbor's gray-values. To extract this information, the rotation-invariant variance measure VAR is introduced:

$$VAR_{P,R} = \frac{1}{P} \sum_{p=0}^{P-1} (g_p - \mu)^2, \quad \text{where } \mu = \frac{1}{P} \sum_{p=0}^{P-1} g_p. \quad (3)$$

Ojala et al. [11] point out that the combination of LBP and VAR turned out to be a powerful feature for texture classification.

2.2   Gradient Calculation and Edge Detection with LBPs

The first question is, why LBPs should be suitable for gradient calculation or edge detection? The answer is given in Fig. 2. For some example images, the LBPs are calculated for three different radii $r \in \{1.0, 2.0, 3.0\}$ and accumulated in LBP histograms. Each histogram has ten bins: nine for the different $LBP_{P,R}^{riu2}$ equivalence classes, which are displayed in Fig. 2 (top left), and one for all other LBPs. Since the accumulation values were highly varying along the bins, the vertical axis is visualized in logarithmic scale. In a first experiment, all pixels of a synthetic image are set to value 127. There are no edges in this image. Four different kinds of artificially generated noise are added subsequently, but not mixed, to find out which LBP distribution appears for which kind of noise. It is obvious that for salt-and-pepper noise mostly spot-like LBPs (classes 0, 1, 8) appear, while the accumulation of class 3 is a result of bilinear interpolation during LBP calculation. The other kinds of noise mainly produce LBPs of classes 0, 1, and 9. If pixel positions of such LBPs are not considered for gradient calculation, noise can be suppressed. For the second experiment, a synthetic image with various edges and well-known natural image *Lena* are considered. The LBPs are calculated for each original image as well as for each image with

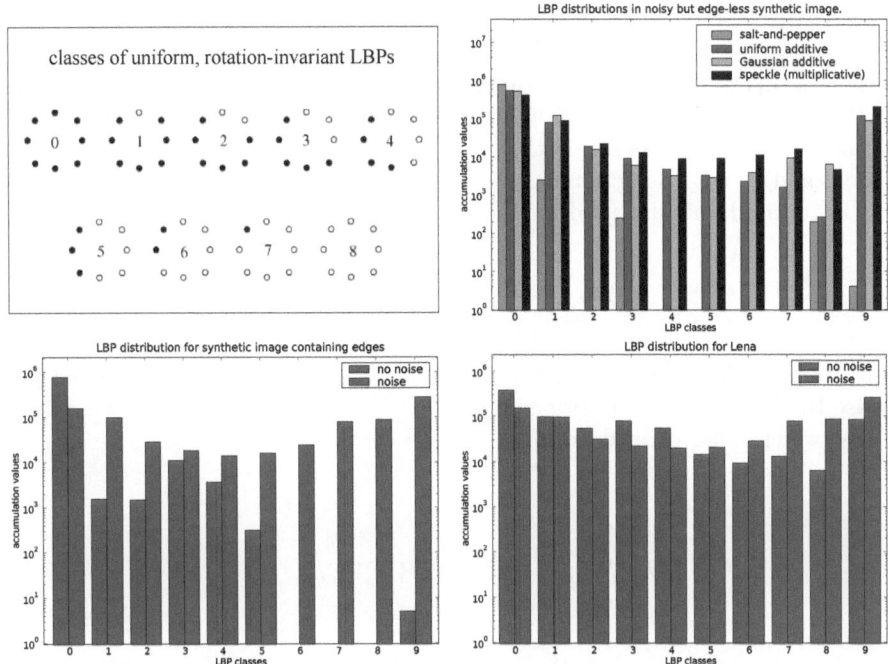

Fig. 2. LBP distributions for an edge-less image with different kinds of noise (top right) and the comparison of LBP distributions for a noise-free and noisy synthetic image with edges (bottom left) and Lena (bottom right)

added combination of noise (Gaussian, speckle, and salt-and-pepper). Class 0 appears most often as it stands for flat, homogeneous image areas. But it also can be seen that classes 3 and 4 mainly represent edges while classes 7 and 8 represent noise.

The basic idea in using LBPs for gradient calculation is to generate a filter rejecting pixel positions of LBPs which are likely to be produced by noise. Thereafter, gradient magnitudes are calculated at the accepted pixel positions using the local variance $VAR_{P,R}$. For easy embedding and testing, the Canny edge detection processing chain is taken as template and modified. The aim of each component is kept, but solved in a different way using LBP. The original Canny algorithm [1] consists of:

1. Noise suppression by smoothing with Gaussian kernel.
2. Calculation of gradient magnitude and direction for each pixel by convolution with a filter matrix.
3. Non-maximum-suppression of the gradient magnitudes in dominant gradient direction.
4. Determination of edge pixels using hysteresis-thresholding.

In the following, the modifications to gain more robustness towards noise are presented step by step.

Noise Suppression: Not only Gaussian noise is the target of noise suppression here, but also salt-and-pepper, uniform additive, and speckle noise. Salt-and-pepper is likely to occur in infrared, while speckle noise is symptomatic for SAR imagery. Especially in applications using such data, another important parameter is used to enhance the robustness of standard LBPs in flat image regions and for edge detection in noisy images: the gray-value threshold T, which was introduced and also used in [14]. Eq. 1 is adapted as follows:

$$LBP_{P,R,T} = \sum_{p=0}^{P-1} s(g_p - g_c - T)2^p. \tag{4}$$

With this modification, a binary decision function f is defined and applied pixelwise to all image pixel positions $c = (x, y)$. f accepts only pixels with related LBPs, which fulfill three criteria:

$$U(LBP_{P,R,T}) = 2 \tag{5}$$

$$LBP_{P,R,T} \neq 2^p, \ p \in \mathcal{P} = \{0, \ldots, P-1\} \tag{6}$$

$$LBP_{P,R,T} \neq 2^P - 1 - 2^p, \ p \in \mathcal{P} \tag{7}$$

This means, only uniform LBPs are allowed, which are not spots (5) or spot-like (6), (7). The assumption is that all non-uniform LBPs and all uniform LBPs violating one of the three criteria are the result of noise. Thus, they should be suppressed before gradients are calculated. This leads to the following formulation of f for each pixel position c:

$$f(c) = \begin{cases} 1, & \text{if (5) and (6) and (7)} \\ 0, & \text{else.} \end{cases} \tag{8}$$

Only pixel positions with $f(c) = 1$ will be considered for the next step.

Gradient Magnitude and Direction: Convolution with Sobel, Prewitt or other filters is an approximation of partial derivatives. Here, the gradients are calculated using $VAR_{P,R,T}$ and $LBP_{P,R,T}$. The gradient magnitudes $G(c)$ for each pixel position c are computed using the equation:

$$G(c) = \begin{cases} \sqrt{VAR_{P,R,T}}, & \text{if } f(c) = 1 \\ 0, & \text{else.} \end{cases} \tag{9}$$

Variance tends to focus too much on bright objects. So, standard-deviation is used instead of variance as it produces more homogeneous edge images. The robustness against noise can be increased significantly using multi-resolution LBPs [11]. In the literature, they are also known as multi-scale LBPs. For the same pixel position c, several LBPs are calculated varying the parameters P and R. In this work, only variations of radius R are considered and P is fixed to

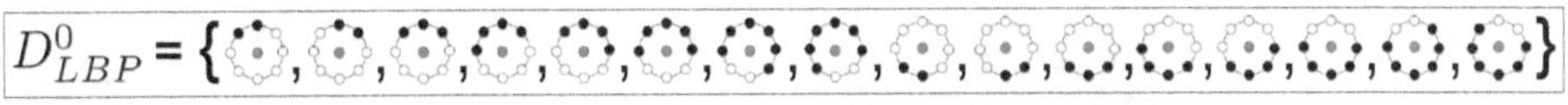

$$D^0_{LBP} = \left\{ \vcenter{\hbox{⬡}}, \vcenter{\hbox{⬡}}, \vcenter{\hbox{⬡}}, \vcenter{\hbox{⬡}}, \vcenter{\hbox{⬡}}, \vcenter{\hbox{⬡}}, \vcenter{\hbox{⬡}}, \vcenter{\hbox{⬡}}, \vcenter{\hbox{⬡}}, \vcenter{\hbox{⬡}}, \vcenter{\hbox{⬡}}, \vcenter{\hbox{⬡}}, \vcenter{\hbox{⬡}}, \vcenter{\hbox{⬡}} \right\}$$

Fig. 3. Set D^0_{LBP} with LBPs of orientation $0°$

$P = 8$. For each LBP accepted by f, $VAR_{P,R,T}$ is calculated and summed up for the gradient magnitude:

$$G(c) = \begin{cases} \sum_{r=R_1}^{R_n} \sqrt{VAR_{P,r,T}}, & \text{if } f(c) = 1 \\ 0, & \text{else.} \end{cases} \tag{10}$$

For each texture primitive in Fig. 1 (b), which is not a spot, eight different rotations are possible. Hence, gradient directions are already available by storing the *not* rotation-invariant $LBP^{u2}_{P,R,T}$ at every pixel position c with $f(c) = 1$.

Non-Maximum-Suppression: In Canny algorithm, the atan2-function is used to calculate gradient directions. These directions are rounded off to only four discretization steps: $0°$, $45°$, $90°$, and $135°$. For each pixel's gradient magnitude, a directed non-maximum-suppression is applied in its gradient direction. The remaining maxima describe a skeleton of gradient magnitudes, from which the edge pixels can be determined by hysteresis-thresholding. In this work, the non-maximum-suppression is performed in exactly the same way, but the direction discretization is already given. All uniform LBPs accepted by f are assigned to one of the four direction sets D^0_{LBP}, D^{45}_{LBP}, D^{90}_{LBP}, or D^{135}_{LBP}. LBPs with an even number of 0s and 1s are ambiguous and, thus, assigned to two sets. The set D^0_{LBP} is shown in Fig. 3. To find out about its direction, the currently considered LBP just has to be re-found in one of the sets.

Determination of Edge Pixels: The last step is to generate a binary edge pixel image B. Therefore, hysteresis-thresholding is used. Two thresholds t_1 and t_2 are determined with $t_1 < t_2$. If a gradient magnitude $G(c)$ exceeds t_2, it is accepted as edge pixel. Then, all pixels are considered, which are connected to this edge pixel, and also accepted if their gradient value is greater than t_1. This approach was adopted with a minor change: a minimum threshold for edge length was introduced. B is the final result of the edge detection approach.

3 Experiments and Evaluation

In this section, the performance of gradient calculation and edge detection with LBPs is evaluated. Before presenting the results, first the experimental setup and the evaluation approach are described and discussed. Finally, the processing of some natural example images coming from standard datasets as well as special surveillance applications is demonstrated.

Fig. 4. Evaluation images with reference edge pixel images

3.1 Experimental Setup

Four parts are needed for the experimental setup: a test image database, a synthetic noise generator with measured signal-to-noise-ratio (SNR), a figure of merit to evaluate the performance, and other edge detection algorithms as competitors. To guarantee repeatability of these experiments, standard approaches have been chosen for each part.

Two test images are used: a synthetic one and Lena. There is a ground-truth edge pixel image for the synthetic image and a sensed-truth for Lena. Sensed-truth means that a humanly sensed good result of Canny edge detection was manually corrected and amended by straightening the typical sinuous lines produced by Canny in case of blurred edges. This is the only way to lay a foundation for a quantitative evaluation on a natural image, which is publicly accessible. The four images have a resolution of 512 × 512 and are shown in Fig. 4.

A random noise generator was implemented to manipulate the original image with salt-and-pepper, uniform additive, Gaussian additive, or speckle (multiplicative) noise. The level of noise was varied and measured by applying two different methods: peak-signal-to-noise-ratio (PSNR) and structural similarity (SSIM). PSNR is widely used to calculate the difference between original and noisy image in decibel (dB) for evaluating noise resistant edge detectors [9,3]. Structural similarity [18] is a rather new approach much more related to the human noise sensing. It is a value between 1 (no noise) and 0 (strong noise) and not depending on the image peak which is a disadvantage of PSNR. Thus, it was decided to present the results using SSIM, although both methods have been applied. However, the results to be shown later were clearly noticeable with both PSNR and SSIM.

As figure of merit F, the proposal of Abdou and Pratt [7] was chosen:

$$F = \frac{1}{\max{(I_I, I_A)}} \sum_{i=1}^{I_A} \frac{1}{1 + \alpha d_i^2}, \tag{11}$$

where I_I and I_A is the number of edge pixels in the ideal and the actually detected edge image, d_i denotes the distance between a detected edge pixel and the nearest edge pixel in the ideal image, and α the penalty constant set to $\frac{1}{9}$ as proposed in [7]. The result is a value between 1 for perfect detection and 0 for poor detection.

The proposed LBP edge detector was compared to three other approaches: two variations of the original Canny algorithm as it is implemented in OpenCV [19] and the edge detector proposed by Korn [2]. Canny in OpenCV doesn't include any smoothing, so two different smoothing strategies are tested with Gaussian and Median filter. Gaussian filter is well-known to be powerful against additive noise, while Median filter is useful against salt-and-pepper and speckle. The Korn edge detector is similar to Canny but using a normalized filter matrix for smoothing and gradient calculation with automatically determined size which is slightly adaptive to varying noise levels.

3.2 Results

Before generating the results, an automatic parameter optimization was applied for each tested algorithm using images with different kinds and levels of noise. It is possible to adjust the parameters for each noise level, of course, but this was not considered since it is not suitable for a real surveillance application.

The OpenCV Canny algorithm has four parameters: size of the smoothing filter, size of the Sobel filter, and the two hysteresis thresholds. For Korn edge detection, filter size factor σ, gradient threshold t, and the two hysteresis parameters have to be set. The LBP edge detector has four parameters as well: LBP radius R, gray-value threshold T, and the two hysteresis thresholds. For all studies in this paper, the number of LBP neighbors was fixed to $P = 8$. Optimization was performed by maximizing the mean figure of merit across the original synthetic and Lena image as well as all different noise level images.

With the best set of parameters, the evaluation was run and its results are shown in Fig. 5. The noise-related plots of F against SSIM show the decreasing performance of edge detection with increasing noise level. G-$Canny$ denotes Canny with Gaussian, and M-$Canny$ with Median smoothing. M-Canny has its advantages only for salt-and-pepper noise. In all other cases, the LBP edge detector is performing similar to the three other approaches for weak noise but better than them for stronger noise starting at a SSIM value of about 0.9. The convergence of Canny and Korn to $F = 0.4$ when processing the Lena image is a weakness of the figure of merit F. Canny and Korn tend to produce false positive edge pixels in case of strong noise, while the LBP edge detector generates false negatives. Since the reference edge pixel image for Lena contains many edge pixels in general, many false positives cause a better result of F than many false negatives in this situation. Without artificial noise, the figure of merit was nearly the same for all four approaches with about 0.95 for the synthetic image and 0.9 for Lena.

It should be mentioned, that no smoothing was necessary for LBP edge detection. This is an advantage as edge detection normally deals with the problem that smoothing with big filter size can suppress desired sophisticated edge structure but small filter size might not be sufficient enough to suppress strong noise.

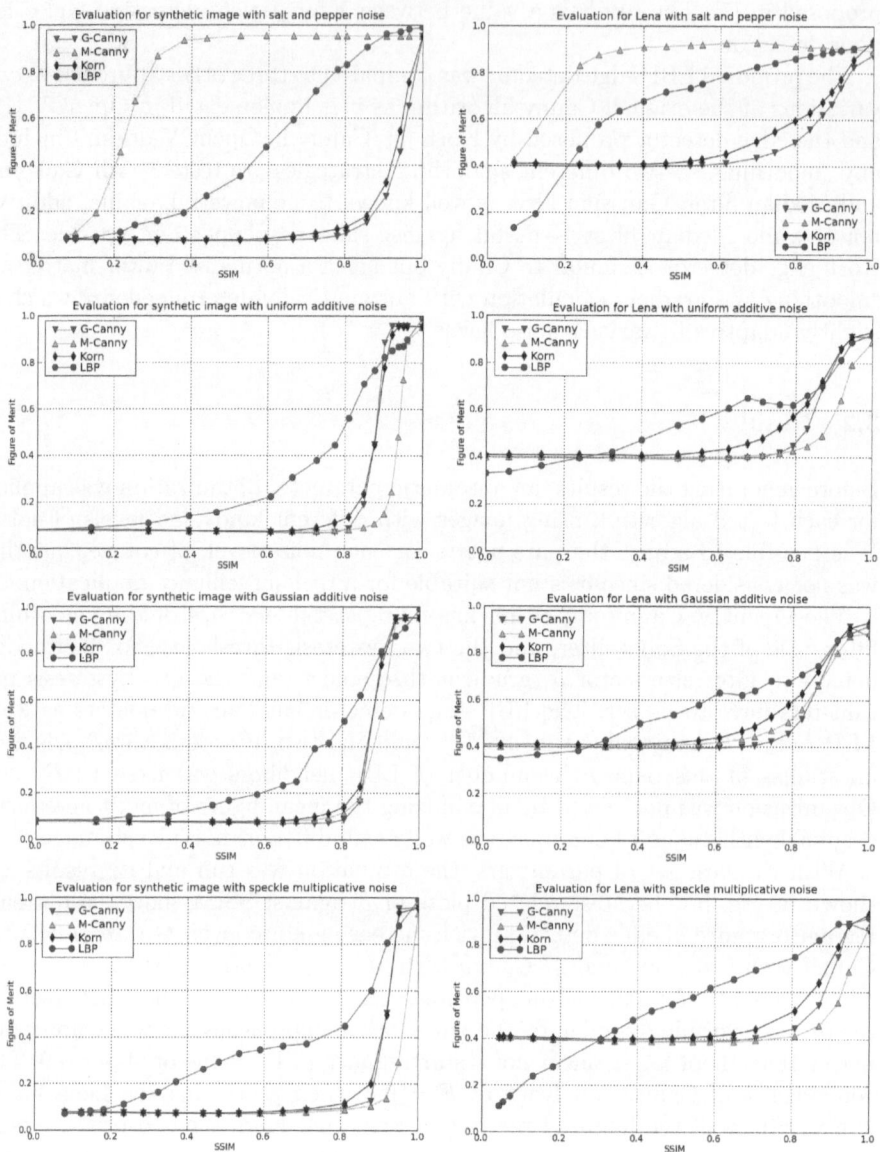

Fig. 5. Evaluation results: figure of merit F plotted against SSIM for synthetic image and Lena with different types and levels of noise

3.3   Examples

For a visual impression of the results, some standard images such as the cameraman or the golf cart have been chosen along with some images from real surveillance applications. They are processed using the LBP approach, OpenCV

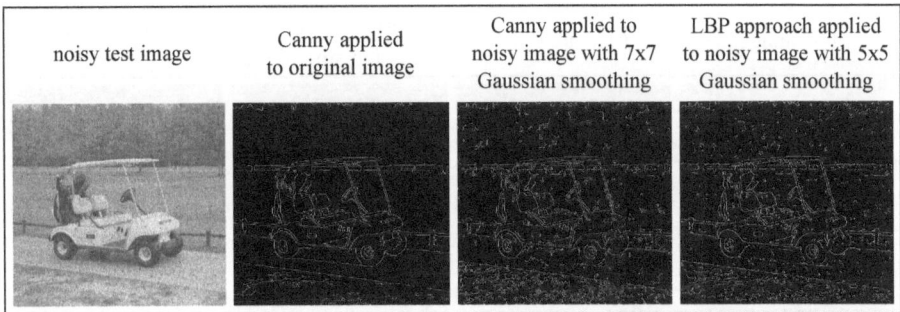

| noisy test image | Canny applied to original image | Canny applied to noisy image with 7x7 Gaussian smoothing | LBP approach applied to noisy image with 5x5 Gaussian smoothing |

Fig. 6. Example for optional Gaussian smoothing in combination with LBP edge detection

Canny with Gaussian smoothing, and the Korn approach. The parameters were adopted from the automatic parameter optimization applied in Section 3.2 with one minor change: since high smoothing parameters were chosen for Canny and Korn, only Canny parameter with a 7×7 filter matrix was directly adopted and σ for Korn filter matrix was lowered to $\sigma = 1.5$ for a better comparison of strong smoothing effects. The examples are visualized in Fig. 7. In the most left column, the original images are located followed by the edge detection results of LBP, Canny, and Korn. In the first row, the original cameraman image is processed. All edge detection results are similar, but due to strong smoothing, some details are lost in the Canny image such as the tower in the background. A combination of Gaussian, speckle, and salt-and-pepper noise is added with a SSIM of 0.75 to the original image and visualized in the second row. The tendency to produce false negative edge pixels rather than false positives is clearly visible in the LBP edge image especially in comparison to the Korn result, which is strongly affected by noise. The Canny edge image is better due to strong Gaussian smoothing. A potential drawback of the LBP approach is shown in the third row. The noisy golf cart has a SSIM of 0.68 to the original image. Nearly all of the edge pixels found by the LBP approach are correct but due to the high false negatives rate, the Canny result visually looks better although it is also affected by noise. If a subsequent algorithm such as line detection with Hough transform is able to handle false negatives better than false positives, this can be an advantage for the LBP approach.

However, it is possible to support the LBP approach with smoothing, too. This is an application dependent alternative method to the proposed one. An example is visualized in Fig. 6. The same noisy golf cart as in Fig. 7 is shown on the left position. The edge detection result of Canny algorithm applied to the original image (without additional noise) and to the noisy image is displayed in the second and third column. Finally, the right image is the result of the LBP approach with previous 5×5 Gaussian smoothing. Visually, the results of Canny and LBP look similar. But when calculating Abdou and Pratt's figure of merit with the Canny result of the original image as reference, the LBP approach

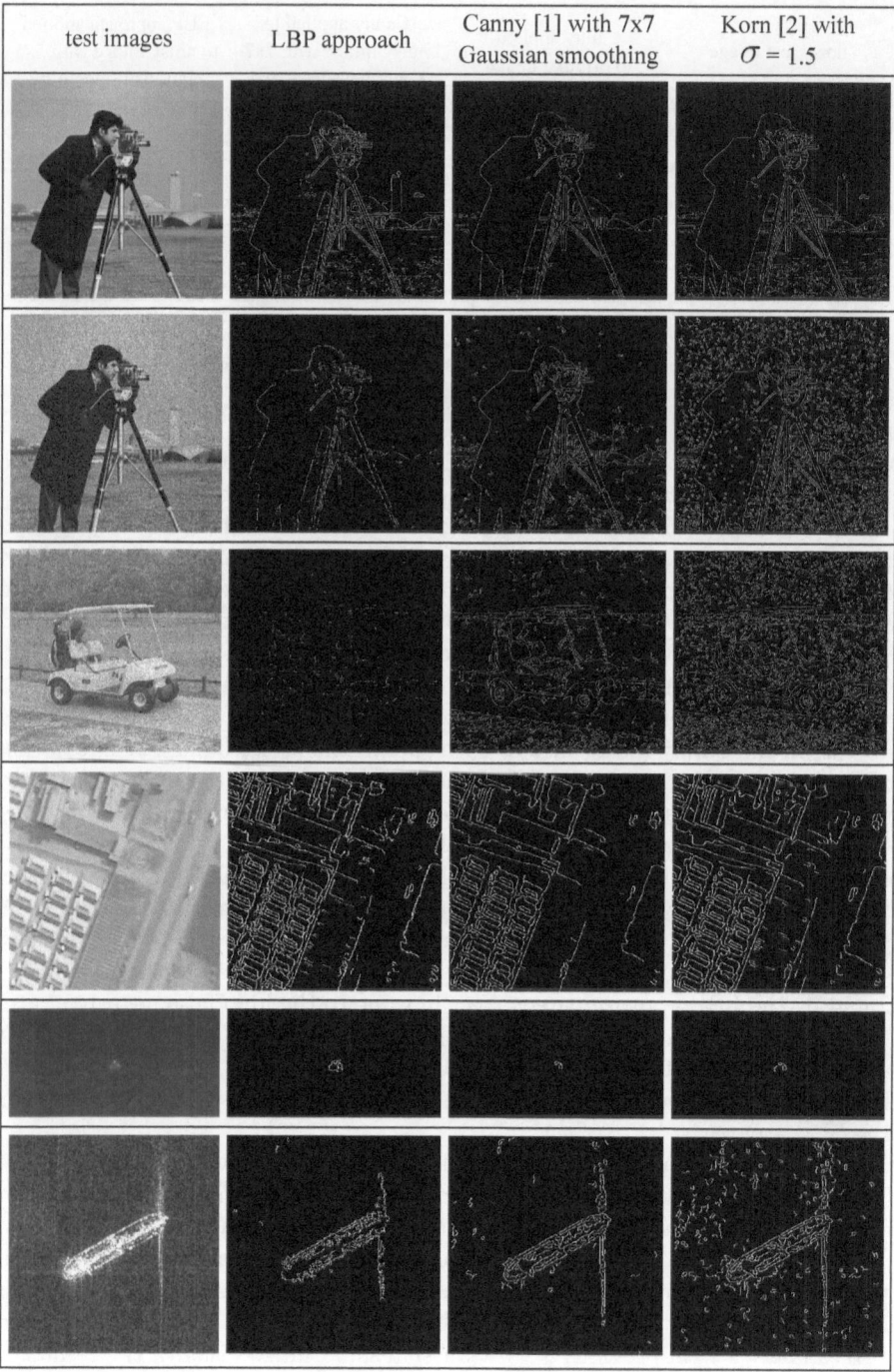

test images	LBP approach	Canny [1] with 7x7 Gaussian smoothing	Korn [2] with $\sigma = 1.5$

Fig. 7. Examples for real data processing: original images and edge detection using LBPs, Canny, and Korn (from left to right column)

performs consistently better with a figure of merit difference between 0.05 and 0.1 along various test images. This result is achieved with a smaller smoothing filter matrix than for Canny and with a Canny result (for original image) as reference for comparison.

The last three examples in Fig. 7 originate from real surveillance applications and were our motivation to develop the proposed approach. The upper one is an image coming from a visual-optical camera mounted on an unmanned aerial vehicle (UAV). In the mid row a ship in a thermal infrared image originating from a buoy camera is shown. Finally, in the lower row a Synthetic Aperture Radar (SAR) image coming from TerraSAR-X satellite with an observed oil tanker is displayed. The vertical smearing effect is typical SAR noise besides the strong speckle. These images are the result of different sensors with different view angles, have different content, different gray-value distribution, as well as different types and levels of noise. LBP edge detection provides the best edge completeness and noise suppression, while both Canny and Korn produce more false negatives in the infrared image and more false positives in the SAR image. This effect was observed in many more example images from different surveillance sensors. Of course, it is possible to find different sets of parameters for Canny and Korn to produce good edge detection results for each of the original images in Fig. 7 separately, but with these examples the robustness of LBP edge detection is demonstrated.

4   Conclusion

A novel approach for gradient calculation and edge detection is presented using Local Binary Patterns (LBPs). The main innovation is the way of noise suppression with a binary decision function f, which only accepts a special subset of LBPs, namely texture primitives, assuming, that all other LBPs are affected by noise. Gradient magnitude is calculated at pixel positions accepted by f with the gray-value variance of LBP neighbors and robustness is achieved by using multi-scale LBPs. For a good testing environment, the LBP approach is embedded to the Canny processing chain consisting of smoothing, gradient calculation, directional non-maximum-suppression, and hysteresis-thresholding. The structure of the processing chain is adopted but implemented using Local Binary Patterns (LBP) instead of Gaussian smoothing and Sobel filter. The evaluation is performed using one synthetic and one natural image with added artificial noise of different types and levels. With increasing noise level, LBP edge detection outperforms the approaches of Canny and Korn concerning Abdou and Pratt's figure of merit. This effect is visually demonstrated with different natural image examples.

In general, inaccuracies occurring at an early stage of a processing chain can significantly affect the performance of all subsequent modules as well as the overall performance. The assumption is, that with higher robustness against noise, not only edge detection but also other algorithms, which rely on gradients instead of edges, can be improved.

References

1. Canny, J.: Computational Approach to Edge Detection. IEEE Transactions on Pattern Analysis and Machine Intelligence 8, 679–698 (1986)
2. Korn, A.: Toward a Symbolic Representation of Intensity Changes in Images. IEEE Transactions on Pattern Analysis and Machine Intelligence 10, 610–625 (1988)
3. Kitanovski, V., Taskovski, D., Panovski, L.: Multi-scale Edge Detection Using Undecimated Wavelet Transform. In: Proceedings of the IEEE International Symposium on Signal Processing and Information Technology, ISSPIT (2008)
4. Agaian, S., Almuntashri, A.: Noise-Resilient Edge Detection Algorithm for Brain MRI Images. In: Proceedings of the Annual International Conference of the IEEE Engineering in Medicine and Biology Society, EMBC (2009)
5. Sun, X., Sun, G.: A New Noise-resistant Algorithm for Edge Detection. In: Proceedings of the Second International Workshop on Education Technology and Computer Science, ETCS (2010)
6. Panetta, K.A., Agaian, S.S., Nercessian, S.C., Almunstashri, A.A.: Shape-dependent canny edge detector. Optical Engineering 50 (2011)
7. Abdou, I.E., Pratt, W.K.: Quantitative design and evaluation of enhancement/thresholding edge detectors. Proceedings of the IEEE 67, 753–763 (1979)
8. Chen, Y., Das, M.: Robust edge and corner detection using noise identification and adaptive thresholding techniques. In: Proceedings of the IEEE International Conference on Electro/Information Technology (2007)
9. Hou, Z.J., Wei, G.W.: A new approach to edge detection. Pattern Recognition 35, 1559–1570 (2002)
10. Chang, C.Y.: Contextual-based Hopfield neural network for medical image edge detection. Optical Engineering 45 (2006)
11. Ojala, T., Pietikäinen, M., Mäenpää, T.: Multiresolution Gray-Scale and Rotation Invariant Texture Classification with Local Binary Patterns. IEEE Transactions on Pattern Analysis and Machine Intelligence 24, 971–987 (2002)
12. Guo, Z., Zhang, L., Zhang, D.: A Completed Modeling of Local Binary Pattern Operator for Texture Classification. IEEE Transactions on Image Processing 19, 1657–1663 (2010)
13. An, K.H., Park, S.H., Chung, Y.S., Moon, K.Y., Chung, M.J.: Learning discriminative multi-scale and multi-position LBP features for face detection based on Ada-LDA. In: Proceedings of the IEEE International Conference on Robotics and Biomimetics (ROBIO), pp. 1117–1122 (2009)
14. Heikkilä, M., Pietikäinen, M.: A Texture-Based Method for Modeling the Background and Detecting Moving Objects. IEEE Transactions on Pattern Analysis and Machine Intelligence 28, 657–662 (2006)
15. Teutsch, M., Saur, G.: Segmentation and Classification of Man-Made Maritime Objects in TerraSAR-X Images. In: Proceedings of the IEEE International Geoscience and Remote Sensing Symposium, IGARSS (2011)
16. Heikkilä, M., Pietikäinen, M., Schmid, C.: Description of Interest Regions with Local Binary Patterns. Pattern Recognition 42, 425–436 (2009)
17. Mäenpää, T.: The Local Binary Pattern Approach to Texture Analysis - Extensions and Applications. Dissertation, University of Oulu, Finland (2003)
18. Wang, Z., Bovik, A.C., Sheikh, H.R., Simoncelli, E.P.: Image quality assessment: From error visibility to structural similarity. IEEE Transactions on Image Processing 13, 600–612 (2004)
19. Bradski, G.: The OpenCV Library. Dr. Dobb's Journal of Software Tools (2000)

Rotation Invariant Co-occurrence among Adjacent LBPs

Ryusuke Nosaka, Chendra Hadi Suryanto, and Kazuhiro Fukui

Graduate School of Systems and Information Engineering,
Department of Computer Science, University of Tsukuba, Japan
{nosaka,chendra}@cvlab.cs.tsukuba.ac.jp,
kfukui@cs.tsukuba.ac.jp

Abstract. In this paper, we propose a new type of local binary pattern (LBP)-based feature, called *Rotation Invariant Co-occurrence among adjacent LBPs* (RIC-LBP), which simultaneously has characteristics of rotation invariance and a high descriptive ability. LBP was originally designed as a texture description for a local region, called a micropattern, and has been extended to various types of LBP-based features. In this paper, we focus on Co-occurrence among Adjacent LBPs (CoALBP). Our proposed feature is enabled by introducing the concept of rotation equivalence class to CoALBP. The validity of the proposed feature is clearly demonstrated through comparisons with various state-of-the-art LBP-based features in experiments using two public datasets, namely, the HEp-2 cell dataset and the UIUC texture database.

1 Introduction

The Local Binary Pattern (LBP) histogram has recently attracted much attention in the area of image recognition. The basic idea behind the LBP histogram is to represent an entire image as a histogram of numerous LBPs, with each LBP extracted from a local region of the image. Many types of LBP-based features [1–5] have been proposed as extensions of the original LBP.

In this paper, we propose a new type of LBP-based feature, which is invariant to rotation of an input image. The proposed feature is an extension of Co-occurrence among Adjacent LBPs (CoALBP) [6], which is an LBP-based feature with a higher descriptive ability than the original LBP. LBP was originally designed as a texture description for a local region, called a micropattern, which consists of binary patterns that represent the magnitude relation between the center pixel of a local region and its neighboring pixels. LBP is obtained by thresholding the image intensity of the surrounding pixels with that of the center pixel. To obtain an LBP histogram feature for use in classification, the binary patterns are converted to decimal numbers as labels, and then a histogram is generated from the labels of all the local regions of an entire image. The main advantage of LBP is its invariance to uniform changes in image intensity over an entire image, making it robust against changes in illumination. This is because LBP considers only the magnitude relation between the center and neighboring pixel intensities. Owing

J.-I. Park and J. Kim (Eds.): ACCV 2012 Workshops, Part I, LNCS 7728, pp. 15–25, 2013.
© Springer-Verlag Berlin Heidelberg 2013

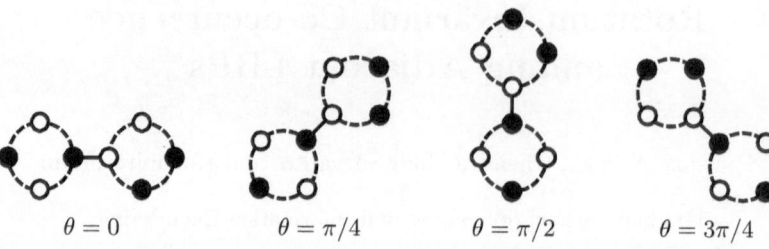

$\theta = 0$ $\theta = \pi/4$ $\theta = \pi/2$ $\theta = 3\pi/4$

Fig. 1. Example of rotation equivalence class of LBP pairs. In this figure, each circle indicates one LBP.

to this characteristic, LBP has become a standard feature for texture recognition, face recognition, and facial expression analysis [1, 5].

To enhance the descriptive ability of LBP, the feature has been extended to CoALBP by introducing the concept of co-occurrence among LBPs so as to extract information related to the more global structures of the input image [6]. However, the CoALBP feature can vary significantly depending on the orientation of the target object. When, for instance, classifying several types of cells with complicated textures, rotation invariance is essential. This is because the orientation of each cell is not relevant to its classification. One could address this problem by preparing all possible LBPs in advance. However, this solution would entail a large memory requirement (to hold the reference patterns) high computational cost.

Several LBP-based features with rotation invariance have already been proposed. They are categorized into two types. The first type focuses on invariance to local rotation of an input image. For example, LBP^{ri} and LBP^{riu2} [4] obtain invariance to local rotation by introducing the concept of rotation equivalence class. The second type focuses on invariance to global rotation. The LBP-HF feature is included in this type. It attains global rotation invariance by applying the discrete Fourier transform to a feature vector of an LBP histogram [7]. Both types can extract distinctive features from an image with rotations. However, these rotation invariant features lack descriptive ability, because they are basically the local features extracted from only micro patterns, without consideration of the relations among micropatterns.

To overcome the problem of low descriptive ability of conventional rotation invariant LBPs, we incorporate the concept of rotation equivalence class into CoALBP. Fig.1 shows an example of the rotation equivalence class of CoALBPs. In this case, we consider that all CoALBPs corresponding to a different angle θ have the same value. Nevertheless, finding such LBP pairs is difficult since the number of possible LBP combinations is huge. To solve this problem, we automatically detect pairs with the same CoALBP value by using a computational algorithm. We call this feature *Rotation Invariant Co-occurrence among adjacent LBPs* (RIC-LBP). RIC-LBP can simultaneously provide a high descriptive ability and invariance to image rotation. The core idea of RIC-LBP is simple yet effective. The validity of RIC-LBP is demonstrated by comparing various

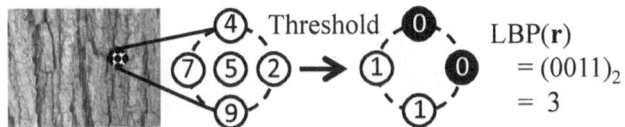

Fig. 2. Flow to obtain LBP from a local region. In this example, the intensity of the center pixel is 5 and those of its neighboring pixels are 2, 4, 7 and 9. Thus, the binary pattern is "0011" and $LBP(\mathbf{r}) = 3$.

state-of-the-art LBP-based features through the experiments using two public datasets, the HEp-2 cell dataset and the UIUC texture database.

The remainder of this paper is organized as follows. In Section 2, we briefly review LBP and co-occurrence among adjacent LBPs. In Section 3, we describe how to impart rotation invariance to CoALBPs. We also explain the RIC-LBP process. In Section 4, we demonstrate the validity of the proposed feature by examining the results of experiments in cell classification and texture recognition using public databases. In the final section, we present our conclusions.

2   LBP and Co-occurrence of Adjacent LBPs

2.1   LBP

LBP[3] is an operator that describes a local region as a binary pattern obtained by thresholding the difference between a center pixel and its neighboring pixels in a local region, as shown in Fig.2. The binary pattern in LBP represents the magnitude relation of intensities, a quantity which is invariant amid uniform changes of image intensity over an entire image. Therefore, LBP is robust against changes in illumination among image patterns, a difficulty commonly found in face and texture images.

Let I be an image intensity and $\mathbf{r} = (x, y)$ be a position vector in I. LBP at $\mathbf{r}$ is defined as follows:

$$LBP(\mathbf{r}) = \sum_{i=0}^{N-1} sgn(I(\mathbf{r} + \Delta \mathbf{s}_i) - I(\mathbf{r}))2^i, \qquad (1)$$

$$sgn(x) = \begin{cases} 1, \text{ if } x \geq 0 \\ 0, \text{ otherwise} \end{cases}, \qquad (2)$$

where N is the number of neighbor pixels. $\Delta \mathbf{s}_i$ is displacement vector from the center pixel to neighboring pixels given by $\Delta \mathbf{s}_i = (s\cos(\theta_i), s\sin(\theta_i))$, where $\theta_i = \frac{2\pi}{N}i$ and s is a scale parameter of LBP.

2.2   Co-occurrence among Adjacent LBPs

The original LBP does not preserve structural information among binary patterns, even though such information may be characteristic of an image. In order

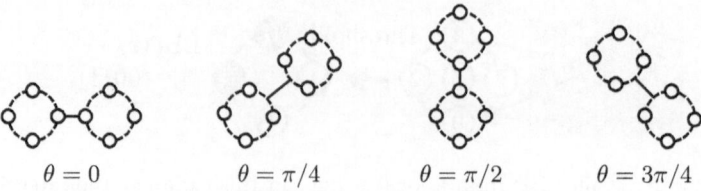

$\theta = 0$ $\qquad\qquad$ $\theta = \pi/4$ $\qquad\qquad$ $\theta = \pi/2$ $\qquad\qquad$ $\theta = 3\pi/4$

Fig. 3. Configurations of an LBP pair

to keep such structural information, we utilize the CoALBP as represented by
LBP pair [6]. The set of CoALBPs over a whole image is converted to a CoALBP
histogram feature. CoALBP (LBP pair) at **r** is written as follows:

$$P(\mathbf{r}, \Delta\mathbf{r}) = (LBP(\mathbf{r}), LBP(\mathbf{r} + \Delta\mathbf{r})), \qquad (3)$$

where $\Delta\mathbf{r} = (r\cos\theta, r\sin\theta)$ is a displacement vector between an LBP pair. The
value of r is an interval between an LBP pair, and $\theta = 0, \pi/4, \pi/2, 3\pi/4$. Fig.3
illustrates the configurations of an LBP pair.

While the LBP produces $2^N (= N_P)$ different output values, the number of
possible combination patterns of an LBP pair $N_P^2 \times 4$ is significantly greater than
that of the LBP itself. That is, an LBP pair can represent a far greater variety of
image patterns than an LBP. The histogram feature generated from these LBP
pairs contains information on the structure of the image, since it describes the
frequency of LBP pairs that are located near each other.

3   Rotation Invariant Co-occurrence among Adjacent LBPs

3.1   Rotation Equivalence Class of LBP Pair

To simultaneously achieve a high descriptive ability and rotation invariance, we
incorporate rotation invariance into CoALBP as represented by an LBP pair.
The simplest way to embed rotation invariance is to attach a rotation invariant
label to each LBP pair. For example, in Fig.4 there are two types of LBP pairs,
each having four configurations. The same label is attached to each of these eight
LBP pairs because each LBP pair is equal to the others in terms of rotation. This
relation among LBP pairs is called rotation equivalence; and a set of rotation
equivalent LBP pairs is called a rotation equivalence class of LBP pairs. Thus,
the LBP pairs in Fig.4 constitute one rotation equivalence class.

As shown in Fig.4, the upper LBP pairs are equivalent to LBP pairs that have
been rotated 180 degrees from the lower LBP pairs. Therefore, for finding the
rotation equivalent LBP pairs, it is necessary to consider only two cases: (i) a
case in which LBP pairs of $\theta = 0, \pi/4, \pi/2, 3\pi/4$ have rotation equivalence and
(ii) a case in which LBP pairs that are rotated by 180 degrees have rotation
equivalence.

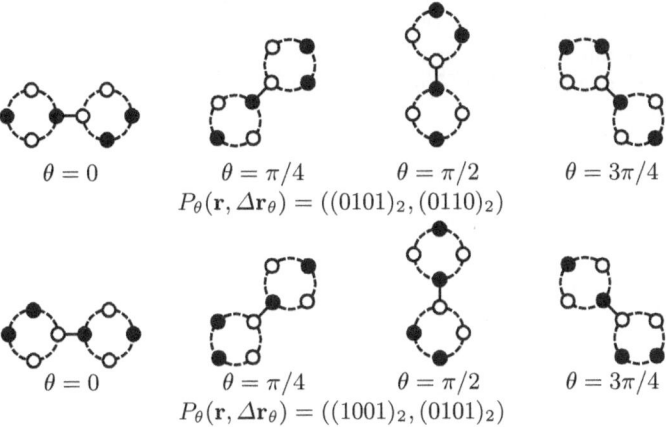

Fig. 4. An example of the rotation equivalence class. The same label is attached to these LBP pairs.

First, in order to consider case (i), we modify the definition of LBP pair. The modified LBP pair is written as follows:

$$P_\theta(\mathbf{r}, \Delta\mathbf{r}_\theta) = (LBP_\theta(\mathbf{r}), LBP_\theta(\mathbf{r} + \Delta\mathbf{r}_\theta)), \tag{4}$$

$$LBP_\theta(\mathbf{r}) = \sum_{i=0}^{N-1} sgn(I(\mathbf{r} + \Delta\mathbf{s}_{i,\theta}) - I(\mathbf{r}))2^i, \tag{5}$$

$$\Delta\mathbf{s}_{i,\theta} = (s\cos(\theta_i + \theta), s\sin(\theta_i + \theta)), \tag{6}$$

where θ serves as the bias of the rotation angle in LBP. Based on the new definition above, the LBP pair of each configuration has the same value in terms of rotation.

Next, we consider case (ii). In this case, we use a rule that an LBP pair that is rotated 180 degrees from $P_\theta(\mathbf{r}, \Delta\mathbf{r}_\theta)$ is equal to $(LBP_{\theta+\pi}(\mathbf{r} + \Delta\mathbf{r}_\theta), LBP_{\theta+\pi}(\mathbf{r}))$. According to this rule, we can consider that these LBP pairs have rotation equivalence. We implement this rule by a mapping table M that has a label for each LBP pair. The mapping table M is generated by using Algorithm 1. In Algorithm 1, "$\gg$" is a circular shift ; also, "$i' = i \gg N/2$" means to rotate LBP i by 180 degree (e.g., $i = (1000)_2$ becomes $i' = (0010)_2$).

By using mapping table M, we define a rotation invariant label for an LBP pair at $\mathbf{r}$ (i.e., RIC-LBP) as follows:

$$P_\theta^{RI}(\mathbf{r}) = M(P_\theta(\mathbf{r}, \Delta\mathbf{r}_\theta)). \tag{7}$$

Finally, an RIC-LBP histogram is generated from $P_\theta^{RI}(\mathbf{r})$ for the entire image.

Since the number of the rotation equivalence classes for the LBP pairs determines the dimension of the RIC-LBP histogram vector, we describe this in more

Algorithm 1. Calculate a mapping table M

Input: N // number of neighbor pixels.
Output: M // mapping table ($N_P \times N_P$ matrix)
   $id \Leftarrow 1, N_P \Leftarrow 2^N$
   for $i = 0, \cdots, N_P - 1$ **do**
     for $j = 0, \cdots, N_P - 1$ **do**
       if $M(i, j) = null$ **then**
         $i' \Leftarrow i \gg N/2, \quad j' \Leftarrow j \gg N/2$
         $M(i, j) \Leftarrow id, \quad M(j', i') \Leftarrow id$
         $id \Leftarrow id + 1$
       end if
     end for
   end for

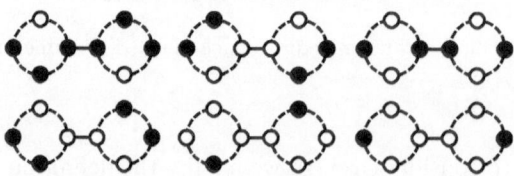

Fig. 5. Examples of symmetric LBP pair

detail as follows. The number of possible LBP pairs is $N_P^2 \times 4$. By considering case (i), the number of possible patterns becomes N_P^2. Moreover, by considering case (ii), the number of possible patterns is halved. Here, we consider a symmetric LBP pair as shown in Fig.5; the number of symmetric LBP pairs is N_P. Therefore, the number of rotation equivalence classes is $N_P(N_P + 1)/2$.

3.2 Process Flow of Generating RIC-LBP Histogram from an Image

We explain how to generate the RIC-LBP histogram from an input image with Eq.(4) and mapping table M.

First, we explain how Eq.(4) and mapping table M work using Fig.6. The example image has four LBPs (Fig.6(a)). The image is decomposed into six LBP pairs (Fig.6(b)). We then have two sets of LBP pairs that have rotation equivalence as indicated by arrows in Fig.6(b). By Eq.(4), the effect of configurations is removed from these LBP pairs, as shown in Fig.6(c). By utilizing mapping table M, these pairs are arranged as shown in Fig.6(d). As we can see, LBP pairs in Fig.6(d) are clearly rotation invariant. By such a process, we obtain an RIC-LBP histogram of the example image, as shown in Fig.6(e).

Next, we explain the overall process flow to obtain a RIC-LBP histogram of an image using Fig.7. Firstly, compute $LBP_\theta(\mathbf{r})$ at every pixel $\mathbf{r}$ throughout the entire input image (Fig.7(a)). Next, compute a histogram of $P_\theta(\mathbf{r}, \Delta \mathbf{r}_\theta)$ (Fig.7(b)). Finally, combine the histogram using mapping table M and obtain a histogram

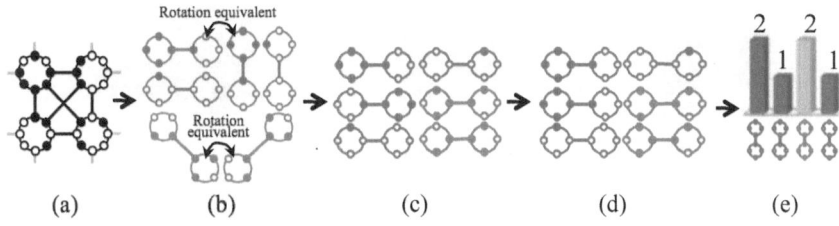

Fig. 6. An example of generating RIC-LBP. (a) Example image. (b) LBP pairs of the example image. (c) Labeling of each LBP pair using Eq.(4). (d) Re-labeling of each LBP pair by applying the mapping table M. (e) RIC-LBP histogram.

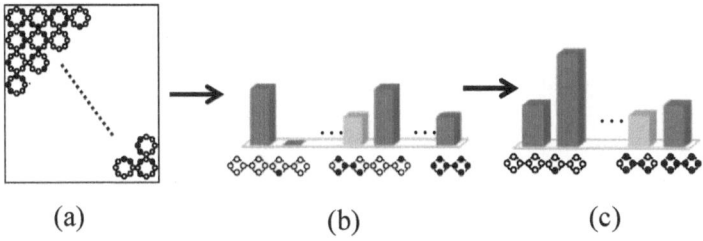

Fig. 7. Process flow of RIC-LBP. (a) Input LBP image. (b) Histogram of $P_\theta(\mathbf{r}, \Delta \mathbf{r}_\theta)$. (c) Histogram of $P_\theta^{RI}(\mathbf{r})$.

of $P_\theta^{RI}(\mathbf{r})$ (Fig.7(c)). Mapping table M is calculated offline. The final histogram is $N_P(N_P + 1)/2$ dimensional vector and is applied to a classifier.

4   Experiments

To evaluate the effectiveness of RIC-LBP, we conducted two types of experiments. The first experiment is for HEp-2 cells classification, an important task to support autoimmune disease diagnosis. Experimental conditions and results are presented in Section 4.1. The second experiment is to apply RIC-LBP to compare its performance relative to other LBP features in general texture recognition, which is described in Section 4.2.

4.1   HEp-2 Cells Classification

Setup. In this experiment, we used the HEp-2 cell dataset from the classification contest at ICPR 2012 [8]. The dataset contains six kinds of antinuclear antibody (ANA) patterns of HEp-2 cell images: *homogeneous, fine speckled, coarse speckled, centromere, cytoplasmatic,* and *nucleolar,* as shown in Fig.8. The total number of images in the dataset is 648. The images are of various sizes.

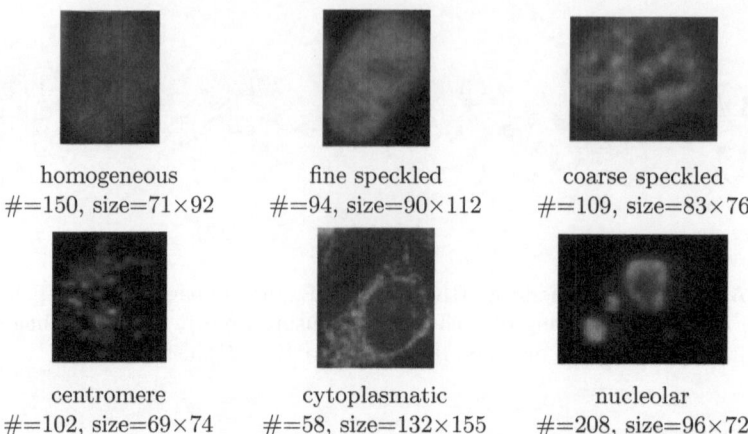

homogeneous
#=150, size=71×92

fine speckled
#=94, size=90×112

coarse speckled
#=109, size=83×76

centromere
#=102, size=69×74

cytoplasmatic
#=58, size=132×155

nucleolar
#=208, size=96×72

Fig. 8. Example images in Hep-2 cell dataset. # is the number of images of each class. "size" is the size of the displayed image; other images not displayed have different sizes.

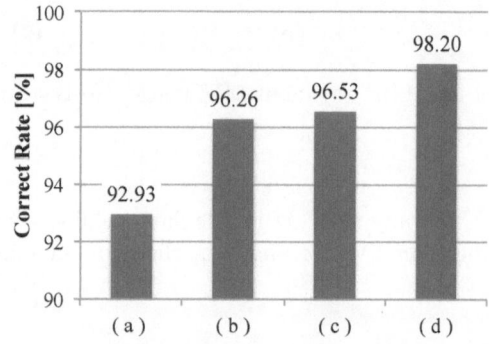

Fig. 9. Performance results. (a) LBP histogram, (b) rotation invariant LBP histogram, (c) LBP pair histogram (CoALBP), (d) rotation invariant LBP pair histogram (RIC-LBP, proposed).

We employed the leave-one-out protocol for evaluation; the correct rate is reported as our experimental result. The parameters of RIC-LBP were set as follows. The radius of LBP was set to $s = 1, 2, 4$ pixels and the intervals of LBP pairs were set to $r = 2, 4, 8$ pixels. Then, the features extracted by each parameter were combined into a final proposed feature vector with dimension of 408 (=136 × 3). The parameters of other methods were also set to produce optimal performance. For classification, the linear SVM was used [9].

Result. First, we show the effectiveness of the proposed feature comparing with various conventional LBP histograms. The baseline result for the original LBP

Table 1. Performance results in HEp-2 cells classification

Method	Correct Rate(%)
LBP^{ri} [4]	96.26
LBP^{riu2}[4]	76.67
LBP-HF [7]	97.23
RIC-LBP (Proposed)	98.20

histogram was 92.93% (Fig.9(a)). When we applied the method with rotation invariance, the correct rate rose to 96.26% (Fig.9(b)). The co-occurrence of adjacent LBPs (i.e. CoALBP) achieved a performance of 96.53% (Fig.9(c)). When we used the proposed RIC-LBP, the performance was further improved to 98.20% (Fig.9(d)). Finally, the proposed method significantly improved the performance of the original LBP by more than 5%. This result demonstrates the effectiveness of both the high descriptive ability of the CoALBPs and the rotation invariance in cell classification.

Next, we compare the results of RIC-LBP with those of other rotation invariant LBP features, as shown in Table 1. As apparent with the experimental results, RIC-LBP outperforms the other methods. These results confirm the significant advantage of RIC-LBP over the conventional methods, especially because the proposed method has not only rotation invariance, but also a high descriptive ability due to CoALBPs.

4.2   Texture Recognition

Setup. We evaluated RIC-LBP for texture recognition using the UIUC texture database [10]. The database contains texture images of 25 classes. Each class consists of 40 images of size 640×480 pixels. Some examples of texture images are shown in Fig.10. The images of each class were randomly split into training and testing sets. This division was repeated 20 times to produce 20 evaluation sets. The average of all correct rates over 20 iterations was defined as the final rate. To increase the difficulty of recognition, we also rotated the texture images by various angles. The parameters of the LBP features were set to the same setting as in the above mentioned experiment.

Results. Experimental results are shown in Table 2. The performance of RIC-LBP was better than that of almost all the other conventional LBP methods, such as LBP^{ri} and LBP^{riu2}. However, the LBP-HF method, which utilizes the discrete Fourier transform, achieved better performance than RIC-LBP. This is because RIC-LBP considers rotation at local regions, whereas LBP-HF considers rotation of the entire image. LBP-HF is thus better suited for this type of texture dataset, which contains global rotation equivalence images. This experimental result indicates that the performance of RIC-LBP for the texture dataset may be further improved by also considering rotation of the entire image by using a method such as the discrete Fourier transform.

bark1   floor1   nkit

wall   water   wood1

Fig. 10. Example images in UIUC texture database

Table 2. Performance results in texture recognition

Method	Correct Rate(%)
LBP[3]	82.55
LBP^{ri} [4]	83.51
LBP^{riu2}[4]	57.33
CoALBP [6]	81.49
LBP-HF [7]	93.60
RIC-LBP (Proposed)	88.27

5   Conclusion

In this paper, we proposed RIC-LBP, a new type of LBP-based feature that simultaneously has the characteristics of rotation invariance and high descriptive ability. Conventional rotation invariant LBP-based features lack descriptive ability. To solve this problem, we focused on CoALBP, which is one effective extension of LBP. Compared with the original LBP, CoALBP has higher descriptive ability since it considers the global relation among LBPs. The proposed RIC-LBP obtained rotation invariance by introducing rotation equivalence class to the CoALBP. The validity of RIC-LBP, in particular, its robustness against local rotations due to transformations of target objects, was confirmed through classification experiments with cells and textures using public databases, specifically the HEp-2 cell dataset and the UIUC texture database.

References

1. Lei, Z., Liao, S., He, R., Pietikäinen, M., Li, S.: Gabor volume based local binary pattern for face representation and recognition. In: Proc. of IEEE International Conference on Automatic Face and Gesture Recognition, pp. 1–6 (2008)

2. Liu, C., Wechsler, H.: Gabor feature based classification using the enhanced fisher linear discriminant model for face recognition. IEEE Trans. on Image Processing 11, 467–476 (2002)
3. Ojala, T., Pietikäinen, M., Harwood, D.: A comparative study of texture measures with classification based on featured distributions. Pattern Recognition 29, 51–59 (1996)
4. Ojala, T., Pietikäinen, M., Mäenpää, T.: Multiresolution gray-scale and rotation invariant texture classification with local binary patterns. IEEE Trans. on Pattern Analysis and Machine Intelligence 24, 971–987 (2002)
5. Zhang, B., Gao, Y., Zhao, S., Liu, J.: Local derivative pattern versus local binary pattern: face recognition with high-order local pattern descriptor. IEEE Trans. on Image Processing 19, 533–544 (2010)
6. Nosaka, R., Ohkawa, Y., Fukui, K.: Feature Extraction Based on Co-occurrence of Adjacent Local Binary Patterns. In: Ho, Y.-S. (ed.) PSIVT 2011, Part II. LNCS, vol. 7088, pp. 82–91. Springer, Heidelberg (2011)
7. Ahonen, T., Matas, J., He, C., Pietikäinen, M.: Rotation Invariant Image Description with Local Binary Pattern Histogram Fourier Features. In: Salberg, A.-B., Hardeberg, J.Y., Jenssen, R. (eds.) SCIA 2009. LNCS, vol. 5575, pp. 61–70. Springer, Heidelberg (2009)
8. Contest on HEp-2 cells classification, http://mivia.unisa.it/hep2contest
9. Fan, R., Chang, K., Hsieh, C., Wang, X., Lin, C.: LIBLINEAR: A library for large linear classification. The Journal of Machine Learning Research 9, 1871–1874 (2008)
10. Lazebnik, S., Schmid, C., Ponce, J.: A sparse texture representation using local affine regions. IEEE Trans. on Pattern Analysis and Machine Intelligence 27, 1265–1278 (2005)

3D LBP-Based Rotationally Invariant Region Description

Jyotirmoy Banerjee[1,2], Adriaan Moelker[1],
Wiro J. Niessen[1,2], and Theo van Walsum[1,2]

[1] Dept. of Radiology, Erasmus MC, Rotterdam, The Netherlands
[2] Dept. of Medical Informatics, Erasmus MC, Rotterdam, The Netherlands
{j.banerjee,t.vanwalsum}@erasmusmc.nl

Abstract. Local binary patterns [LBP] [1] are popular texture descriptors in many image analysis tasks. One of the important aspects of this texture descriptor is their rotational invariance. Most work in LBP has focused on 2D images. Here, we present a three dimensional LBP with a rotational invariant operator using spherical harmonics. Unlike Fehr and Burkhardt [2], the invariance is constructed implicitly, without considering all possible combinations of the pattern. We demonstrate the 3D LBP on phantom data and a clinical CTA dataset.

1 Introduction

Visual tasks such as detection, localization, categorization, and recognition are important subjects of study in computer vision and image analysis. These tasks are often difficult due to apparent within-class inhomogeneity or variability. Part of this within-class variability may be due to the image formation process. Invariant image descriptors extract information from images which is invariant to the variability introduced due to the imaging process, such as noise, distortions, illumination, scale changes, occlusion, etc. One class of such descriptors is texture patterns. Texture have received considerable attention [3] [4] [5] with application in areas of medical imaging [6] [7], image retrieval, remote sensing and object recognition [8]. The local binary patterns [LBP], introduced by Ojala et al. [9], is an efficient method for texture description in 2D. The aim of our work is to extend the conventional LBP and its rotational invariant property mentioned in [1], to a 3D paradigm.

LBP - LBP is a simple and computationally efficient way to describe local image content, with impressive texture discriminative properties. Applications of LBP descriptors are evident in texture classification and face analysis [10]. Though it encapsulates textural information, the conventional LBP operator has a number of limitations which are discussed by Liu et al. [11]. The prominent disadvantages are: weak spatial support and sensitivity to noise. Ojala et al. [1] addressed the first issue by introducing a multi-resolution framework. The sensitivity to noise was addressed by grouping the noisy patterns into one bin and defining the remainder of the patterns as "uniform", corresponding to binary label sequence that has no more than two transitions between "0" and "1"

J.-I. Park and J. Kim (Eds.): ACCV 2012 Workshops, Part I, LNCS 7728, pp. 26–37, 2013.
© Springer-Verlag Berlin Heidelberg 2013

among all pairs of the adjacent binary labels. However, in practice this is an oversimplifying assumption. The uniform LBPs extracted from texture images having more complicated shapes may not necessarily be the patterns dominating the texture. Liao et al. [12] proposed a method that makes use of the most frequently occurring patterns to capture textural information. The frequently occurring or dominant patterns are estimated from training examples. An adaptive framework was proposed by Guo et al. [13] to obtain most discriminative patterns.

Complementary measures - To boost the descriptive power of LBP, several complementary measures were proposed. Ojala et al. [1] included local contrast. Guo et al. [14] and Liu et al. [11] incorporated intensity information, considering the intensities of the center pixel and those of its neighbors. Orientation information was incorporated by Chen et al. [15]. Nanni et al. [6] considered different shapes for neighborhood calculations.

Rotational Invariance - Rotational invariance was originally described by Ojala et al. [1], where the pixel pattern is circularly bit-wise right shifted and the unique identifier is minimum of the generated patterns. Guo et al. [13] consider a rotation-invariant strategy from nonrotation-invariant histograms of LBPs. The method keeps the original rotation-variant features but finds a matching strategy to deal with the rotation. Invariance is globally constructed in Zhao et al. [16] for the whole region by histogramming noninvariant LBPs. Unlike Ojala et al. [1], they achieve rotational invariance implicitly in the Fourier domain, without considering all possible combinations of the patterns. Their frequency domain representation of LBP histograms is a band-limited representation which ignores higher frequencies. This is shown to be robust to other histogram-based invariant texture descriptors, which normalize rotation locally. However, smaller footprints or regions would lead to sparse histograms. Fourier representation of sparse signals is not conducive to similarity measures.

3D LBP - Recently there has been interest in dynamic texture analysis. LBP descriptors were proposed to deal with rotations and view variations in video. They essentially analyze dynamic texture in 2D time series. Zhao et al. [16] have designed invariance for 2D images and extended to 2D time series using bi-planes. Extending LBP to full 3D volume presents few challenges. A circle in 2D would translate to a sphere in 3D. Equidistant sampling on a sphere is not as trivial as on a circle. The notion of ordering is lost in 3D because of the dimensionality, which was an essential step in calculating rotational invariance in 2D. Fehr and Burkhardt [2] proposed a rotationally invariant LBP on volume data. For each LBP computation, correlation between the gray values of all points on the neighborhood sphere with radius R and the weight factor which is a volume representation in an arbitrary but fixed order binomial factors $\{2^0, \ldots, 2^{P-1}\}$, is performed in the spherical harmonic domain. Similar to Ojala et al. [1], rotational invariance is achieved from the computation of the minimum over all angles.

Our method - In this work, we present a rotationally invariant 3D LBP, where unlike Fehr and Burkhardt [2], the invariance is constructed implicitly,

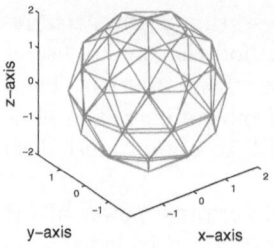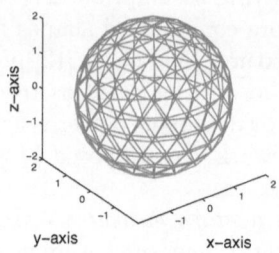

Fig. 1. Icosahedron Spherical Sampling. Left Fig.: 42 Sample points (P_1), Right Fig.: 162 Sample points (P_2).

without considering all possible combinations of the pattern. Spherical harmonics is the mathematical foundation behind our computation [17]. The theory of spherical harmonics states that any rotation of a spherical function does not change its L_2-norm [18]. These features capture invariance to rotation, however with some ambiguity. There will be different signals having similar L_2-norm [19]. The ambiguity is due to loss of phase information. Fehr [19] additionally used bispectrum to address the issue. We choose a simple statistical measure, which encodes the phase angle of a signal. Gluckman [20] has shown that the phase information in an image has relationship with the non-Gaussian statistics, such as kurtosis.

Similar to the Fourier representation of Zhao et al. [16], our harmonic representation of LBP, increases spatial support. However, unlike Zhao et al. [16], where invariance is calculated globally, our method estimates it locally. This is useful in describing regions with small footprints. Our method is a complete three dimensional rotationally invariant modeling of LBP.

2   Method

We present a method for rotationally invariant description of landmarks or regions in 3D using LBP. The LBP in 3D requires a spherical sampling, which is represented in a spherical harmonics framework [17]. The framework helps in obtaining rotation invariant representation. Further, the region information is collected to a set of histograms that are invariant to rotation. The similarity between any two regions can be computed using the Chi-square distance measure [21] between the corresponding set of histograms.

2.1   Spherical Harmonics

Spherical harmonics (SH) is a mathematical framework, generally used to describe a function on a sphere [17]. They are essentially a spherical analog to the Fourier basis. Spherical harmonic functions are defined on imaginary numbers.

We are interested in approximating real functions over the sphere, so we will use the real basis of spherical harmonics. The real spherical harmonic function $Y_\ell^m(\theta, \phi)$ of degree ℓ and order m is given by

$$Y_\ell^m(\theta, \phi) = \begin{cases} \sqrt{2}K_\ell^m \cos(m\phi)P_\ell^m(\cos\theta) & m > 0 \\ \sqrt{2}K_\ell^m \sin(-m\phi)P_\ell^{-m}(\cos\theta) & m < 0 \\ K_\ell^0 P_\ell^0(\cos\theta) & m = 0 \end{cases} \tag{1}$$

where θ, ϕ are polar, azimuthal angles respectively. P is the Associated Legendre polynomials and $K_\ell^m = \sqrt{\frac{(2\ell+1)}{4\pi} \frac{(\ell-|m|)!}{(\ell+|m|)!}}$.

Projecting spherical harmonic functions into spherical harmonic coefficients is straight forward. To calculate a coefficient c_ℓ^m with degree ℓ and order m, we integrate the product of the function f and the spherical harmonic function Y_ℓ^m, in effect projecting how much the function is like the basis function:

$$c_\ell^m = \oint f(\theta, \phi)Y_\ell^m(\theta, \phi)d\Omega$$

where Ω represents the sphere and $(\theta, \phi) \in \Omega$.

The function can be reconstructed to a band-limited approximation (n bands), by reversing the above step, i.e. $\tilde{f}(\theta, \phi) = \sum_{\ell=0}^{n-1} \sum_{m=-\ell}^{m=\ell} c_\ell^m Y_\ell^m(\theta, \phi)$.

2.2   Three Dimensional Rotational Invariant LBP

For convenience we use the similar notation as used by Ojala et al. [1]. Texture representation f^T in a local neighborhood of a monochrome volume is defined as the joint distribution of the binary values of P voxels:

$$f^T \approx t(s(g_0 - g_c), s(g_1 - g_c), \ldots, s(g_{P-1} - g_c))$$

where

$$s(x) = \begin{cases} 1 \text{ if } x \geqslant 0, \\ 0 \text{ if } x < 0. \end{cases}$$

and gray value g_c corresponds to the gray value of the center voxel of the local neighborhood and $g_p(p = 0, \ldots, P-1)$ correspond to the gray values of P equally spaced sample points on a sphere of radius R, around the center voxel.

Spherical harmonics can be used to obtain a rotation invariant representations [18] in 3D. As any rotation of a spherical function does not change the L_2-norm, a set of equally spaced pixels on a sphere of radius R can be represented using an index invariant to rotation.

We define the rotationally invariant local binary pattern per voxel as

$$LBP_{P,R}^{ri3D} = \{\|f_0\|, \|f_1\|, \ldots, \|f_{(n-1)}\|\} \tag{2}$$

where f_ℓ are the frequency components [18] of a function f^T, given by $f_\ell(\theta, \phi) = \sum_{m=-\ell}^{m=\ell} c_\ell^m Y_\ell^m(\theta, \phi)$ and $\|f_\ell\| = \left(\sum_\theta \sum_\phi |f_\ell(\theta, \phi)|^2\right)^{1/2}$, where (θ, ϕ) is an element of the spherical sampling scheme Ω_S.

However, as mentioned by Fehr [19], the power spectrum features are ambiguous, i.e. two different signals may have the same power spectrum. The ambiguity is due to loss of phase information. Variance cannot be used for discrimination, as same power spectrum shares the same variance. We propose the use of kurtosis to address the ambiguity. Gluckman [20] in his work has reported relationship between the phase angle of a signal and the non-Gaussian statistics, kurtosis. It is shown that both local and global correlations in the phase angle lead to many of the statistical regularities, such as kurtosis.

Kurtosis measures how sharply peaked a distribution is, relative to its width, and is defined as $\kappa = ((\mu_4/\mu_2{}^2) - 3)$, where μ_i denotes the ith central moment and in particular, μ_2 is the variance. The kurtosis is normalized to zero for a Gaussian distribution.

The 3D rotationally invariant texture feature per voxel is then described as

$$\kappa LBP_{P,R}^{ri3D} = \{LBP_{P,R}^{ri3D}, \kappa\}. \tag{3}$$

Note that the spherical harmonics are performed on the binary texture pattern f^T and kurtosis κ is estimated over the gray level intensity distribution obtained from the spherical sampling.

2.3   Spherical Sampling in 3D

Three dimensional LBP construction requires sampling over a sphere of radius R. It is non-trivial to have an equidistant sampling over a sphere. To approximate this we use the icosahedron. Icosahedron structure is used to sample the surface of the sphere. To make an icosahedron approximate a sphere more closely, the triangles making up the icosahedron can intuitively be subdivided by splitting the edges of the triangle and then making the new split edges into more triangles [22]. The frequency component f represents how many times the struts of the base icosahedron have been subdivided. Icosahedrons of frequency 2 and 4 as shown in Figure 1 have $10f^2 + 2$ number of vertices, i.e. 42 and 162 sample points, respectively. In the voxel grid, trilinear interpolation is used to estimate the gray value at the vertices of the icosahedron.

2.4   Histogram Matching

The 3D rotationally invariant texture feature is a set of variables (see Equation 3) per voxel, where the number of variables is equivalent to the number of spherical harmonic bands n plus one. The additional term is due to the inclusion of kurtosis. One histogram from each of the variables can be built to represent the texture region. Each histogram consisting of a set of bins measuring the count of an event falling into a given range of a variable. The minimum and maximum values of a variable, required for histogram normalization is empirically estimated. Similarity/dissimilarity between two regions can be estimated using the distance measure between the histograms. The Bhattacharyya measure (or coefficient) and Chi-square measure [21] are two popular measures of similarity between two distributions. The final matching score is derived by adding scores from all the histogram pairs.

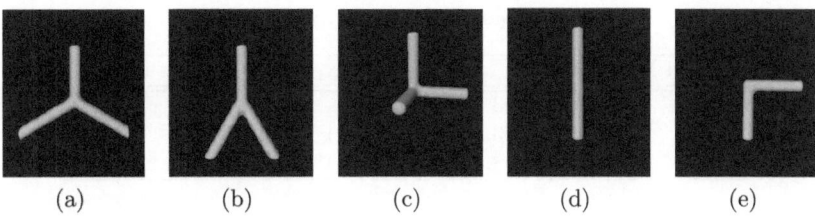

Fig. 2. Surface rendered images of 3D vessel-like structures - (a) All angles are 120°, (b) One of the angle is 60°, other angles are 150° each, (c) All angles are 90°, (d) Straight and (e) Angle is 90°. The volume has binary intensity values. The vessel shapes have voxel intensity 1 and the rest of region in the volume have voxel intensity 0. The shapes are constructed out of cylindrical branches with diameter D each. Shapes in (a), (b) and (c) are made of three cylindrical arms, shape in (d) is made of one cylindrical arm and shape in (e) is made of two cylindrical arms.

3   Experiments and Results

We investigated the properties of both the rotationally invariant features and their application in histogram-based region descriptors, using phantom data. Additionally, we demonstrate the application of our method to the localization of landmarks in medical imaging data.

3.1   Evaluation of Rotational Invariant Features

Purpose of the first experiment is to investigate to what extent our proposed features are indeed rotationally invariant, and can be used to discriminate between various 3D patterns. To this end, we compute the correlation of the rotationally invariant features (Equation 3) on several rotated versions of phantom volumes with different embedded vessel-like structures (see Figure 2).

Phantom volumes with different embedded vessel-like structures - The images shown in Figure 2 are the surface rendered images of five cubic digital phantoms with different embedded vessel-like structures. Vessels shown in Figure 2 constitute foreground while the rest is background. The volumes have binary intensity levels, with foreground and background intensity levels as 1 and 0, respectively. The centroids of the different shapes are located at $(63, 63, 63)$, in their respective volumes of size $[128 \times 128 \times 128]$. The vessel-like structures are constructed from cylindrical branches with diameter D each.

In Figure 2(a) all angles between the vessel branches are 120°. In Figure 2(b) one of the angle is 60°, other angles are 150° each. In Figure 2(c) all angles between the vessel branches are 90°. Figure 2(d) vessel structure has no branches and is linearly oriented. Figure 2(e) the vessel structure is bent at 90°. Shapes in Figures 2(a), 2(b) and 2(c) are made of three cylindrical arms, shape in Figure 2(d) is made of one cylindrical arm and shape in Figure 2(e) is made of two cylindrical arms.

Table 1. Rotation Matrix

Rotation Matrix	Center	Axis (x, y, z)	Angle (in radians)
M_A	$(63.5, 63.5, 63.5)$	-	-
M_B	$(63.5, 63.5, 63.5)$	$(0, 0, 1)$	$\pi/2$
M_C	$(63.5, 63.5, 63.5)$	$(1, 1, 1)$	$\pi/3$

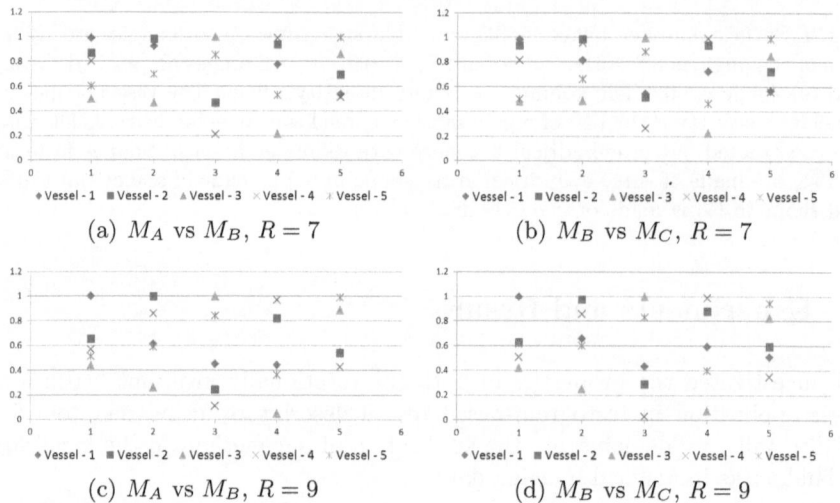

(a) M_A vs M_B, $R = 7$

(b) M_B vs M_C, $R = 7$

(c) M_A vs M_B, $R = 9$

(d) M_B vs M_C, $R = 9$

Fig. 3. Voxel level texture feature correlation between different vessel-like structures in Figure 2, and their rotated versions. In the X and Y axis we have the vessel-like structures and the correlation scores, respectively. Parameters - Diameter of the cylindrical arms $D = 11$ (in Figure 2); $P_2 = 162$ sample points; Sampling Radius R.

Rotation matrix - The rotation matrices used to rotate the phantoms shown in Figure 2 are described in Table 1. The rotation center coincides with the center of the vessel structure. Rotation matrix M_A denotes no rotation. Matrix M_B represents rotation along z-axis by $\pi/2$ radian angles. Matrix M_C represents rotation along $(1, 1, 1)$ axis by $\pi/3$ radian angles. The rotation matrices when applied to the five volumes shown in Figure 2, rotate the shapes around the center of the vessel structure. The voxel location and the rotation center in our evaluation are $(63, 63, 63)$ and $(63.5, 63.5, 63.5)$, respectively.

We evaluate the rotational invariant property of the texture feature (see Equation 3), evaluating it per voxel location. Various intensity patterns can be obtained by sampling around any given voxel from the different volumes in Figure 2. For evaluation purpose we choose the centroids of the vessel structures appearing in these volumes. The results are shown in Figure 3 and Figure 4. They show the correlation between the κLBP index per voxel location obtained from the five different volumes in Figure 2, and their rotated versions. Various intensity patterns can be generated from Figure 2 by varying the sampling radius R or the

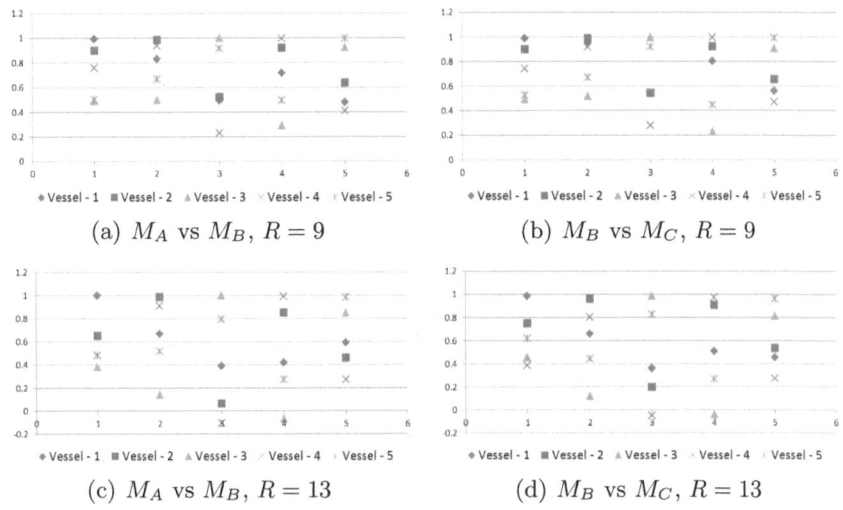

(a) M_A vs M_B, $R = 9$ (b) M_B vs M_C, $R = 9$

(c) M_A vs M_B, $R = 13$ (d) M_B vs M_C, $R = 13$

Fig. 4. Voxel level texture feature correlation between different vessel-like structures in Figure 2, and their rotated versions. In the X and Y axis we have the vessel-like structures and the correlation scores, respectively. Parameters - Diameter of the cylindrical arms $D = 15$ (in Figure 2); $P_2 = 162$ sample points; Sampling Radius R.

diameter D of the cylindrical arms of the vessel-like structures. In Figure 3 we use a sampling radius R of 7 and 9, keeping the diameter D at 11. In Figure 4 we use a sampling radius R of 9 and 13, changing the diameter D to 15. To ensure different intensity patterns with each sampling, the sampling radius is chosen greater than the radius of the cylindrical arms in Figure 2. In Figure 3 and Figure 4 the κLPB feature of a voxel from the volumes in Figure 2, correlates very well with its rotated version. Correlation ranges from -1.00 to $+1.00$. A correlation of 1.00, is a perfect correlation, while a correlation of 0 means there's no relationship between the two variables.

3.2   Evaluation of Region Descriptors

In the next experiment, we investigate the rotationally invariant property of the histogram-based region descriptors. To evaluate the rotational invariance property of the histogram-based region descriptors, the region ought to be spherical. The radius of spherical region is the descriptor radius K. We select one spherical region from each of the five volumes in Figure 2. The center of the regions coincides with the centroid of the vessel-like structures. Since the rotationally invariant texture feature (see Equation 3) is a set of variables, the region description is accumulated to multiple histograms, each histogram corresponding to one of the texture variables. In Figure 5(a) and Figure 5(b) the number of sample points P are 42 and 162, respectively. The descriptor radius K is set to 5. Increase in sampling rate, improves the spherical function approximation, which translates to improved discriminative ability of the descriptor (see Figure 5). We

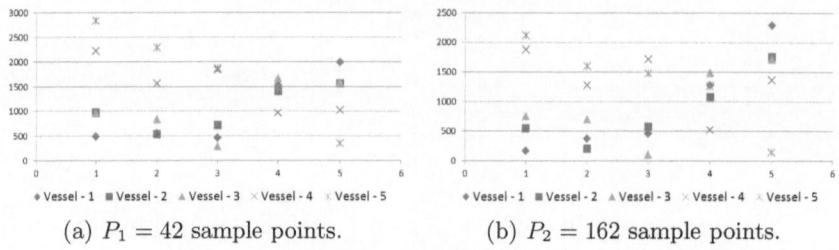

(a) $P_1 = 42$ sample points.   (b) $P_2 = 162$ sample points.

Fig. 5. Chi-square distance measure between five regions from Figure 2, and their rotated versions (Rotation Matrix M_C). In the X and Y axis we have the vessel-like structures and the Chi-square distances, respectively. Parameters - Diameter of the cylindrical arms D (in Figure 2) = 11; Sampling Radius $R = 9$; Descriptor Radius $K = 5$; 20 Bins per histogram.

use the Chi-square distance [21] to measure the difference between regions from all the volumes and their rotated version, as shown in Figure 5. The lower the Chi-square score, the more similar are the histograms. Figure 5 shows that the regions correlate very well with its rotated version, and weekly correlates with other regions.

3.3   Clinical Examples / Evaluation

To investigate the use of LBP as a 3D texture descriptor in medical image analysis (e.g. [6], [7]), we test our approach to describe a region or landmark in liver CTA volume. The liver CTA volume shown in Figure 6(a), is rotated on axis $(0,0,1)$ by $\pi/3$ angle to obtain the new volume, shown in Figure 6(b). The parameters, sampling radius is set to $R = 5$; descriptor radius is set to $K = 5$; and sample points is set to $P_1 = 42$. A landmark location is selected in the left-hand side volume in Figure 6(a). The same landmark is then searched in the rotated volume, and as shows in the Figure 6(b) we are able to retrieve back the landmark location. The Chi-square distance map for the axial, sagittal and coronal planes are shown in Figure 6(c), Figure 6(d) and Figure 6(e), respectively. The distance maps have a peak at the selected landmark location. Figure 7 show the histograms of the descriptor at various locations in the liver CTA volume like thin vessel structure, large vessel structure, liver tissue and liver boundary. The histograms show the first three elements of the rotationally invariant feature from the Equation 2. The histogram-based region descriptor has different signature for different structural location in the liver CTA volumes, which show its discriminative ability in visual classification tasks.

4   Discussion and Conclusion

Volumetric data, like their 2D counterpart have an inherent textural property. The textural property of a 3D region can aid in its region description. LBP

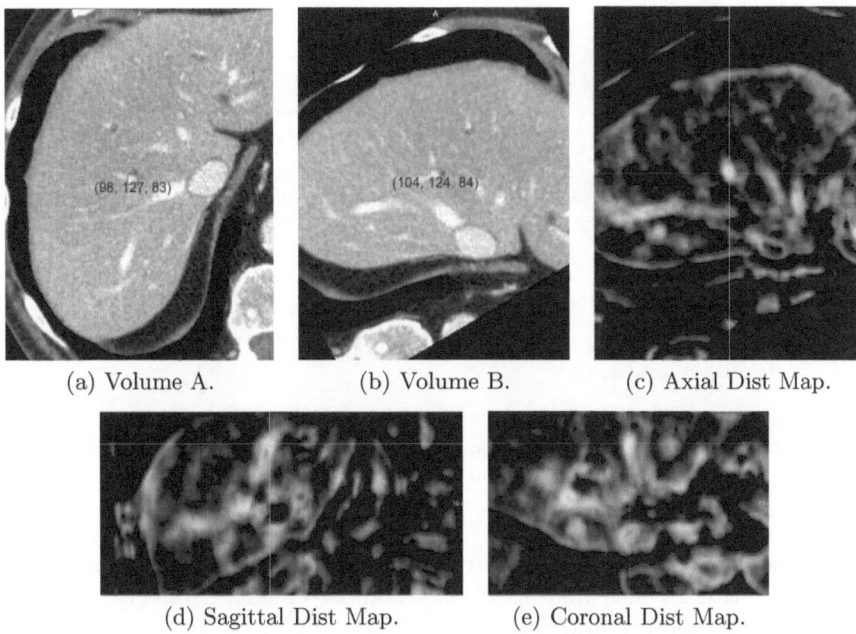

(a) Volume A.   (b) Volume B.   (c) Axial Dist Map.

(d) Sagittal Dist Map.   (e) Coronal Dist Map.

Fig. 6. Liver CTA volume: A landmark location is selected in Volume A. The same landmark location is then searched in the Volume B; a) Volume A, b) Volume B: Obtained by rotating Volume A on axis $(0, 0, 1)$ by $\frac{\pi}{3}$ radians, c) The axial Chi-square distance map, d) The sagittal Chi-square distance map and e) The coronal Chi-square distance map. Parameters - Sampling Radius $R = 5$; Descriptor Radius $K = 5$; Sample points $P_1 = 42$; 10 Bins per histogram.

which was conventionally designed for image texture description is extended in this work to region description of volumetric data. The rotational invariance property helps in view invariant region detection. We show the application of our histogram-based region descriptors in distinguishing various vessel-like structures in phantom data and landmark detection in medical imaging data.

The method presented has several parameters that need to be determined appropriately. The LBP parameters, spherical sample points P and sampling radius R are related as the spherical neighborhood corresponding to a given R contains a limited number of non-redundant sample points. To capture the vessel-like structures well (Figure 2) in the phantom experiments, the sampling radius R is chosen greater than the radius of the cylindrical arms. The 3D region descriptor covers a spherical region of radius K and is described in terms of multiple histograms. Histogram bins provide an estimate of the number of corresponding texture patterns. It is relevant to choose the appropriate number of bins and their range that appear in the histogram carefully, as they directly affect the distance measure. If we use too few bins, the histogram doesn't really portray the data very well. If we have too many bins, we get a broken comb look, which also doesn't give a sense of the distribution. Care should be taken

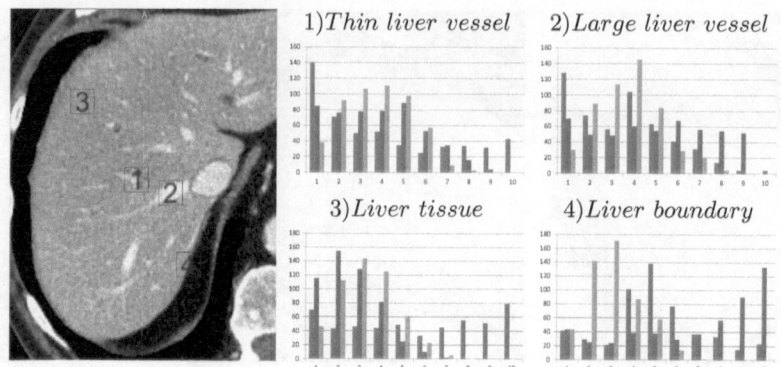

Fig. 7. Histogram descriptors of various locations in liver CTA volume. Histograms show the first three features from the Equation 2. Parameters - Sampling Radius $R = 5$; Descriptor Radius $K = 5$; Sample points $P_1 = 42$; 10 Bins per histogram.

so that there are enough bins and are utilized well. In our phantom experiments histograms with 20 bins were used. The number of bins is directly related to the size of the spherical descriptor region. A descriptor radius K of 5 is used in most of our experiments, which corresponds to 515 entries in the histogram, i.e. approximately 27 entries per bin. The number of histograms is determined by the number of feature variables in Equation 3. As we work with a band limited approximation of a spherical function, the spherical harmonics of bands n equal to 6 in Equation 2 was empirically decided. Hence, the number of histograms per descriptor is $(n + 1)$ equal to 7. The additional histogram is representative of the kurtosis term in the formulation.

To measure the similarity between two regions we use the Chi-square histogram distance metric, since it is popular in previous works [16] [13]. During experimentations we considered the Bhattacharya distance; however from our pilot experiments we found Chi-square metric to be more discriminative.

In our work we focus on a rotationally invariant representation of the 3D LBP. Many of the proposed complementary measures [14] [11] [15] are independent of the LBP representation, and thus could easily be integrated in our 3D approach.

Concluding, we presented a method for rotationally invariant 3D LBP, using spherical harmonic decomposition. We applied the method on vessel-like phantom data and a clinical dataset, with encouraging results. More in-depth analysis and integration of complementary measures is part of future work.

References

1. Ojala, T., Pietikäinen, M., Mäenpää, T.: Multiresolution gray-scale and rotation invariant texture classification with local binary patterns. IEEE Trans. Pattern Anal. Mach. Intell. 24, 971–987 (2002)
2. Fehr, J., Burkhardt, H.: 3d rotation invariant local binary patterns. In: ICPR, pp. 1–4 (2008)

3. Varma, M., Zisserman, A.: A statistical approach to texture classification from single images. International Journal of Computer Vision 62, 61–81 (2005)
4. Tuceryan, M., Jain, A.K.: Handbook of pattern recognition & computer vision, pp. 235–276. World Scientific Publishing Co., Inc., River Edge (1993)
5. Gonzalez, R.C., Woods, R.E.: Digital Image Processing, 2nd edn. Addison-Wesley Longman Publishing Co., Inc., Boston (2001)
6. Nanni, L., Lumini, A., Brahnam, S.: Local binary patterns variants as texture descriptors for medical image analysis. Artif. Intell. Med. 49, 117–125 (2010)
7. Sørensen, L., Shaker, S.B., de Bruijne, M.: Quantitative analysis of pulmonary emphysema using local binary patterns. IEEE Trans. Med. Imaging 29, 559–569 (2010)
8. Chen, C.H., Pau, L.F., Wang, P.S.P. (eds.): Handbook of Pattern Recognition and Computer Vision, 2nd edn. World Scientific Publishing Co., Inc., River Edge (2000)
9. Ojala, T., Pietikäinen, M., Harwood, D.: A comparative study of texture measures with classification based on featured distributions. Pattern Recognition 29, 51–59 (1996)
10. Jun, B., Kim, T., Kim, D.: A compact local binary pattern using maximization of mutual information for face analysis. Pattern Recognition 44, 532–543 (2011)
11. Liu, L., Zhao, L., Long, Y., Kuang, G., Fieguth, P.W.: Extended local binary patterns for texture classification. Image Vision Comput. 30, 86–99 (2012)
12. Liao, S., Law, M.W.K., Chung, A.C.S.: Dominant local binary patterns for texture classification. IEEE Transactions on Image Processing 18, 1107–1118 (2009)
13. Guo, Z., Zhang, L., Zhang, D.: Rotation invariant texture classification using lbp variance (lbpv) with global matching. Pattern Recognition 43, 706–719 (2010)
14. Guo, Z., Zhang, L., Zhang, D.: A completed modeling of local binary pattern operator for texture classification. IEEE Transactions on Image Processing 19, 1657–1663 (2010)
15. Chen, J., Shan, S., He, C., Zhao, G., Pietikäinen, M., Chen, X., Gao, W.: Wld: A robust local image descriptor. IEEE Trans. Pattern Anal. Mach. Intell. 32, 1705–1720 (2010)
16. Zhao, G., Ahonen, T., Matas, J., Pietikäinen, M.: Rotation-invariant image and video description with local binary pattern features. IEEE Transactions on Image Processing 21, 1465–1477 (2012)
17. Green, R.: Spherical harmonic lighting: The gritty details. Archives of the Game Developers Conference (2003)
18. Kazhdan, M.M., Funkhouser, T.A., Rusinkiewicz, S.: Rotation invariant spherical harmonic representation of 3d shape descriptors. In: Symposium on Geometry Processing, pp. 156–165 (2003)
19. Fehr, J.: Local rotation invariant patch descriptors for 3d vector fields. In: ICPR, pp. 1381–1384 (2010)
20. Gluckman, J.: Kurtosis and the phase structure of images. In: 3rd International Workshop on Statistical and Computational Theories of Vision (in conjunction with ICCV 2003) (2003)
21. Cha, S.H.: Comprehensive survey on distance/similarity measures between probability density functions. International Journal of Mathematical Models and Methods in Applied Sciences 1, 300–307 (2007)
22. Fuller, R.B.: U.s. patent no. 2,682,235 (building construction) (1954)

Dynamic Texture Synthesis in Space with a Spatio-temporal Descriptor

Rocio A. Lizarraga-Morales[1], Yimo Guo[2],
Guoying Zhao[2], and Matti Pietikäinen[2]

[1] Universidad de Guanajauto, DICIS. Salamanca, Guanajuato, Mexico
[2] Center for Machine Vision Research, Department of Electrical and Information
Engineering, University of Oulu. PO Box 4500, FI-90014, Finland

Abstract. Dynamic textures are image sequences recording texture in
motion. Given a sample video, the goal of synthesis is to create a new sequence enlarged in spatial and/or temporal domain, which looks percep-
tually similar to the input. Most synthesis methods are mainly focused
on extending sequences only in the temporal domain. In this paper, we
propose a dynamic texture synthesis approach for spatial domain, where
we aim to enlarge the frame size while preserving the aspect and motion
of the original video. For this purpose, we use a patch-based synthesis
method based on LBP-TOP features. In our approach, 3D patch regions
from the input are selected and copied to an output sequence. Usually,
in other patch-based approaches, the selection of the patches is based
only in the color, which cannot capture the spatial and temporal information, causing an unnatural look in the output. In contrast, we propose
to use the LBP-TOP operator, which implicitly represents information
about appearance, dynamics and correlation between frames. The experiments show that the use of the LBP-TOP improves the performance of
other methods giving a good description of the structure and motion of
dynamic textures without generating visible discontinuities or artifacts.

1   Introduction

Texture synthesis (TSyn) has generated considerable research interest in recent
years, since it is an essential element in many computer graphics applications.
Given a sample texture, the goal is to synthesize a new texture that looks per-
ceptually similar to the input, with an arbitrary size specified by the user. TSyn
is a practical alternative way to create textures for a given surface, instead of the
more traditional ways like hand drawing or scanning pictures [13]. One primary
advantage of TSyn lies on the storage requirements, because it only needs to
store a small sample of the texture, regardless of the size of the surface to cover.

Dynamic textures (DTs) are video sequences that are spatially repetitive and
temporally stationary [5]. Basically, DTs are textures in motion. Analogously to
the definition of TSyn, dynamic texture synthesis (DTSyn) consists in creating
an infinite sequence, either in space or time domains, using a video exemplar

J.-I. Park and J. Kim (Eds.): ACCV 2012 Workshops, Part I, LNCS 7728, pp. 38–49, 2013.

as input. The time domain comprises the duration of the video, while the spatial domain consists of enlarging the image size. Both domains must preserve a natural appearance and motion in the outputs.

Different methods for DT synthesis have been proposed. These approaches can be separated into two categories: parametric and non-parametric. Parametric methods are applied to model the behavior of a given phenomenon as a linear dynamic system [3,4,8,15], and typically, they are focused on the two domains at the same time. Even though these methods are able to obtain an output similar to the input, the visual quality is not realistic enough. On the other hand, non-parametric, or exemplar-based, methods are based on taking small parts from an input sample as elements to build the output. Results of non-parametric methods look more natural and realistic than the parametric methods, in view of that these approaches reuse the information of the input.

Non-parametric approaches have been used to synthesize dynamic textures in both time and space domains. Considerable work has been developed for DTSyn along the time domain, for example, in [6,11,7]. The idea behind these techniques for extending the duration of the video, is to find sets of matching frames in the input video and then, jump between these frames during playback. On the other hand, in order to enlarge the frame size, but keeping the duration of the video with non-parametric methods, two main approaches have been followed: pixel-based and patch-based methods. The essential difference between these two methods is in how the information is transferred to the output. As their name says, the pixel-based methods transfer one pixel at a time. The value of each pixel in the output is determined by comparing its spatial neighborhood with all neighborhoods in the input texture. Some pixel based techniques that have been applied for DTSyn are those introduced by Bar-Joseph et al. [1] and Wei and Levoy [14]. By contrast, patch-based techniques select and copy whole neighborhoods each time to the output. With these methods, the speed and quality of synthesis can be improved. However, the problem of how to avoid mismatches between adjacent patches arises. The patch is pasted on the output with a portion of overlapped volume with the already synthesized portion. The patch can be just blended, or an optimal cut can be found for seaming the two patches. In these methods, each patch must be carefully selected depending on a given visual feature. Typically, only the color of the pixels is considered. A significant number of patch-based approaches for static texture synthesis has been proposed, while dynamic textures synthesis has not received the same attention. One representative method for DTSyn in space is the proposed by Kwatra et al. [7], using dynamic programming to find an optimal path to cut through the overlapped regions considering only the color of pixels. Even though pixel-based and patch-based approaches have obtained good results, the influence of the visual features used for DT description for DTSyn remains unexplored. According to Chetverikov and Péteri [2], fundamental issues regarding the DT description include the combination of appearance with motion features. This issue cannot be achieved by only using the intensity of pixels and must be considered for DTSyn implementations.

In this paper, we propose the use of local binary patterns from three orthogonal planes [16] as a reference feature in a non-parametric patch-based method for DTSyn in space. This operator can capture the structure of local brightness variations in three orthogonal planes, and therefore, describe appearance and motion based on the local spatial and temporal patterns. The use of this operator gives to our approach an advantage in comparison to those based only on the color. In addition, it is not intricate since we do not need an optimization of the boundary zone between adjacent patches. Experiments carried out on different dynamic textures show that the use of LBP-TOP features allows a better description of DT patches and preserves the structure and dynamics without generating visible discontinuities between regions. It is also shown that our method can achieve better or at least similar performance to previously proposed methods.

This paper is organized as follows: in Section 2, the LBP-TOP operator and the synthesis algorithm are presented. Experiments and results are presented in Section 3, and concluding remarks are given in Section 4.

2   Dynamic Texture Synthesis in Space Using LBP-TOP Features

The proposed method for DTSyn in spatial domain is carried out by using a spatio-temporal descriptor as visual feature, which allows a better perceptual representation of DT. Details of the implementation are given below.

2.1   Spatio-Temporal Descriptor

The local binary pattern from three orthogonal planes (LBP-TOP) [16], is a spatio-temporal descriptor for dynamic textures. The LBP-TOP considers the co-occurrences in three planes XY, XT and YT, capturing information about the space-time transitions, as shown in Fig. 1.

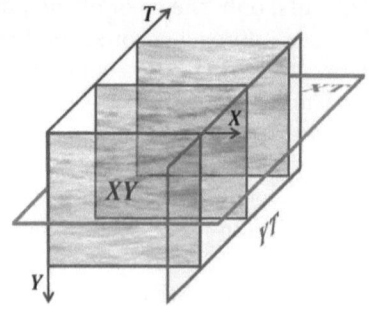

Fig. 1. The LBP-TOP feature is obtained by extracting the LBPs from three orthogonal planes

The LBP-TOP is an extension of the Local Binary Patterns (LBP) presented by Ojala *et al.* [9]. As it is known, the LBP is a theoretically simple, yet efficient approach, to characterize the spatial structure of local texture. Basically, the operator labels a given pixel of an image by thresholding its neighbors in function of the pixel intensity and summing the thresholded values weighted by powers of two. According to Ojala, a static texture T in a local neighborhood of a monochrome texture image is defined as the joint distribution of the gray levels of $P(P > 1)$ image pixels $T = t(g_c, g_0, \ldots, g_{P-1})$, where g_c is the gray value of the center pixel and $g_p(p = 0, 1, \ldots, P - 1)$ are the gray values of P equally spaced pixels on a circle radius $R(R > 0)$, that form a circularly symmetric neighbor set. If the coordinates of g_c are (x_c, y_c), then the coordinates of g_p are $(x_c - R \sin(2\pi p/P), y_c + R \cos(2\pi p/P))$. The LBP code for the pixel g_c is defined as

$$LBP_{P,R}(g_c) = \sum_{p=0}^{P-1} s(g_p - g_c)2^p \qquad (1)$$

where the thresholding function $s(\cdot)$ is defined in equation 2. More details can be further consulted in [9].

$$s(t) = \begin{cases} 1, t \geq 0 \\ 0, \text{otherwise} \end{cases} \qquad (2)$$

For the spatio-temporal extension of the LBP, named as LBP-TOP, the local patterns are extracted from the XY, XT and YT planes. Each code is denoted as XY-LBP for the space domain, and XT-LBP and YT-LBP for space-time transitions [16]. In the LBP-TOP approach, the three planes intersect in the center pixel and three different patterns are extracted in function of that central pixel. The local pattern of a pixel from XY plane, contains information about the appearance and, in the local patterns from XT and YT planes, statistics of motion in horizontal and vertical directions are included. In this case, the radii in axes X,Y and T are R_X, R_Y and R_T respectively and the number of neighboring points in each plane are defined as P_{XY}, P_{XT}, P_{YT}. Supposing that the coordinates of the center pixel $g_{t_c,c}$ are (x_c, y_c, t_c), the coordinates of the neighbors $g_{XY,p}$ in the plane XY are given by $(x_c - R_X \sin(2\pi p/P_{XY}), y_c + R_Y \cos(2\pi p/P_{XY}), t_c)$. Analogously, the coordinates of $g_{XT,p}$ in the plane XT are $(x_c - R_X \sin(2\pi p/P_{XT}), y_c, t_c - R_T \cos(2\pi p/P_{XT}))$, and the coordinates of $g_{YT,p}$ on the plane YT are $(x_c, y_c - R_Y \cos(2\pi p/P_{YT}), t_c - R_T \sin(2\pi p/P_{YT}))$. Consequently, every pixel in the input video is represented by 3 codes, one for each orthogonal plane.

For the implementation proposed in this paper, each pixel of the input sequence V_{in} is analyzed with the LBP-TOP operator, in such a way that we obtain an LBP-TOP-coded sequence $V_{LBP-TOP}$. Each pixel in the $V_{LBP-TOP}$ sequence is coded by three values, comprising each of the space-time patterns of the local neighborhood, as can be seen in Fig. 2. As we said before, in patch-based methods each patch must be carefully selected depending on a given visual feature, then, the patch is positioned with some overlapped area with the already synthesized portion. To accomplish this task, we use $V_{LBP-TOP}$ as a temporary sequence for the patch description in the selection process.

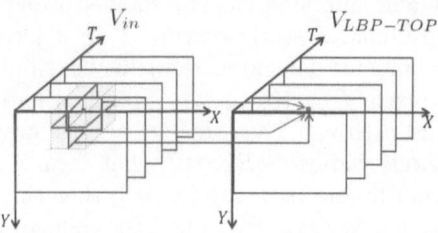

Fig. 2. Each pixel in the corresponding LBP-TOP sequence is obtained by extracting the LBPs from the three orthogonal planes in the input sequence

2.2   Dynamic Texture Synthesis in Space Domain

In this paper, we propose the use of LBP-TOP features [16] in a non-parametric patch-based method for DTSyn in space. As mentioned, non-parametric algorithms basically select patches, or blocks from the input as elements to build an output. The use of LBP-TOP features, allows us to consider the spatial and temporal relations among pixels and, therefore, obtain more information about a given block and its possible neighbor blocks.

Our method is cyclical, in each step we select a block B_k from the input sample video V_{in} and copy it to the output video V_{out}. To avoid discontinuities or artifacts between blocks, we must carefully select B_k based on the blocks already pasted $\{B_0, \ldots, B_{k-1}\}$ in V_{out}. At the beginning, a block B_0, of $W_x \times W_y \times W_t$ pixels size, is randomly selected from the input V_{in}, and copied to the upper left corner of the output V_{out}. The following blocks needed to fulfill the output, are positioned in raster scan order in such a way that they are partially overlapped with previously pasted blocks. The overlapped volume between two blocks is of size $O_x \times O_y \times O_t$ pixels. In Fig. 3 an example of a video block, the boundary zone where two blocks should match and an example of the overlapped volume between two blocks are illustrated. In Figure 3(b), the selected block B_k has a boundary zone E_{B_k} and the previously pasted volume in V_{out} has a boundary zone E_{out}. According to our method and in order to avoid discontinuities, E_{B_k} and E_{out} should match.

The appropriate description and selection of each block becomes a key issue in our method. In the block selection step, we consider the similarity of the spatio-temporal features on the boundary zones. For this, we first build a set of candidate blocks A_B of V_{in}, which are considered to match with the previously pasted volumes in V_{out}. Then we select one block randomly from the set. The random selection is performed to keep a good diversity on the blocks selected. Two blocks are considered to match if the distance in the corresponding overlapping volume is lower than a distance tolerance, specified by the user. We

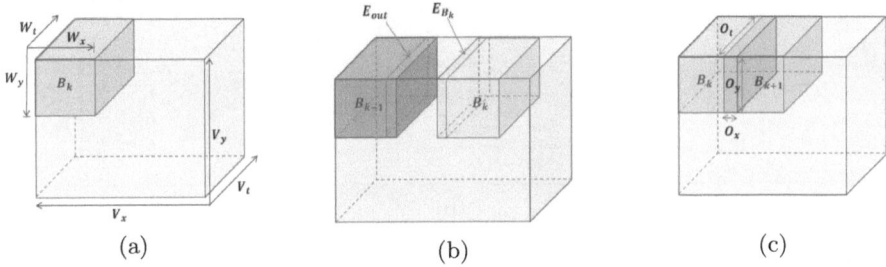

Fig. 3. Examples of (a) a video block, (b) the boundary zone of two different blocks and (c) the overlapped volume between two video blocks. The boundary zones should match.

construct a set of all the potential blocks $B_{(x,y,t)}$ to be considered to match with E_{out}. Let $B_{(x,y,t)}$ be the block whose upper left corner is at (x,y,t) in V_{in}. We construct

$$A_B = \{B_{(x,y,t)}|d(E_{B_{(x,y,t)}}, E_{out}) < d_{max}, B_{(x,y,t)\in V_{in}}\} \tag{3}$$

where $E_{B_{(x,y,t)}}$ is the boundary zone of $B_{(x,y,t)}$ and d_{max} is the distance tolerance between two boundary zones. Details on how to compute $d(\cdot)$ are given later. When we have determined all the potential blocks, we pick one randomly from A_B to be the kth block B_k to be pasted on V_{out}. The size of A_B depends on how many blocks satisfy the similarity constraints. With a high value of d_{max} the output will have a better quality but, few blocks would be considered to be part of A_B. By contrast, with a low tolerance a big number of blocks will be part of the set and there will be more options to select, but the quality of the output will be compromised. For a given d_{max}, the set A_B could be empty. In such case, we choose B_k to be the block $B_{(x,y,t)}$ in V_{in} with the smallest distance to the boundary zone of the output E_{out}. In Fig. 4 the three possible configurations of the overlapping zones between the already pasted zones E_{out} and the new patch B_k are shown. The first possibility, shown in Fig. 4(a), is when B_k is on the first row and goes after B_0. The second is when B_k is the first block in the second or subsequent rows (Fig. 4(b)). The third is when B_k is not the first on the second or subsequent rows as shown in Fig. 4(c), here the total distance is the addition of the above and left boundaries distances.

The algorithm to pursue for synthesizing dynamic texture in space can be described as follows:

1. Let V_{in} be an input DT sample of $V_x \times V_y \times V_t$ pixels size. Set the synthesis block size as $W_x \times W_y \times W_t$, and the size of the overlapped volume of two adjacent blocks as $O_x \times O_y \times O_t$. Consider $V_t = W_t = O_t$.
2. Obtain the $V_{LBP-TOP}$ sequence of V_{in}.
3. Transfer the first block B_0 from V_{in} to the upper left corner of the output V_{out} by random selection. Set $k = 1$.
4. Synthesize next block in raster scan order:

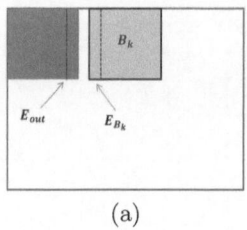 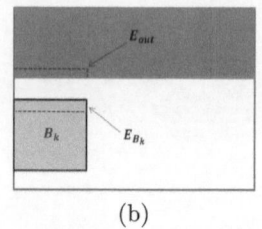 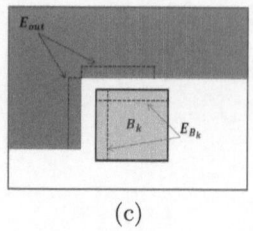

(a)   (b)   (c)

Fig. 4. Three possible overlapping zones between the output E_{out} and the new block E_k. (a) B_k is in the first row, but after B_0. (b) B_k is the first block in the second or subsequent rows and (c) B_k is not the first on the second or subsequent rows.

a) Select a set of candidate blocks A_B from V_{in}, such that for each block in A_B, the boundary zone satisfies the overlap constraints (above and left) with the previously pasted blocks, with certain tolerance distance between the blocks, computed using the LBP-TOP features.
b) Pick one block randomly from A_B to be B_k and paste it from V_{in} to V_{out}. Set $k = k + 1$. Perform blending in the boundary zones.
c) Repeat until V_{out} is completely synthesized.

On the overlapped volume, in order to obtain smooth transitions and minimize artifacts between two adjacent blocks, we blend the volumes using a feathering algorithm [12]. This algorithm set weights to the pixels for attenuating the intensity around the blocks' boundaries using a ramp style transition. As a result, the possible discontinuities are avoided, and uniform transitions are achieved.

The sizes of a given block and the overlapped volume are dependent on the properties of a particular DT, hence, in our algorithm they can be adjusted by the user. This characteristic makes our algorithm flexible and controllable. The boundary zone should be large enough to avoid mismatching features across the borders but at the same time, it should be small to be tolerant to the border constraints. Usually, the overlap volume is a small fraction of the block size, 1/6 of the total volume in our experiments.

In this approach, the overlap distance between the boundary zones of a given block $E_{B_{(x,y,t)}}$ and the output E_{out} is estimated by using the L2 norm among the LBP values of each orthogonal plane. This error is defined as:

$$d(E_{B_{(x,y,t)}}, E_{out}) = \frac{1}{V} \sum_{i=1}^{V} \sum_{j=1}^{3} \left[p_{B_{(x,y,t)}}^{j}(i) - p_{out}^{j}(i) \right]^2 \qquad (4)$$

where V is the number of pixels in the overlapped volume. $p_{B_k}^{j}(i)$ and $p_{out}^{j}(i)$ represent the LBP values of the ith pixel in the overlapping zones on the jth orthogonal plane, respectively. For color DTs, we compute the LBP-TOP code for each color channel. In this paper, we use the RGB color space, the final overlapping distance is the sum of the errors in each color component. The matching estimation between two blocks is computed based on their LBP values from the $V_{LBP-TOP}$ sequence. As a result, spatial and temporal features are considered simultaneously for the block description.

3 Experiments and Results

In this section, we present two series of tests that have been accomplished in order to evaluate the performance of our method. At first, a visual evaluation of performance is made on a variety of dynamic textures. Afterwards, comparisons between the proposed approach with other state-of-the-art methods are made to validate the application of it. All the resulting videos are available on the website: dl.dropbox.com/u/13100121/LBP2012Results.zip

3.1 Performance on a Variety of DTs

In the first experiment, a set of videos was selected for evaluating our approach performance on different types of dynamic textures. The videos were selected from the DynTex database [10], which provides a comprehensive range of high-quality DTs and can be used for various research purposes. In Figs. 5(a)-(f), a frame (176×120 pixels size) taken from the original videos is shown. The selected sequences correspond to videos that show: spatio-temporal stationarity (a-c), a scene with a variety of textures and colors with different kind of dynamics (d) and a scene composed by structured objects (e-f).

In Figs. 5(g)-(l), the results of the synthesized outputs enlarged to 200×200 pixels size are presented. Spatial dimensions of the block $W_x \times W_y$ used for synthesis are shown below each image. As we said before, the size of the block is a user-specifiable parameter and should be proportional to the size of the spatial or temporal texture patterns. Here, the size of the overlapped volume $O_x \times O_y$ is $1/6$ of $W_x \times W_y$. As we can observe in Figs. 5(g)-(i), our method preserves the spatio-temporal stationarity of the input and the borders between blocks are almost invisible. It is worth mentioning that in our method, we do not need to do an additional optimal seam on the borders to achieve smooth transitions, such as the graph cut used in [7]. This soft transition is achieved because of the selection of the blocks, based on the LBP-TOP features. The corresponding output for the video shown in Fig. 5(d) is presented in Fig. 5(j), where the same variety of colors and the diversity of surfaces is maintained. In this video, the transitions between blocks are also invisible. Sequences shown in Figs. 5(e)-(f) are different in the sense that they are composed of structured objects. Therefore, it is crucial that the structure of these objects can be maintained in the output, where we aim to generate an array of these objects. The synthesized results, seen in Figs. 5 (k)-(l), exhibit such arrangement showing that our method can keep the shape and structure of the given object without adding any discontinuity.

(g) 20 × 20   (h) 30 × 30   (i) 55 × 55   (j) 20 × 20   (k) 60 × 60   (l) 90 × 90

Fig. 5. Results of spatial synthesis. (a-f) A frame taken from the original sequence. (g-l) the corresponding synthesis result with the video block size used. The block size is proportional to the size of the spatial or temporal patterns.

3.2 Performance Comparison

The second experiment consist of a comparison with other state-of-the-art methods. We have compared our approach with the methods proposed by Wei and Levoy [14], Bar-Joseph *et al.* [1] and Kwatra et al. [7]. The firsts two are pixel-based approaches, while the third is a patch-based method.

We have borrowed the sequences named OCEAN and SMOKE (frame of 150 × 112 pixels size) used by Wei and Levoy in their experiments and made a comparison of the quality of the results. In Fig. 6 a frame extracted from the original sample, from the result of Wei and Levoy and from our result are presented. Here, it is observed that the videos obtained by Wei and Levoy (frame of 150 × 112 pixels size) are blurred, while the videos generated by our method (frame of 170 × 170 pixels size) keeps a natural appearance and motion of the two phenomena.

A second comparison is made with the results obtained by Bar-Joseph *et al.* [1]. We have used the sequences named as CROWD and JELLY FISH (frame of 256 × 256 pixels size). In Fig. 7, a frame from each resulting sequence is presented. Here, it is observed that the videos obtained by Bar-Joseph (frame of 256 × 256 pixels size) have some artifacts, blurred spots and discontinuities, while the videos generated by our method (frame of 280 × 280 pixels size) keep a natural look.

In a third comparison, we synthesized the sequence named RIVER (frame of 176 × 112 pixels size), provided by Kwatra *et al.* [7] for spatial extension. As it can be seen from both results (frame of 200 × 150 pixels size) shown in Fig. 8, Kwatra has generated good results of DT synthesis, which can be taken as the baseline to compare with. From the experimental results, we found that our method also achieves a good performance. It is observed that the appearance and dynamics of the water are preserved.

We have found that our method obtains very good results with sequences that present some spatial homogeneity, however we have detected limitations of our method on a very specific type of dynamic textures. Our approach does not work very well when a moving object occupies a big portion of the scene and thus,

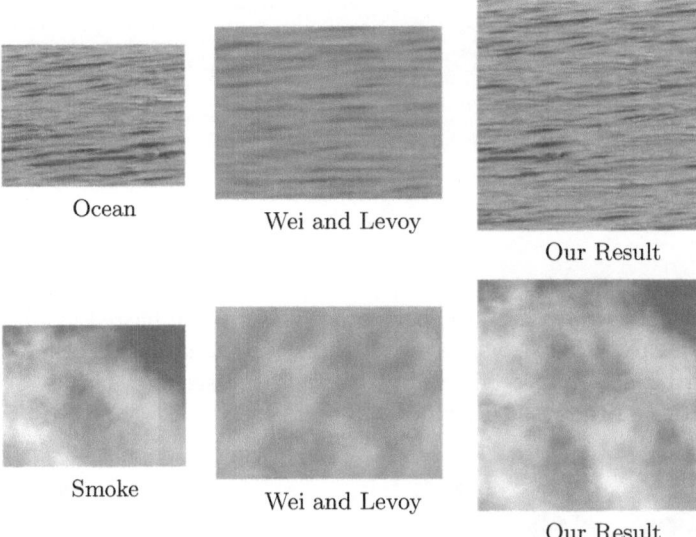

Fig. 6. Comparisons between the proposed approach and the method proposed by Wei and Levoy [14]

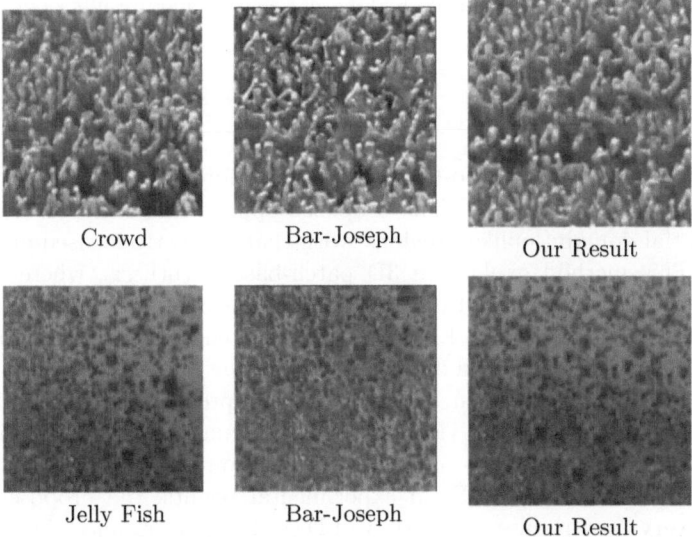

Fig. 7. Comparison between the proposed approach and the method proposed by Bar-Joseph *et al.* [1]

River   Kwatra   Our Result

Fig. 8. Comparison between the proposed approach and the method proposed by Kwatra *et al.* [7]

there is not enough diversity to choose the blocks to be pasted on the output. Examples of this, using sequences form the DynTex database are shown in Fig. 9 where there is certain repeatability between the selected blocks, leading to discontinuities on the resulting videos.

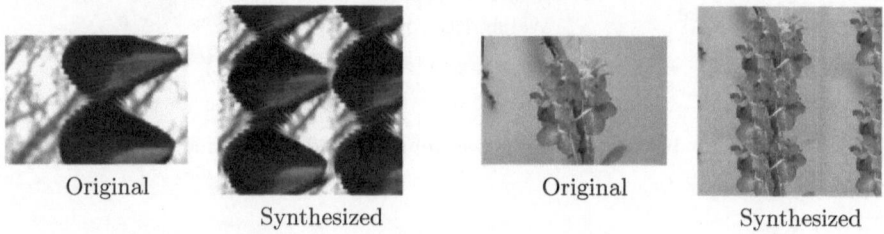

Original   Synthesized   Original   Synthesized

Fig. 9. Two examples where our method did not achieve good results as in the general case

4   Conclusions

In this paper, the use of spatio-temporal features for dynamic textures synthesis in space has been considered. The proposed approach is centered on synthesis in the spatial domain, unlike previous work that is mostly focused on temporal domain. This method explores a 3D patch-based synthesis, where the patch selection is accomplished by taking LBP-TOP features, instead of just making use of the intensity of pixels. LBP-TOP features can enhance the capability of describing the appearance and dynamics of DTs due to the local spatio-temporal patterns extracted. The main advantage of the presented approach is that it preserves on the output the visual similarity, dynamics and continuity of the input. Furthermore, no additional seam optimization is needed to achieve smooth transitions between blocks. From experimental results, the proposed method produces very good results on a variety of DTs. The performance of the proposed method has shown to be better than, or at least equal to other methods. In future work, the inclusion of a temporal domain synthesis approach will be considered.

Acknowledgement. The authors would like to thank the Academy of Finland, the Finnish CIMO and the Universidad de Guanajuato for the financial support.

References

1. Bar-Joseph, Z., El-Yaniv, R., Lischinski, D., Werman, M.: Texture mixing and texture movie synthesis using statistical learning. IEEE Trans. on Visualization and Computer Graphics 7, 120–135 (2001)
2. Chetverikov, D., Peteri, R.: A brief survey of dynamic texture description and recognition. In: Proc. of the CORES 2005, vol. 30, pp. 17–26 (2005)
3. Constantini, R., Sbaiz, L., Susstrunk, S.: Higher order SVD analysis for dynamic texture synthesis. IEEE Trans. on Image Processing 17, 42–52 (2008)
4. Doretto, G., Jones, E., Soatto, S.: Spatially Homogeneous Dynamic Textures. In: Pajdla, T., Matas, J(G.) (eds.) ECCV 2004. LNCS, vol. 3022, pp. 591–602. Springer, Heidelberg (2004)
5. Ghanem, B., Ahuja, N.: Phase PCA for dynamic texture video compression. In: Proc. of the IEEE ICIP 2007, vol. 3, pp. 425–428 (2007)
6. Guo, Y., Zhao, G., Chen, J., Pietikäinen, M., Xu, Z.: Dynamic texture synthesis using a spatial temporal descriptor. In: Proc. of the IEEE ICIP 2009, pp. 2277–2280 (2009)
7. Kwatra, V., Schodl, A., Essa, I., Turk, G., Bobick, A.: Graphcut textures: Image and video synthesis using graph cuts. ACM Trans. on Graphics 22, 277–286 (2003)
8. Liu, C.B., Lin, R.S., Ahuja, N., Yang, M.H.: Dynamic textures synthesis as non-linear manifold learning and traversing. In: Proc. of the BMVC 2006, pp. 859–868 (2006)
9. Ojala, T., Pietikäinen, M., Mäenpää, T.: Multiresolution gray-scale and rotation invariant texture classification with local binary patterns. IEEE Trans. on Pattern Analysis and Machine Intelligence 24, 971–987 (2002)
10. Peteri, R., Fazekas, S., Huiskes, M.J.: DynTex: A comprehensive database of dynamic textures. Pattern Recognition Letters 31, 1627–1632 (2010)
11. Schodl, A., Szeliski, R., Salesin, D., Essa, I.: Video textures. In: Proc. of the ACM SIGGRAPH 2000, pp. 489–498 (2000)
12. Szeliski, R., Shum, H.Y.: Creating full view panoramic image mosaics and environment maps. In: Proc. of the ACM SIGGRAPH 1997, pp. 251–258 (1997)
13. Wei, L.Y., Lefebvre, S., Kwatra, V., Turk, G.: State of the art in example-based texture synthesis. In: Eurographics 2009, EG-STAR, pp. 93–117 (2009)
14. Wei, L.Y., Levoy, M.: Fast texture synthesis using tree-structured vector quantization. In: Proc. of the ACM SIGGRAPH 2000, pp. 479–488 (2000)
15. Yuan, L., Wen, F., Liu, C., Shum, H.-Y.: Synthesizing Dynamic Texture with Closed-Loop Linear Dynamic System. In: Pajdla, T., Matas, J(G.) (eds.) ECCV 2004. LNCS, vol. 3022, pp. 603–616. Springer, Heidelberg (2004)
16. Zhao, G., Pietikäinen, M.: Dynamic texture recognition using local binary patterns with an application to facial expressions. IEEE Trans. on Pattern Analysis and Machine Intelligence 29, 915–928 (2007)

Adaptive Kernel Size Selection for Correntropy Based Metric

Ying Tan, Yuchun Fang, Yang Li, and Wang Dai

School of Computer Engineering and Science, Shanghai University, Shanghai, China

Abstract. The correntropy is originally proposed to measure the similarity between two random variables and developed as a novel metrics for feature matching. As a kernel method, the parameter of kernel function is very important for correntropy metrics. In this paper, we propose an adaptive parameter selection strategy for correntropy metrics and deduce a close-form solution based on the Maximum Correntropy Criterion (MCC). Moreover, considering the correlation of localized features, we modify the classic correntropy into a block-wise metrics. We verify the proposed metrics in face recognition applications taking Local Binary Pattern (LBP) features. Combined with the proposed adaptive parameter selection strategy, the modified block-wise correntropy metrics could result in much better performance in the experiments.

1 Introduction

In many research fields such as pattern recognition, data mining and machine learning, metrics plays an essentials role in applications such as image retrieval, object recognition and clustering. Metrics has significant influence to the performance of many classic algorithms such as K-means and nearest-neighbor classifier.

Previous research on metrics cares about the performance of metrics for specific applications. Rubner et al. [1] made a comparison of nine general image measurements based on many computational experiments and classified them into four categories: heuristic histogram distances, non-parametric test statistics, information-theoretic divergences and ground distance measures. Others learned metrics based on models such as hidden markov models [2] or information bottleneck theory [3]. Normally, the selection of metrics is decided by the specific problems. According to the image representation with two dimensional principal component analysis (2D-PCA), Zuo et al. [4] proposed assembled matrix distance metric. Under specific assumptions, Liu [5] proposed two new metrics - Probability Reasoning Model Whitened Cosine (PWC) metric and Within-Class Whitened Cosine (WWC) metric according to combining Bayes decision rule with whitened cosine similarity. Zhu et al. [6] proposed the Rank-Order Distance to deal with the situation of incomplete distribution of samples in feature space for face tagging.

Among the research topics on metrics, metric learning has become a hot topic recently. Xing et al. [7] used a Mahalanobis learning distance with side information in clustering. Jain et al. [8] learned the parameterization of Mahalanobis

J.-I. Park and J. Kim (Eds.): ACCV 2012 Workshops, Part I, LNCS 7728, pp. 50–60, 2013.
© Springer-Verlag Berlin Heidelberg 2013

metric and encoded the learned information into randomized hash functions for image retrieval. The motivation of metric learning is to achieve an appropriate distance from training examples to accurately reflect the underlying relationships among components of features.

Stimulated by the idea of metric learning, we propose a learning method to adaptively select parameters for correntropy metrics. The concept of correntropy was introduced firstly for blind deconvolution [9] and later on developed into a kernel metrics [10]. Similar to other kernel methods, the kernel parameter is crucial for correntropy metrics. However, there is no available solution for this problem. In this paper, we propose a learning method and deduce a close-form solution to obtain kernel parameters for correntropy metrics based on the Maximum Correntropy Criterion (MCC). Moreover, considering the correlation among features, we propose a modified correntropy metrics. We verify the proposed metrics in face recognition applications taking use of Local Binary Pattern (LBP) features. Combined with the adaptive parameter selection method, the proposed modified block-wise correntropy results in prominently improved performance compared with other popular block-wise metrics for LBP features in face recognition.

The paper is structured as follows. Section 2 presents a review about correntropy. In Section 3, we show the basic theory about MCC and describe the close-form solution to calculate the kernel size. A modified correntropy is proposed as a block-wise metrics in Section 4. Section 5 shows comparison of our method with the general metrics in face recognition experiment. Section 6 draws the conclusions.

2   Correntropy

As a novel metrics, correntropy originates from the framework of Information Theoretic Learning (ITL). Combining the Renyís quadratic entropy with the Parzen estimation[10], the correntropy metrics is defined in Eqn.(1).

$$V_\sigma = E(\kappa_\sigma(X - Y)) \tag{1}$$

where X and Y are two random variables, $\kappa_\sigma(\cdot)$ is a kernel function, and $E(\cdot)$ denotes the expectation operator. The basic idea of correntropy is to measure the similarity of two random variables with the expectation of the diversity from couples of samples.

In practice, it is hard to learn the joint PDF between X and Y. Liu et al.[10] estimate the correntropy with a finite number of samples according to Eqn.(2).

$$\widetilde{V}_\sigma(X, Y) = \frac{1}{N} \sum_{i=1}^{N} \kappa_\sigma(X_i - Y_i) \tag{2}$$

where X_i and Y_i are the i-th sample of X and Y respectively. The Gaussian kernel is usually adopted in calculation of correntropy as shown in Eqn.(3).

$$\kappa_\sigma(x) = \frac{1}{\sqrt{2\pi}\sigma} \exp(-\frac{x^2}{2\sigma^2}) \tag{3}$$

where σ is the size of kernel.

Further, Liu et al. [11] advance it to measure the distance of two discrete vectors on the dimension level. Namely it is correntropy induced metric (CIM). And we can learn it following the Eqn.(4):

$$CIM(A, B) = \sqrt{\kappa_\sigma(0) - \frac{1}{N} \sum_{i=1}^{N} \kappa_\sigma(a_i - b_i)} \tag{4}$$

where A and B are discrete vectors $A = (a_1, a_2, \ldots, a_N)$ and $B = (b_1, b_2, \ldots, b_N)$, $\kappa_\sigma(x)$ is the Gaussian kernel with kernel size σ.

For a given kernel size σ, $\kappa_\sigma(0)$ is constant. So we use an equivalent form for Eqn.(4) in this paper and name it as correntropy for dimension (CD):

$$D_{dim}(A, B) = \frac{1}{N} \sum_{i=1}^{N} \kappa_\sigma(a_i - b_i) \tag{5}$$

Apparently, the kernel size is a variable and important parameter for correntropy based metrics. However, there is no general solution to the selection of σ, which is the first problem we concentrate on in this paper.

3   Adaptive Kernel Size Seletion with MCC

3.1   The Contribution of Kernel Size to CD

Let $X_f = (x_1, x_2, \ldots, x_n)$ and $Y_f = (y_1, y_2, \ldots, y_n)$ be the feature vectors of two samples. For each dimension of the feature vector, we define:

$$D_i = |x_i - y_i| \qquad i = 1, 2, \ldots, n \tag{6}$$

Taking face recognition as example, we adopt the classic Uniform Local Binary Pattern (ULBP) feature [12] and build histogram of D_i, $i = 1, 2, \ldots, n$ to estimate the distribution of genuine and imposter as shown in **Fig.1 (a)**.

From **Fig.1 (a)**, we can learn that the proportion of genuine is higher than imposter in the interval [0,3]. Based on such observation, we increase the weight of the dimensions with smaller distance to enhance the separability between genuine and imposter. Such idea can be realized to use the CD metrics with a appropriate parameter σ as shown in Eqn.(7).

$$D_{dim}(X_f, Y_f) = \frac{1}{n} \sum_{i=1}^{n} \kappa_\sigma(x_i - y_i) \tag{7}$$

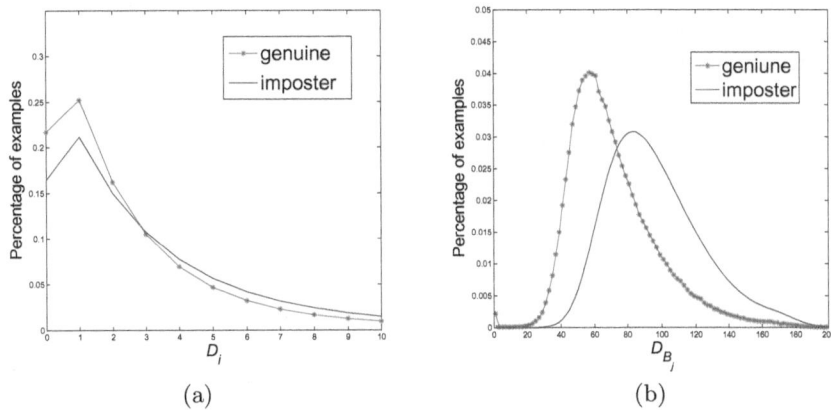

Fig. 1. (a) The distribution of D_i from genuine and imposter. **(b)** The distribution of D_{B_j} from genuine and imposter. The horizontal axis of two sub-figure denote the variation of D_i, D_{B_j} respectively. The vertical axis denotes the percentage of genuine and imposter.

In the ideal case, it only has one point of intersection between histograms. Actually, it will appear multi-points in the practical application. More detailed discussion in Section 3.3.

3.2 Maximum Correntropy Criterion

To solve the problem of kernel size selection especially, we employ MMC to deal with it. In this section, we will have a review about MMC.

MCC is a cost function defined to estimate parameters [11]. For different problems, the physical meaning and number of parameter are different. For our problem, there is one parameter kernel size needs to be estimated as formulated in Eqn.(8):

$$\hat{\theta} = \max_{\theta} \frac{1}{N} \sum_{i=1}^{N} \kappa_\sigma(e_i) \tag{8}$$

where $\kappa_\sigma(\cdot)$ is a Gaussian kernel, σ is the kernel size and e_i denotes the error of the i-th components of the feature vector.

It can be learned that MCC maximizes the weighting factor of the smaller error components and weakens the weighting factor of the larger error components. Since the the kernel size σ denotes the scale of error of the components. Hence, in feature matching, it has a prominent influence to the final distance.

3.3 Close-Form Solution for Kernel Size Selection

In Eqn.(7), the kernel size actually decides the weight of each dimension. Based on MCC, we propose a learning method of kernel size. The two histograms of genuine and imposter in **Fig.1.(a)** are denoted respectively as h_s, h_d . $h(i)$ means the value of i-th bin in the histogram h and B is the bin number of histogram. It follows the properties below:

$$\sum_{i=0}^{B} h_s(i) = 1 \quad \sum_{i=0}^{B} h_d(i) = 1 \tag{9}$$

$$\exists 0 < t < B, \quad st \begin{cases} h_s(x) > h_d(x) & x \leq t \\ h_s(x) < h_d(x) & x > t \end{cases} \tag{10}$$

When there is one point of intersection between histograms e.g. Fig.1 (a), it is fairly straightforward to learn t. However, histograms cross each other at two or more points in fact. In the case of multiple points of intersection, we acquire the approximate effect with calculating t by Eqn.(11):

$$t = arg \max_{m} (\sum_{i=0}^{m} (h_s(i) - h_d(i))) \quad 0 < m < B \tag{11}$$

Applying the idea of MMC for this problem, we can obtain an objective function as follow:

$$F = \max_{\sigma} \sum_{i=0}^{B} (h_s(i) - h_d(i)) \kappa_\sigma(i) \tag{12}$$

In Eqn(12), based the basic point that the smaller error term owns the larger weight factor, we set a gaussian kernel to ensure it. According to maximizing the score of positive samples, it will be beneficial for our task.

Selecting the kernel size is equivalent to selecting the bandwidth. In such case, σ plays a role of low pass filter to preserve the main energy. Moreover, with i increasing, the value of $\kappa_\sigma(i)$ will decreases much faster. So it dose not matter to remove the second part in Eqn.(13).

$$\sum_{i=0}^{B} (h_s(i) - h_d(i)) \kappa_\sigma(i)$$

$$= \sum_{i=0}^{t} (h_s(i) - h_d(i)) \kappa_\sigma(i) - \sum_{j=t+1}^{B} (h_d(i) - h_s(i)) \kappa_\sigma(i) \tag{13}$$

Thus the objective function is reformed as

$$\tilde{F} = \max_{\sigma} \sum_{i=0}^{t} (h_s(i) - h_d(i)) \kappa_\sigma(i) \tag{14}$$

On one hand, there is no the analytical solution for the object function in Eqn.(14). On the other hand, the enumeration method is time consuming and impractical to solve this problem. Hence, we approximate the objective function by the following form to obtain an acceptable solution.

$$\widetilde{F} = \max_{\sigma} \sum_{i=0}^{t} \ln((h_s(i) - h_d(i))\kappa_\sigma(i)) \tag{15}$$

This sub-optimal solution can simplify the solving process and prevent overfitting. Another advantage of this solution is that it can be easily applied in the incremental learning and real-time identification.

The solving process is as follows:

$$\sum_{i=0}^{t} \ln((h_s(i) - h_d(i))\kappa_\sigma(i))$$

$$= \sum_{i=0}^{t} \ln((h_s(i) - h_d(i)) \cdot \frac{1}{\sqrt{2\pi}}) + \sum_{i=0}^{t} \ln\frac{1}{\sigma} + \sum_{i=0}^{t} -\frac{i^2}{2\sigma^2} \tag{16}$$

Due to $\sum_{i=0}^{t} \ln((h_s(i) - h_d(i)) \cdot \frac{1}{\sqrt{2\pi}})$ is constant, Eqn.(16) derivation of σ :

$$\frac{\partial}{\partial \sigma}(\sum_{i=0}^{t}(\ln(h_s(i) - h_d(i))\kappa_\sigma(i)))$$

$$= \sum_{i=0}^{t} -\frac{1}{\sigma} + \sum_{i=0}^{t} \frac{i^2}{\sigma^3}$$

$$= -\frac{t+1}{\sigma} + \frac{t(t+1)(2t+1)}{6\sigma^3} \tag{17}$$

So it takes the maximum with $\sigma = \sqrt{\frac{t(2t+1)}{6}}$

4   Modified Correntropy

Though in orthogonal spaces the components of feature are supposed to be independent, there are more features whose components are correlated in practical applications such as images and video analysis etc. Hence, there exist such subsets of components of the feature that there is correlation inside the subset but no correlation among subsets. It is possible that the distance on the level of subsets is more powerful. Based on the above analysis, we develop the classic correntropy into a modified block-wise metrics.

Assume that the feature is divided into M unoverlapped subsets or blocks $R_1, R_2, \ldots, R_M$ and O_i denotes the dimension number of the i-th block. So the feature subset correspond to the i-th block is denoted as $R_i = (x_{i,1}, x_{i,2}, \ldots, x_{i,O_i})$. Let $R_i^X = (x_{i,1}, x_{i,2}, \ldots, x_{i,O_i})$ and $R_i^Y = (y_{i,1}, y_{i,2}, \ldots, y_{i,O_i})$ denote the feature subset of the i-th block from two samples X_s and Y_s.

We define the distance of the i-th block from two samples X_s and Y_s as follows:

$$D(R_i^X, R_i^Y) = \sum_{j=1}^{O_i} |x_{i,j} - y_{i,j}| \qquad j = 1, 2, \ldots, M \qquad (18)$$

Let $X_f = (x_1, x_2, \ldots, x_n)$ and $Y_f = (y_1, y_2, \ldots, y_n)$ be features of two images. Suppose the feature contains M blocks. n is the dimension of image feature. The dimension of the j-th block is O_j. The distance of the j-th block is defined as

$$D_{B_j} = \sum_{i=1}^{O_j} |x_{A_{j-1}+1} - y_{A_{j-1}+1}| \qquad j = 1, 2, \ldots, M \qquad (19)$$

where $A_{j-1} = \sum_{i=1}^{j-1} O_i$.

The block-wise correntropy (BC) is defined as

$$D_{block}(X_f, Y_f) = \frac{1}{M} \sum_{i=1}^{M} \kappa_\sigma(D_{B_j}) \qquad (20)$$

Like in Section 3.1, we build a block-wise histogram to learn the distribution of D_{B_j}, $j = 1, 2, \ldots, M$. An example with the ULBP feature is shown in **Fig.1 (b)**. For ULBP feature, the blocks are just the histogram of the local region of images. **Fig.1.(b)** shows that the block-wise separability is more prominent than that of the dimension level. Hence, the block-wise metrics with adaptive kernel size will be more powerful.

Setting weights for every local region to improve the performance is very popular. Ahonen et al.[12] adopt the weighted distance to measure LBP and different weights are set to every block. The weights are assigned based on the recognition rate of the blocks. So the weight is obtained in the local level. The block-wise correntropy metrics can also be regarded as learning adaptive weights for every blocks. However, the proposed metrics works on a global level, in the sense that the kernel size is the same to all blocks.

5   Experimental Analysis

5.1   Experimental Settings

To validate the proposed metrics, we perform face recognition experiments with the LBP features. The LBP features are sequences of histograms of LBP of blocked face regions. Hence, it is very suitable for verifying the block-wised metrics. We adopt two most popular and successful LBP features, i.e. the ULBP and the Local Gabor Binary Pattern (LGBP) feature. The latter is the benchmark feature in face recognition [13].

The tests are performed on frontal facial subsets of two benchmark face databases, i.e. the Facial Recognition Technology (FERET) and the Face Recognition Grand Challenge (FRGC 2.0) as listed in Table 1. We take 4 images per subject from FERET and 6 images per subject from FRGC so that half of the samples can be used to learn the parameters. From both databases, the number of subjects is the maximum with available 4 or 6 frontal face images. Face images of the two subsets contains different expressions and illumination variations. All the face images have been registered and preprocessed to the size of 140*160. Some examples of the two subsets are shown as in Fig.2.

Table 1. Face Databases

	Subject Number	Image/ Subject	Size of Training Set	Size of Testing Set
FERET	256	4	2	2
FRGC	459	6	2	4

(a)Examples of FERET subset

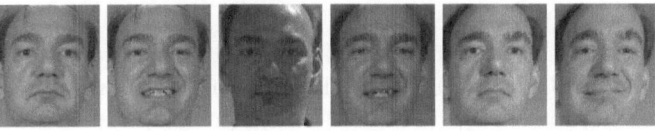

(b)Examples of FRGC subset

Fig. 2. Examples of Experimental Data

5.2 Comparative Results

We compare the proposed metrics with several most frequently adopted distances for LBP feature including L_1 distance, χ^2 distance, Cosine distance. For LBP features, the results of block-wise L_1 , χ^2 distance are the same as the non-block version. For cross validation, we select 10 groups of training set and testing set randomly and calculate the average recognition rate with the Nearest Neighbor classifier. For both ULBP and LGBP, we take the LBP histogram obtained from the same sub-image as the blocks for BC. The results are summarized in **Table 2** and **Table 3**. The recognition rate in **Table 2** and **Table 3** demonstrates the effectiveness of both BC and CD.

Table 2. The comparison of different metrics on FERET

Type of Metric	Metric	$ULBP_{2,8}(7*8)$	$ULBP_{2,8}(14*16)$	$LGBP$
Accordance	L_1	0.8765	0.8982	0.8988
	χ_2	0.8716	0.8943	0.8947
Normal	Cosine	0.8168	0.84102	0.8891
	CD	**0.8869**	**0.8554**	**0.8943**
block-wise	Cosine-block	0.8589	0.8822	0.8949
	BC	**0.9044**	**0.9146**	**0.9136**

Table 3. The comparison of different metrics on FRGC

Type of Metric	Metric	$ULBP_{2,8}(7*8)$	$ULBP_{2,8}(14*16)$	$LGBP$
Accordance	L_1	0.6057	0.6486	0.6704
	χ_2	0.6075	0.6467	0.6825
Normal	Cosine	0.56879	0.6192	0.6663
	CD	**0.6126**	**0.6242**	**0.6748**
block-wise	Cosine-block	0.5882	0.6352	0.6816
	BC	**0.6259**	**0.6662**	**0.6922**

$ULBP_{R,P}(m*n)$ denotes ULBP feature with sampling radius R, sampling density P and $m*n$ sub-images, and Cosine-block denotes calculating the Cosine distance on the level of block.

The CD performs better than the other normal metrics with the low dimensional features $ULBP_{2,8}(7*8)$. For the high-dimensional features $ULBP_{2,8}(14*16)$, $LGBP$, such advantage exists but not as prominent as in the case of low dimensional features. This is due to the separability of distance of each dimension is not so significant for higher dimensional features as to lower dimensional feature. Especially for LBP features, since each dimension is a the bin value of a histogram. The higher the dimension, the lower the variance in each dimension for images with the same blocking strategy and resolution. Hence, the MCC based kernel size selection is sensitive to the disparity between positive samples and negative samples. Another reason for the phenomenon is that the higher dimensional feature contains much noise, which will deteriorate the reliability of metric learning. After all, the size of training sets is not enlarged with respect to the dimension.

However, the recognition rate of the proposed BC is the highest for all three different features and two face image sets. It achieves much higher accuracy compared with all the other normal metrics and block-wise metrics. Moreover, the proposed BC metrics is robust in the case of higher dimensional features. Such prominent advantage shows that adopting block distance discards the influence of noise and enhances the separability between genuine and imposter.

As to the computation cost, the proposed metrics need an off-line stage to obtain the kernel size. In on-line usage, according to Eqn.(7), we need to calculate

N times Gaussian kernels for the CD metrics. However, with BC metrics, only M Gaussian kernels need to be calculated as shown in Eqn.(20). In fact, since $M \ll N$, the computation cost drops vastly. Comparing to the simple metrics such L_1, BC and CD may need more extra computation. But compared with the other metrics, the computation complexity of BC is the same or even lower. After all, the BC metrics outperforms the other metrics nearly 2% in accuracy in each group of tests.

6   Conclusion

In this paper, we propose an adaptive kernel-size selection solution and a modified correntropy metrics for feature matching. Through optimizing the objective function based on MCC, we derive a close-form solution of the kernel size based on learning the separability of subsets of feature components. We extend the correntropy into a block-wise metrics BC, which takes advantage of block distance to decrease the influence of noise. The proposed metrics is tested in LBP feature spaces of face recognition experiments. Comparison with the other most frequently adopted metrics demonstrates the advantages of the proposed block-wise correntropy metrics. The experimental results also validates the proposed kernel size selection. The proposed metrics and kernel-size selection strategy can also be applied in other computer vision tasks such as image retrieval and object recognition.

Acknowledgments. The research is funded by the National Natural Science Foundation of China (No.61170155) and the Shanghai Leading Academic Discipline Project (No.J50103).

References

1. Rubner, Y., Puzicha, J., Tomasi, C., Buhmann, J.: Empirical evaluation of dissimilarity measures for color and texture. Computer Vision and Image Understanding 84, 25–43 (2001)
2. Smyth, P.: Clustering sequences with hidden markov models. In: Advances in Neural Information Processing Systems, NIPS (1997)
3. Slonim, N., Tishby, N.: Agglomerative information bottleneck. In: Advances in Neural Information Processing Systems, NIPS (1999)
4. Zuo, W., Zhang, D., Wang, K.: An assembled matrix distance metric for 2dpca-based image recognition. Pattern Recognition Letters 27, 210–216 (2006)
5. Liu, C.: The bayes decision rule induced similarity measures. IEEE Transactions on Pattern Analysis and Machine Intelligence 29, 1086–1090 (2007)
6. Zhu, C., Wen, F., Sun, J.: A rank-order distance based clustering algorithm for face tagging. In: Computer Vision and Pattern Recognition, CVPR (2011)
7. Xing, E.P., Ng, A.Y., Jordan, M.I., Russell, S.J.: Distance metric learning with application to clustering with side-information. In: Advances in Neural Information Processing Systems, NIPS (2002)

8. Jain, P., Kulis, B., Grauman, K.: Fast image search for learned metrics. In: Computer Vision and Pattern Recognition, CVPR (2008)
9. Santamaria, I., Pokharel, P.P., Principe, J.C.: Generalized correlation function: definition, properties, and application to blind equalization. IEEE Transactions on Signal Processing 54, 2187–2197 (2006)
10. Liu, W., Pokharel, P.P., Principe, J.C.: Correntropy: A localized similarity measure. In: International Joint Conference on Neural Networks (2006)
11. Liu, W., Pokharel, P.P., Principe, J.C.: Correntropy: Properties and applications in non-gaussian signal processing. IEEE Transactions on Signal Processing 55, 5286–5298 (2007)
12. Ahonen, T., Hadid, A., Pietikainen, M.: Face recognition with local binary patterns. IEEE Transactions on Pattern Analysis and Machine Intelligence 28, 2037–2041 (2006)
13. Zhang, W., Shan, S., Gao, W., Chen, X., Zhang, H.: Local gabor binary pattern histogram sequence (lgbphs): a novel non-statistical model for face representation and recognition. In: International Conference on Computer Vision (2005)

Vitality Assessment of Boar Sperm Using an Adaptive LBP Based on Oriented Deviation

Oscar García-Olalla, Enrique Alegre,
Laura Fernández-Robles, and María Teresa García-Ordás

University of León
{ogaro,enrique.alegre,l.fernandez,mgaro}@unileon.es

Abstract. A new method to describe sperm vitality using a hybrid combination of local and global texture descriptors is proposed in this paper. In this regard, a new adaptive local binary pattern (ALBP) descriptor is presented in order to carry out the local description. It is built by adding oriented standard deviation information to an ALBP descriptor in order to achieve a more complete representation of the images and hence it has been called ALBPS. Regarding semen vitality assessment, ALBPS outperformed previous literature works with an 81.88% of accuracy and it also yielded higher hit rates than the LBP and ALBP base-line methods. Concerning the global description of sperm heads, several classical texture algorithms were tested and a descriptor based on Wavelet transform and Haralick feature extraction (WCF13) obtained the best results. Both local and global descriptors were combined and the classification was carried out with a Support Vector Machine. Therefore, our proposal is novel in three ways. First, a new local feature extraction method ALBPS is introduced. Second, a hybrid method combining the proposed local ALBPS and a global descriptor is presented outperforming our first approach and all other methods evaluated for this problem. Third, vitality classification accuracy is greatly improved with the two former texture descriptors presented. F-Score and accuracy values were computed in order to measure the performance. The best overall result was yielded by combining ALBPS with WCF13 reaching a F-Score equals to 0.886 and an accuracy of 85.63%.

1   Introduction

Sperm assessment is an essential task for porcine industry. The huge demand of alimentary products based on pork meat has resulted in lots of companies around the world trying to obtain high quality goods at the lower price available. In most cases, artificial insemination is used for creating new litter of pigs and, for this reason, the semen used must have a quality as higher as possible. Assessing the semen and checking that a sample has a high proportion of alive spermatozoa is a routine procedure in almost every single Semen Production Center. Nowadays, the automatic vitality assessment is carried out using fluorescent stains what it is time consuming and expensive due to the required equipments. In this paper we present a method that allows to assess the semen vitality using phase contrast images, without fluorescence, with very promising results.

J.-I. Park and J. Kim (Eds.): ACCV 2012 Workshops, Part I, LNCS 7728, pp. 61–72, 2013.
© Springer-Verlag Berlin Heidelberg 2013

Several works have addressed some of the problems related to the semen analysis using digital image processing. Most of them uses CASA (Computer-Assisted Semen Analysis) systems for evaluating the sperm motility [1] or for studying motility patterns among sperm cell, morphology and boar fertility [2,3]. Other research lines are focused on implementing algorithms to characterize the spermatozoa shape by using spectral approaches [4,5], or they have been looking for subpopulations using shape descriptors over the head of the spermatozoa [6]. But there are not CASA systems which deal with vitality assessment using only phase contrast images.

Texture analysis and a number of classification methods have been used successfully in the literature applied to a wide range of fields but there are few computer vision works which deal with boar sperm analysis. In general, computer-based systems designed for semen analysis tasks should reliably segment the heads of the spermatozoa [7], extract the patterns which characterise them and finally classify those patterns in order to estimate how many dead spermatozoa are present in the sample. There are some works using texture or shape analysis to classify the spermatozoon acrosome as intact or damaged. Those approaches evaluate the acrosome integrity in different ways, sometimes using complex descriptors such as the Curvelet transform [8] and, other times, evaluating a broad range of texture and based-moments descriptors [9]. As the final goal is not to correctly classify each espermatozoon but to know the right proportions of both classes, some authors have proposed new methods for estimating class proportions in boar semen analysis using the Hellinger Distance [10] and even they have applied the former methods to more general problems [11].

But there are few works that have evaluated the vitality of a sample classifying the spermatozoa heads as dead or alive. The more recent works are the ones proposed by Alegre et al. [12,13] that obtained a 76.80% of hit rate using texture descriptors when testing images captured at 100× and, more recently, Garca-Olalla et al. in [14] achieved a 78,67% of hit rate combining LBP and NCSR (n concentric squares resized).

In this paper we proposed a new algorithm called ALBPS based on the Adaptive Local binary pattern combined with the oriented standard deviation vector. These algorithm has been evaluated using a sperm head database obtaining the best results ever achieved.

The rest of the paper is organized as follows: Section 2 describes the followed methodology. The selected dataset, experiment setup and results achieved with the different descriptors are presented in section 3. Finally, section 4 shows our conclusions.

2   Methodology

2.1   Global Texture Descriptors

In a first approach the used dataset, composed of images showing boar spermatozoa heads, has been characterized by several global texture descriptors.

Seven classical texture descriptors have been computed for each spermatozoa head. The first one, was a features vector made up of four statistical measures of texture taken from the gray scale original image. This vector contains the average gray level, the average contrast, a measure of uniformity and the image entropy. Other descriptors gathered the image texture information using normalize or affine moments. Specifically, the seven Hu's moments [15] and the six invariant affine moments proposed by Flusser have been obtained [16]. Furthermore, two more features vectors based on moments have been used. The first one contains nine values coming from the Legendre polynomials corresponding with the five first moments from order 0 to 2 and the third last are moments of order third and fourth. The fifth features vector for global texture is a twenty seven dimensional descriptor, made up of the Zernike orthogonal moments [17] up to fourth order, what makes nine features, and including the real, imaginary and absolute values, it sums up to 27 values.

With a different focus, the two last global texture descriptors evaluated use Haralick's features [18] obtained from the GLCMs (Grey Level Co-occurrence Matrix) that is computed on the original image and also on the first level decomposition of the wavelet transform with a Haar mother function. Using the five matrices, the original image and the four coefficients matrices from the wavelet decomposition, the two last features vectors have been computed. Therefore, the sixth global descriptor is a 65 dimensional vector which contains 13 out of the 14 features proposed by Haralick, leaving out just the maximal correlation coefficient. The last global texture description has been carried out computing just four Haralick features, such as the Energy, Contrast, Correlation and the Inverse Different Moment on the original gray scale image and the first wavelet decomposition, yielding a twenty dimensional features vector.

2.2   Local Binary Pattern

Local Binary Pattern (LBP) [19] is a gray-scale texture descriptor that extracts the local spatial structure of an image. Given a pixel, a pattern code is computed by comparing this pixel with the value of its neighbours:

$$LBP_{P,R} = \sum_{p=0}^{P-1} s(g_p - g_c)2^p \ , \ s(x) = \begin{cases} 1 \text{ if } x \geq 0 \\ 0 \text{ if } x < 0 \end{cases} \tag{1}$$

where g_c is the value of the central pixel, g_p is the value of its neighbour p, P are the number of neighbours and R is the radius of the neighbourhood.

After LBP is obtained for each pixel, in this work, a histogram is built in order to describe the whole image using $P + 2$ bins, yielding the features vector of the image. The pattern extraction process for one pixel is shown in Figure 1.

2.3   Adaptive Local Binary Pattern

In [20], Guo et al. proposed an adaptive descriptor based on Local Binary Pattern motivated by the lack of information about the orientation in the local

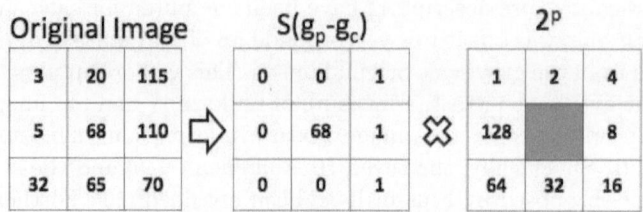

Fig. 1. Local Binary Pattern process over one gray scale pixel with P=8 and R=1. LBP code assigned to the central pixel is calculated by multiplying the output of the threshold function by the term 2^p for each neighbour pixel and then summing all those values.

binary pattern approach. Their method takes into account the oriented mean and standard deviation of the local absolute difference in order to make the matching more robust against local spatial structure changes. To minimize the variations of the mean and standard deviation of the directional differences, Guo et al. proposed a scheme that minimizes the directional difference $|g_c - w_p * g_p|$ along different orientations adding the parameter w.

The objective function is defined as follows:

$$w_p = arg_w \min \left\{ \sum_{i=1}^{N} \sum_{j=1}^{M} |g_c(i,j) - w \cdot g_p(i,j)|^2 \right\} \tag{2}$$

where w_p is the weight element used to minimize the directional difference and N and M are the number of rows and columns in the image respectively. Each weight w_p is estimated along one orientation $2p\pi/P$ for the whole image.

To solve this, Guo et al. used LSE (Least Squared Estimation) technique to optimize the w weight parameter vector.

Finally, ALBP method is:

$$LBP_{P,R} = \sum_{p=0}^{P-1} s(g_p - w_p \cdot g_c)2^p \ , \ s(x) = \begin{cases} 1 \text{ if } x \geq 0 \\ 0 \text{ if } x < 0 \end{cases} \tag{3}$$

2.4 Proposed Method: ALBPS

In [20], the oriented mean and standard deviation were used in the matching algorithm to improve the classification performance. However, the ALBP method proposed by Guo et al. does not take into account these statistical values to compute the image descriptor, instead they were only applied to minimize the directional difference along the different orientations using the weight parameter w_p. Our proposal includes the standard deviation information not in the matching method but in the descriptor algorithm and it is called ALBPS on that account. In addition, whereas a 1 by 1 matching technique was proposed by Guo et al., our scheme uses a Support Vector Machine algorithm in order to classify a descriptor. This is a huge advantage in most cases because, in this way, it is possible to use a fast and powerful classifier that will perform very well when the training set is big enough.

The standard deviation vector σ is obtained using the equation 4.

$$\sigma_p = \sqrt{\sum_{i=1}^{N}\sum_{j=1}^{M}(g_c(i,j) - g_p(i,j) - \mu_p)^2/(M \cdot N)} \tag{4}$$

where N and M are the numbers of rows and columns respectively, $g_c(i,j)$ is the center pixel at position (i,j), $g_c(i,j)$ is neighbourhood of $g_c(i,j)$ lying along orientation $2p\pi/P$ with radius R and μ_p the oriented mean obtained using:

$$\mu_p = \sum_{i=1}^{N}\sum_{j=1}^{M}|g_c(i,j) - g_p(i,j)|/(M \cdot N) \tag{5}$$

Our proposed descriptor is obtained by concatenating the $P+2$ bins histogram values of LBP approach together with the P-dimensional standard deviation vector, yielding a descriptor of $2P+2$ features with P the size of the neighbourhood.

We have selected the standard deviation because this statistic can reflect the high difference of homogeneity seen in the dead heads (heterogeneous texture) in contrast with the alive ones (homogeneous texture). We preferred this statistic rather than the mean because sometimes dead heads present black and white dots that can be counteracted when calculating the mean value. Go to the first row of figure 3 to see examples of dead and alive spermatozoa heads.

3   Experiments

3.1   Dataset

The lack of publicly available databases of dead and alive boar sperm images forced us to collect an image dataset. This set of images has been captured in CENTROTEC, an Artificial Insemination Center that is a University of Leon spin-off. The sperm was obtained from boars of three different races: Piyorker, Large White and Landrace. 450 pairs of images have been captured using a Nikon Eclipse microscope and a Baster A312f camera of progressive scan. Each of these pairs contains an image in positive phase contrast and a fluorescent image obtained using two different stains: propidium iodide (PI) that dyes dead spermatozoa as red and dichlorofluorescein (DCF) for turning green the alive spermatozoa. We encourage the reader to see more about the sample preparation in [21]. We have captured the phase contrast images for developing and testing the texture descriptors evaluated on the proposed method. The fluorescent images were used to obtain the ground truth in order to label all the heads in the data set. Examples of this captures can be shown in figure 2.

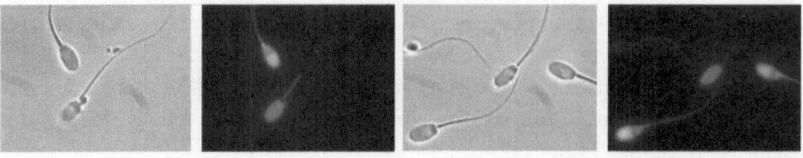

Fig. 2. Two pairs of images captured. In the left of each pair the gray scale images and in the right the fluorescent ones.

After labelling all the images, each head has been automatically registered in order to assure scale and rotation invariance. First of all, the heads have been rotated to its vertical position. This is performed by relating an sperm head with an ellipse and correcting the orientation of the major axis to achieve verticality. Then, the image has been right and left cropped leaving head's pixels untouched. Afterwards, the tail coordinates has been detected. Evaluating if the tail is placed in the bottom half or in the top half of the image will let us know if the spermatozoon has its head up or down respectively. In the second case, the image has been flipped, leading to equal orientations. Then, the image has been up and down cropped leaving head's pixels intact.

Finally, a 3×3 texture range filter has been applied over the whole dataset in order to reduce the non-informative areas and therefore facilitate the subsequent dataset description and classification. Figure 3 shows gray scale dead and alive heads and their filtered outputs.

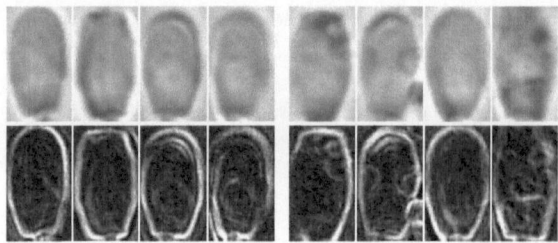

Fig. 3. The first row shows registered gray scale heads and the second row their range filtered outputs. The first four images are examples of alive heads and the last four of dead ones.

3.2   Experimental Setup

Once the 450 images are range filtered and the ground truth vector is obtained using the fluorescent images, a cross validation algorithm has been implemented in order to avoid biased results. First of all, a 20% random subset of the total number of images is kept back in order to get the test results. With the rest of the images, a 10-fold crossvalidation has been carried out. Classification was accomplished using Support Vector Machine (SVM) with Least Squares (LS) training algorithm.

Since this dataset is skewed due to the high number of alive heads in contrast with the low number of dead ones, F-Score has been used as one of the quality metrics, and it was applied over the skewed class, in this case, the dead images subset.

F-Score has been computed as $F-Score = 2 \cdot Precision \cdot Recall/(Precision + Recall)$, where Precision has been computed as $Precision = TP/(TP + FP)$ and Recall as $Recall = TP/(TP + FN)$, being TP the number of true positives in the classification, TN the number of true negatives and, FP and FN the false positives and false negatives respectively. Note that the positive class is the one with less elements in the training set, in this instance the positive class corresponds to the dead heads.

F-Score results are in the range $[0, 1]$ where values near 0 indicate a poor classification and values close to 1 show a good performance.

As the process of selecting the training and cross validation subsets is a random one, sometimes the number of images in each class could be quite balanced. For this reason, the accuracy measure has been taken into account, and computed as $Accuracy = (TP + TN)/(TP + FP + TN + FN)$

Therefore, we obtained F-Score and accuracy measures for all 10-folds combinations of training and cross validation datasets. Afterwards, the classifier which outperformed the others both in terms of accuracy and F-Score has been selected as the best classifier and its parameters were used to classify the test set in order to get a more reliable performance. By using a cross validation set instead of directly a test set we avoid that the decision about the best classifier is influenced by the random cross validation set and the classifier cannot generalize well to future test samples.

In figure 4 a scheme of this process is showed step by step.

3.3   Experimental Results

In this subsection, we show the performance evaluation results on the proposed description method in terms of the F-Score and accuracy using our own dataset. All experiments are carried out using Support Vector Machine with Least Squares and a linear kernel. Experiments using different kernels have been performed obtaining worse results than with the linear one.

Performance Evaluation Using Global Texture Descriptors
As it was explained in section 2.1, in our first approach several well-known global texture descriptors have been evaluated. In table 1 (left) it is possible to see the F-Score, Precision, Recall and Accuracy achieved with these descriptors whereas in figure 5 (left) F-Score and accuracy results and how they are directly related are shown graphically.

As it can be noticed, using WCF13 the performance improved compared to the rest of global descriptors, yielding both the best F-Score value and the best accuracy. In contrast, values from Hu, Zernike and Flusser moments are quite low, Hu just obtained a 65% of accuracy which is an unacceptable result for this

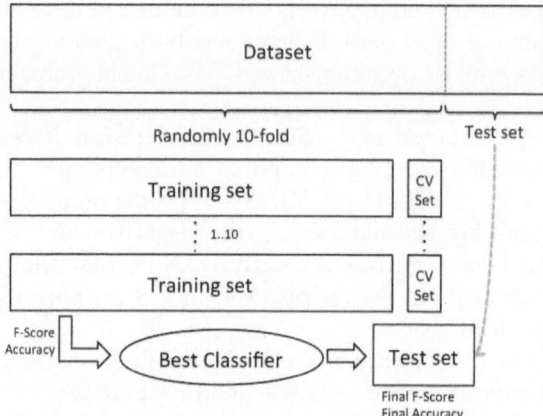

Fig. 4. Scheme of the experimental setup

Table 1. Performance of global (left) and local (right) texture descriptors LBP and ALBP using $R = 1$, $P = 8$ and $R = 2$, $P = 16$ as neighbourhoods

	WCF13	WCF4	Statistical	Legendre	Flusser	Zernike	Hu	ALBP2,16	ALBP1,8	LBP1,8	LBP2,16
F-Score	**0.800**	0.795	0.783	0.780	0.739	0.736	0.720	**0.737**	0.683	0.674	0.603
Precision(%)	**80.43**	79.54	78.72	71.00	75.58	71.38	66.24	**73.26**	67.47	67.05	60.00
Recall(%)	**79.57**	79.54	78.10	86.59	72.65	76.29	78.76	**74.12**	69.14	67.82	60.67
Accuracy(%)	**76.88**	76.75	75.69	75.00	70.88	70.19	65.44	**71.88**	67.50	64.38	55.63

problem. Therefore, we can conclude that global texture descriptors offer poor results for assessing the vitality of boar semen samples.

Performance Evaluation Using Local Texture Descriptors

In this experiment, we used the local texture descriptors LBP and the adaptive version ALBP proposed by Guo et al. [20].

Two different neighbourhoods, $R = 1$, $P = 8$ and $R = 2$, $P = 16$ have been used, to measure F-Score and accuracy in both cases. Their performance can be seen in table 1 (right) and figure 5 (right).

ALBP behaves better than LBP in all cases. ALBP2,16 obtains similar results to global descriptors, with a F-Score equals to 0.737 and an accuracy of 71.88%, outperforming in more than a 15% the accuracy of the classical LBP2,16. Nevertheless, global WCF13 descriptor outperforms ALBP approach.

ALBPS Compared with Previous Local Texture Descriptors

In this experiment, the performance using our proposed method, ALBPS, which includes oriented standard deviation information for the images description has been assessed. To measure the performance of our proposal, we compared it with LBP and ALBP methods. Results when concatenating LBP histogram from the original LBP method with oriented standard deviation vector (LBPS) have also been obtained.

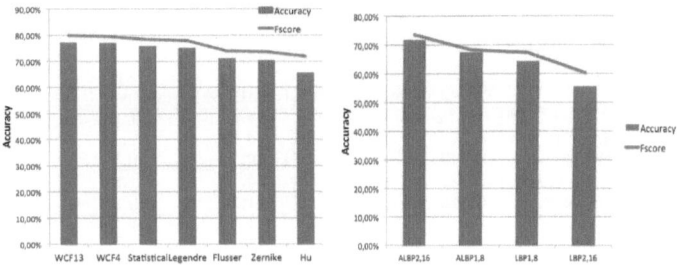

Fig. 5. Performance of different global texture descriptors (left) and performance of local texture descriptors LBP and ALBP using $R = 1$, $P = 8$ and $R = 2$, $P = 16$ as neighbourhoods (right). The F-Score is extended to the range $[0-100]$ in order to preserve the legibility of the graphic.

In table 2 and figure 6 (left), a comparison between ALBP2,16 that is the best previous local texture descriptor and our proposed descriptors is shown. Adding oriented standard deviation to local texture descriptors, LBPS and ALBPS, improves both original LBP and ALBP methods which verifies the effectiveness of our proposal. The best overall result is achieved with ALBPS with $R = 2$, $P = 16$ (ALBPS2,16) yielding a F-Score equals to 0.842 and a 81.88% of accuracy which means an improvement of 14.25% in F-Score and of 13.91% in accuracy over the base method, ALBP2,16. Moreover, we would like to highlight that ALBPS2,16 also outperforms global texture description, specifically, by a 5.25% in F-Score and by a 6.5% in accuracy with regard to WCF13. It is also noticeble that ALBPS2,16 also outperforms the previous related works found in the literature since the best approach [14] obtained a hit rate of just 78.67%. Therefore, in the carried out experiments is clear that our proposed method outperforms global traditional descriptors, previous local texture descriptors based on LBP and previous related works.

A New Improvement: Combining Global and Local Features

Our last experiment consisted of combining the analysed local descriptors with the best outperforming global features into a new hybrid features vector. We intended to introduce global context to resolve ambiguities that can occur locally when an image has multiple similar regions. Consequently, WCF13 and WCF4 were merged with the studied local descriptors yielding the results shown in table 3 and figure 6 (right). The best overall result was achieved when combining WCF13 and ALBPS2,16 reaching a F-Score of 0.886 and a 85.63% of accuracy outperforming the results obtained with local and global texture separately. Particularly, WCF13+ALBPS2,16 improves F-Score value in a 5.23% and accuracy in a 4.58% with regard to our individual local descriptor ALBPS2,16.

It is important to note the high value of recall (89%) obtained, which means that the algorithm detects a high percentage of dead heads. Specifically, it only misclassifies a 11% of them as alive ones. Since there are many more images of alive heads than dead ones in our dataset, predicting that a new test image is

Table 2. Performance of our proposed texture descriptors compared with the best of previous local texture descriptors ALBP2,16

	F-Score	Precision (%)	Recall (%)	Accuracy (%)
ALBPS2,16	**0.842**	**83.70**	**84.62**	**81.88**
ALBPS1,8	0.753	76.50	74.27	72.13
LBPS2,16	0.747	73.61	76.17	70.50
ALBP2,16	0.737	73.26	74.12	71.88
LBPS1,8	0.710	71.07	71.10	69.25

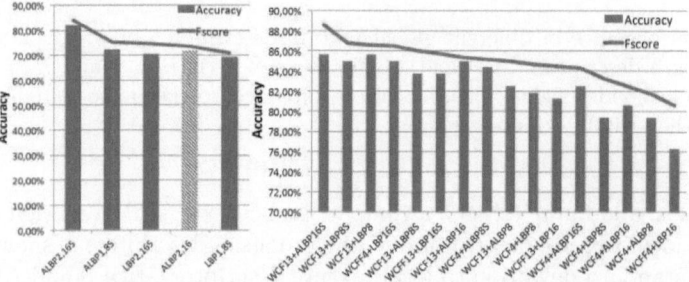

Fig. 6. Performance of our proposed texture descriptors and the best of previous local texture descriptors ALBP2,16 (dotted bar) (left) and performance of hybrid global+local texture descriptors (right). The F-Score is extended to the range [0-100] in order to preserve the legibility of the graphic.

Table 3. Performance of hybrid global+local texture descriptors

	F-Score	Precision (%)	Recall (%)	Accuracy (%)
WCF13+ALBPS2,16	**0.886**	**88.12**	**89.00**	**85.63**
WCF13+LBPS1,8	0.867	89.77	84.04	85.00
WCF13+LBP1,8	0.865	93.67	80.43	85.63
WCF4+LBPS2,16	0.865	87.50	85.56	85.00
WCF13+ALBPS1,8	0.860	84.21	87.91	83.75
WCF13+LBPS2,16	0.857	80.41	91.76	83.75
WCF13+ALBP2,16	0.854	86.42	84.34	85.00
WCF4+ALBPS1,8	0.852	87.80	82.76	84.38
WCF13+ALBP1,8	0.850	85.87	84.04	82.50
WCF4+LBP1,8	0.847	86.02	83.33	81.88
WCF13+LBP2,16	0.845	82.83	86.32	81.25
WCF4+ALBPS2,16	0.843	84.27	84.27	82.50
WCF4+LBPS1,8	0.832	81.19	85.42	79.38
WCF4+ALBP2,16	0.825	82.02	82.95	80.63
WCF4+ALBP1,8	0.818	76.29	88.10	79.38
WCF4+LBP2,16	0.806	79.80	81.44	76.25

alive has a higher probability of chance than otherwise. A value of recall equals to zero would imply that the algorithm is just classifying all images as alive (no skewed class) and therefore it is obtaining a high accuracy next to 100% without being useful for the goal of our task. As a consequence, our approach is correctly classifying the database even though the existence of a skewed class.

Besides, we want to highlight that this hybrid WCF13+ALBPS2,16 descriptor classified with SVM obtains better results than the previous works found in the literature. In section 1, we found hit rates of 76.80% in [12,13] and 78.67% in [14]. Therefore, the approach presented in this paper obtains about a 8.85% of improvement over previous works.

4 Conclusions

In this paper, we proposed a new local texture descriptor ALBPS by adding an oriented standard deviation term to the ALBP descriptor. It also has been proved that adding this new term to the classical LBP its performance also increases. In addition, we have combined the local proposed descriptor, ALBPS2,16 with the global WCF13 descriptor obtaining a features vector which contains local and global information. The experimental results showed that the hybrid features extracted by the proposed method provide a better performance than previous works when using a robust SVM classification. Also, we were able to ascertain that the skewed class, the dead one, was successfully classified reaching a recall of 89%. A F-Score equals to 0.886 and an accuracy of 85.63% were yielded by WCF13+ALBPS2,16 which is a very interesting result for classifying the vitality of boar spermatozoa heads as dead or alive.

Aknowledgements. This work has been supported by grants DPI2009-08424, PR2009-0280 and via the pre doctoral FPU fellowship program from the Spanish Government.

The authors would like to thank CENTROTEC for providing us the semen samples and for their collaboration in the acquisition of the images.

References

1. Contri, A., Valorz, C., Faustini, M., Wegher, L., Caluccio, A.: Effect of semen preparation on casa motility results in cryopreserved bull spermatozoa. Theriogenology 74, 424–443 (2010)
2. Didion, B.: Computer-assisted semen analysis and its utility for profiling boar semen samples. Theriogenology 70, 1374–1376 (2008)
3. Verstegen, J., Iguer-Quada, M., Onclin, K.: Computer assiisted semen analyzers in andrology research and veterinry practice. Theriogenology 57, 149–179 (2002)
4. Beletti, M., Costa, L., Viana, M.: A spectral framework for sperm shape characterization. Computers in Biology and Medicine 35, 463–473 (2005)
5. Severa, L., Machal, L., Svabova, L., Mamica, O.: Evaluation of shape variability of stallion sperm heads by means of image analysis and fourier descriptors. Animal Reproduction Science 119, 50–55 (2010)
6. Thurston, L., Watson, P., Mileham, A., Holt, W.: Morphologically distinct sperm subpopulations defined by fourier shape descriptors in fresh ejaculates correlate with variation in boar semen quality following cryopreservation. Journal of Andrology 22, 382–394 (2001)
7. Gonzalez-Castro, V., Alegre, E., Morala-Arguello, P., Suarez, S.: A combined and intelligent new segmentation method for boar semen based on thresholding and watershed transform. International Journal of Imaging 2, 70–80 (2009)

8. Curvelet-based texture description to classify intact and damaged boar spermatozoa, Aveiro (2012)
9. Alegre, E., González-Castro, V., Aláiz-Rodríguez, R., García-Ordás, M.T.: Texture and moments-based classification of the acrosome integrity of boar spermatozoa images. Computer Methods and Programs in Biomedicine (2012)
10. González-Castro, V., Aláiz-Rodríguez, R., Guzmán-Martínez, R., Alegre, E.: Estimating class proportions in boar semen analysis using the hellinger distance. In: XIII The Twenty Third International Conference on Industrial, Engineering & Other Applications of Applied Intelligent Systems, IEA-AIE 2010 (2010)
11. González-Castro, V., Aláiz-Rodríguez, R., Alegre, E.: Class distribution estimation based on the hellinger distance. Information Sciences (2012)
12. Alegre, E., García-Olalla, O., González-Castro, V., Joshi, S.: Boar Spermatozoa Classification Using Longitudinal and Transversal Profiles (LTP) Descriptor in Digital Images. In: Aggarwal, J.K., Barneva, R.P., Brimkov, V.E., Koroutchev, K.N., Korutcheva, E.R. (eds.) IWCIA 2011. LNCS, vol. 6636, pp. 410–419. Springer, Heidelberg (2011)
13. Alegre, E., García-Ordás, M.T., González-Castro, V., Karthikeyan, S.: Vitality Assessment of Boar Sperm Using N Concentric Squares Resized (NCSR) Texture Descriptor in Digital Images. In: Vitrià, J., Sanches, J.M., Hernández, M. (eds.) IbPRIA 2011. LNCS, vol. 6669, pp. 540–547. Springer, Heidelberg (2011)
14. Garcia-Olalla, O., Garcia-Ordas, M.T., Garcia-Ordas, D., Fernandez-Robles, L., Alegre, E.: Vitality assessment of boar sperm using n concentric squares resized and local binary pattern in gray scale images. In: XXXIII Jornadas de Automatica (2012)
15. Hu, M.K.: Visual pattern recognition by moment invariants. IRE Transactions on Information Theory 8, 179–187 (1962)
16. Flusser, J., Suk, T.: Affine moment invariants: a new tool for character recognition. Pattern Recognition Letters 15, 433–436 (1994)
17. Zernike, F.: Diffraction theory of the cut procedure and its improved form, the phase contrast method. In: Physica, pp. 689–704 (1934)
18. Haralick, R.M., Shanmugam, K., Dinstein, I.: Textural features for image classification. IEEE Transactions on Systems, Man and Cybernetics 3, 610–621 (1973)
19. Ojala, T., Pietikäinen, M., Harwood, D.: Performance evaluation of texture measures with classification based on kullback discrimination of distributions. In: Proceedings of the 12th IAPR International Conference on Pattern Recognition, ICPR 1994 (1994)
20. Guo, Z., Zhang, L., Zhang, D., Zhang, S.: Rotation invariant texture classification using adaptive lbp with directional statistical features. In: 2010 17th IEEE International Conference on Image Processing (ICIP), pp. 285–288 (2010)
21. Sanchez, L., Petkov, N., Alegre, E.: Statistical approach to boar semen evaluation using intracellular intensity distribution of head images. Cellular and Molecular Biology 52, 38–43 (2006)

Background Subtraction
Based on Multi-channel SILTP

Fan Ma and Nong Sang

Institute for Pattern Recognition and Artificial Intelligence
Huazhong University of Science and Technology
Wuhan, Hubei 430074, China

Abstract. Background subtraction is the first step in many video
surveillance systems, its performance has a decisive influence on the
result of the post-processing. An effective background subtraction algorithm should distinguish foreground from the background sensitively,
and adapt to the variation of background scenes robustly, such as illumination changes or dynamic scenes. In this paper, a novel pixel-wise background subtraction algorithm is introduced. First, we propose a novel
texture descriptor named Multi-Channel Scale Invariant Local Ternary
Pattern(MC-SILTP). The pattern is cross-calculated in RGB color channels with the Scale Invariant Local Ternary Pattern operator. This descriptor does not only show an excellent performance in abundant texture
regions, but also in flat regions. Secondly, we model each background
pixel with a codebook rather than estimating the probability density
functions. The codebook is consisted of many MC-SILTP samples actually observed in the past. A lot of experiments have been done over the
proposed approach, results indicates that this approach is well balanced
in sensitivity and robustness. It can handle the tricky problem of illumination changes robustly while detecting complete objects in flat areas
sensitively. Comparison between the proposed one and several popular
background subtraction algorithms demonstrates that it outperforms the
state-of-the-art.

1 Introduction

Detecting foreground in video sequences captured by a stationary camera is a
fundamental processing in video surveillance systems, whose output will be the
groundwork of the higher-level process, such as object tracking or counting. A
popular approach to discriminate foreground objects in the scenes is background
subtraction. The basic idea of background subtraction is to build an appropriate
distribution of the features extracted from images to represent the background,
and then compare each new observation with the distribution to classify it to
the background or not.

The comprehensive application of background subtraction in diverse scenes
makes it a hot topic in computer vision. The most popular background subtraction is Mixture of Gaussians [1], it adopts more than one Gaussian distributions

J.-I. Park and J. Kim (Eds.): ACCV 2012 Workshops, Part I, LNCS 7728, pp. 73–84, 2013.
© Springer-Verlag Berlin Heidelberg 2013

to represent the intensity value of a background pixel. This approach can adapt to dynamic scenes with repetitive moving background, but it does not work very well in complex situations, such as the illumination changes gradually or suddenly, cast shadows moving along with objects, dynamic background moving with different frequency.

To deal with the tricky problems in complex scenes, a lot of effort on background subtraction has been done over the last decades. Generally speaking, improvement mainly manifests in two aspects. The first one is to introduce advanced probabilistic models to represent the background, such as the non-parametric kernel density estimation approach proposed in [2], joint domain-range density estimation in [3]. The other line is to employ a better feature representation by discovering a new robust feature descriptor [4,5,6,7] or combining two different features together [8,9,10]. The background subtraction presented in this paper belongs to the second line.

The common used features in background subtraction are intensity value [1] and RGB values [11],they are both the direct reflection of the visual information in the scenes. But there is a common drawback of them that they are too sensitive to adapt to illumination changes, and cause a lot of misclassification. To tackle this problem, some robust texture features are discovered [4,6,7]. In [4], discriminative texture feature LBP is first proposed to background subtraction, each pixel is modeled as a group of LBP histograms calculated over the neighborhoods around. This method pioneers the use of texture descriptor to handle illumination variation problems in feature level, and it is more computational efficient than [12] that employs a special algorithm to detect moving shadows. εLBP [6] and SILTP [7] are the texture features recently developed from LBP used in background subtraction, they exceed LBP in computational efficiency and tolerance to noises. Despite that both of them are very robust to illumination variations, they perform poorly in flat areas and results in some "holes" in objects. Some authors combine different features to benefit from both. For example, color and edge information is used to model the background in [13]. Authors utilize shape and color information in [8], and a multi-layer background subtraction based on color and texture described by LBP is proposed in [9]. The idea of employing a fusion of two features can be useful to a limited extent. They can not always be mutually complementary, sometimes they conflict, and shortcomings of the two features still have an influence on the background subtraction.

In this paper, we introduce a newly discovered feature called Multi-Channel Scale Invariant Local Ternary Pattern(MC-SILTP) to the background subtraction algorithm, where the scale means color scale pixel value. MC-SILTP is improved from SILTP [7] proposed by S. Liao with the idea to combine texture descriptor with color information, and it extends from spatial space to spatial and feature spaces. MC-SILTP cross-calculates the RGB SILTPs of each channel to get a more precise description of the texture. Experiments demonstrate that MC-SILTP shows all the great properties that SILTP owns, and it can do a great job especially in flat areas. As to the background modeling procedure, a quantization/clustering technique is employed, we model the background pixel with a codebook consisted of real MC-SILTP samples observed, it can deal with dynamic scenes and outperform the state-of-the-art.

The paper presents the background subtraction algorithm in the following order. In section 2, we give a detail introduction about Multi-Channel Scale Invariant Local Ternary Pattern, and compare the performance of MC-SILTP and SILTP in some situations. In Section 3, the framework for the proposed background subtraction algorithm is discussed. And section 4 shows the experimental results, and a comprehensive comparison between the proposed approach and several other background subtraction algorithms is done. We end up with a conclusion in section 5.

2   Texture Description with Multi-channel SILTP

Scale Invariant Local Ternary Pattern is a gray scale invariant texture primitive statistic, and it is a newly developed texture descriptor used in background subtraction [7]. It converts the pixels of an image into the form of binary by thresholding the gray value of the center pixel with its neighborhoods. Given a pixel located at (X_c, Y_c) , the SILTP calculates as follows:

$$SILTP_{N,R}^{\tau}(x_c, y_c) = \bigoplus_{k=0}^{N-1} S_\tau(I_c, I_k), S_\tau = \begin{cases} 01, I_k > (1+\tau)I_c \\ 10, I_k < (1-\tau)I_c \\ 00, otherwise \end{cases} . \qquad (1)$$

where I_c and $\{I_k\}_{k=0...N-1}$ correspond to the gray intensity values of the center pixel and its N neighborhood pixels. The sign $\oplus$ indicates concatenation operator of binary strings. τ is a scale factor affecting the tolerant range. The most important properties of SILTP are its computational efficiency, its tolerance against illumination changes and local image noises within a range. However, there is a common drawback of SILTP that it does not work very robustly on flat image areas, where the gray values of pixels are similar, and it is hard to describe the texture only in spatial space.

In this paper, we propose a novel texture descriptor called Multi-Channel Scale Invariant Local Ternary Pattern, which extends to feature space,and operates on the three channels of RGB images rather than the only channel of gray images to get the texture patterns, given by:

$$SILTP_R^{\tau}(x_c, y_c) = \oplus_{k=0}^{N-1} S_\tau(I_R, I_{B,k})$$
$$SILTP_G^{\tau}(x_c, y_c) = \oplus_{k=0}^{N-1} S_\tau(I_G, I_{R,k}) \ S_\tau = \begin{cases} 01, I_k > (1+\tau)I_c \\ 10, I_k < (1-\tau)I_c \\ 00, otherwise \end{cases} . \qquad (2)$$
$$SILTP_B^{\tau}(x_c, y_c) = \oplus_{k=0}^{N-1} S_\tau(I_B, I_{G,k})$$

where I_R, I_G and I_B correspond to the RGB values of the center pixel, respectively; $I_{R,K}, I_{G,K}$ and $I_{B,K}$ correspond to the RGB values of the neighborhoods. The thresholding method of MC-SILTP is as same as the original SILTP, but

MC-SILTP cross-computes the RGB SILTPs, and thresholds a pixel's color value of one color channel with the other color channel of its neighborhood pixels. In this way, it can really make great progress with SILTP. Firstly, MC-SILTP extends to feature space to make full use of color information and gets a more precise texture description. Especially in flat areas where the gray values are similar, and the gray values of spatial neighborhoods can not supply enough information. It is significant to introduce feature space to calculate the texture. Secondly, MC-SILTP enhances the relevance of each channel and gets a pattern which makes every channel closely linked to each other. This pattern can give a better reflection of the visual information. See Fig.1 for a detail illustration of the MC-SILTP operator.

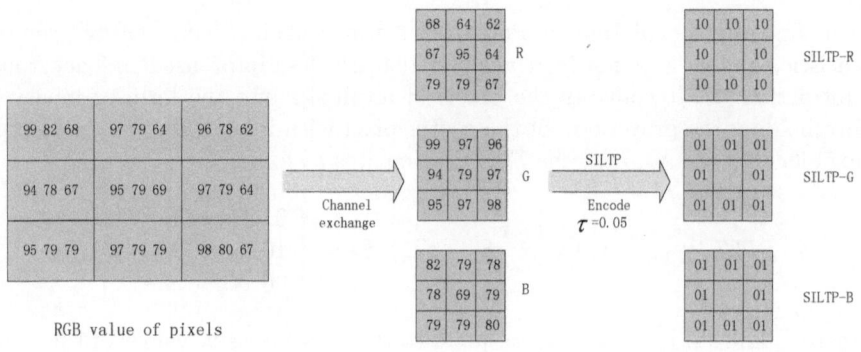

Fig. 1. Calculating the binary patterns

Take real videos "Lobby" and "ShoppingMall" from the open data set I2R [14], and apply SILTP and MC-SILTP on them to do some research and comparison. As shown in Fig.2, there are three fixed pixel positions labeled out in the scenes in the first column. Among them, the blue one is a pixel changing from background to foreground, the red one is a background pixel with illumination changes, and the green one is a background pixel with moving shadows sometimes. To the convenience of statement in illustration, We apply $\oplus$ on SILTP-R, SILTP-G and SILTP-B to get a combined MC-SILTP based on Equ.3.

$$SILTP_{RGB}^{\tau}\left(x_c, y_c\right) = SILTP_R^{\tau}\left(x_c, y_c\right) \oplus SILTP_G^{\tau}\left(x_c, y_c\right) \oplus SILTP_B^{\tau}\left(x_c, y_c\right) .$$
(3)

Both of the operators take 8 neighborhoods of the center pixel into account, so there are 16 bits in SILTP and 48 bits in MC-SILTP. We calculate the distance of two patterns by summing up the number of bits they differ from each other. For example, there are two SILTPs: 01000110 00101010 and 00010110 10101010, the distance between them is 3. An effective way to get the distance of two patterns is to make use of XOR operator.

Observe the distribution curves of SILTP and MC-SILTP of the three positions throughout the videos, we can figure out that MC-SILTP has the advantage of SILTP in the complex scenes with moving cast shadows and illumination changes, as the red and blue curves show in Fig.2 The small scaling threshold factor in SILTP can handle some illumination problems in abundant texture regions. In flat areas, the gray values decrease and become more and more similar as the shadows cast or light dims, it tenders to result in zero SILTP consisted of all zero bits. This zero pattern cannot describe the texture well and differs from the original pattern. This explains for the reason why the red and blue distribution curves of SILTP are up-and-down during the experiment. As to MC-SITLP cross-calculated on the RGB channels, although the values of the same channel approximate to each other in these situations as the gray values do, but the difference between two channels still keeps, so MC-SITLP can stay invariant by threshoding the values from different channels, and perform excellently in flat areas as the curves in Fig.2 show. Besides, MC-SITLP encodes the pattern by the same thresholding method of SILTP with an adaptive tolerative range, it shows all the great properties that SILTP owns, such as computational efficiency and robustness to local image noises.

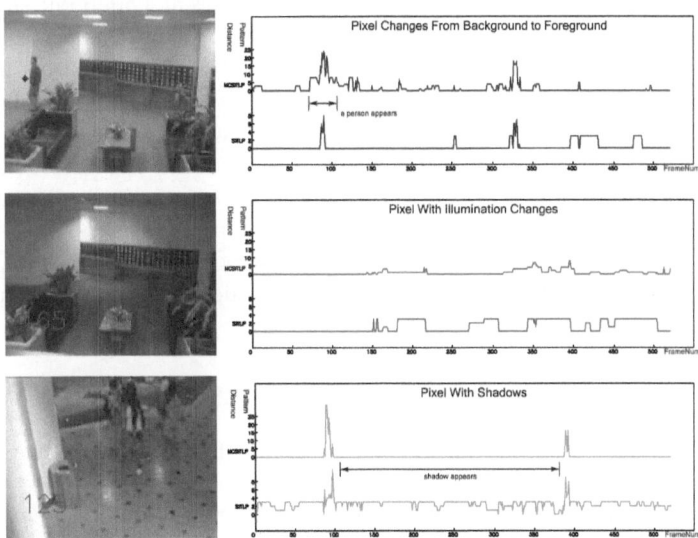

Fig. 2. Comparison of SILTP and MC-SILTP operators on three different pixels. First column: the real scenes chosen for experiments. Second column: the distribution curves of the pixels labeled out in the scenes on the first column along with time.(x axis is the frame number, y axis is the distance between the new patterns and the pattern got at frame 0).

3   Background Subtraction Based on MC-SILTP

In this section, we employ MC-SILTP to the statistical model of background in pixel-wise manner, and give a detail description of the framework for background subtraction algorithm, including background model construction and foreground detection.

Most of the mainstream background subtraction algorithms rely on probability density functions(pdfs)[1] or statistical parameters[2,3]. The innovative mechanism presented in ViBe[15] indicates that already observed samples would have a higher probability to appear again, and it makes more sense to model the background pixel with a group of already observed values than an explicit pixel distribution such as pdfs. So we construct the background model for each pixel with a codebook consisting of MC-SILTPs already being observed.

3.1   Background Model Construction

Given a color video sequence, apply MC-SILTP operator on it to get the pixel process [1] of a single pixel during the training period T, which can be defined as $\{p1, p2, ...p_T\}$. Let $M_t = \{cb_1, cb_2, ...cb_L\}$ be the codebook representing the background pixel at time t. Each pixel has a different codebook size according to the variations of the samples and the longest length is set to be L_{Max} . $cb_{i(i=1...L)} = \{p_i, c_i\}$ is the codeword, in which p_i is consisted of three RGB SILTPs $p_i = \{p_i^r, p_i^g, p_i^b\}$, and c_i is the weighting coefficient reflecting the contribution that the pattern has made to the background.

In the initial period, the codebook is empty, and add the observed patterns to the codebook with a small weighting coefficient ω_0. ω_0 is a constant for initialization, and it is set to be $1/T$.

In the train period, calculate the distance between the pattern observed p_{new} and the codewords in the codebook, then decide which codeword does the pattern match to(if there is one) .The distance of two MC-SILTPs is calculated as Equ.4

$$d(p_i, p_j) = d(p_i^r, p_j^r) + d(p_i^g, p_j^g) + d(.p_i^b, p_j^b) . \tag{4}$$

Where $d(p_i, p_j)$is define as the number of different bits between the two local patterns p_i and p_j.

A match is found if the distance between the new observation p_{new} and a pattern of the codeword is zero, and then adjust the weighting coefficients of the codewords as follows:

$$c_{k,t} = c_{k,t-1} + M\omega_0 . \tag{5}$$

Where M is 1 for the codeword matched,and 0 for the remaining codewords in the codebook.

If no match is found, add a new codeword cb_{L+1} to the codebook, and $cb_{L+1} = \{p_{L+1} = p_{new}, c_{L+1} = w_0\}$. To improve the speed of the algorithm, we relocate the codewords by the weighting coefficients in descending order.

3.2   Foreground Detection

In foreground detection procedure, first determine the number of background codewords by Equ.6

$$M = \arg\min_{r}(\sum_{i=1}^{r} c_i > T_b) \ . \tag{6}$$

Where T_b is a sampling threshold deciding how many codewords may be regarded as background. A bigger value for T_b can make the model contain more repetitive background patterns. A small value for T_b may make the most frequency pattern to represent the background.

Afterwards, compute the distances between the new pattern p_{new} and the background codewords, and then choose the smallest one d_s to be the distance of the pattern and background model. Then the pattern is classified to background or foreground by thresholding the distance d_s with the detection parameter T_s. The value of T_s can be got by experiments.

Then update the background model to keep it up with the changes in the dynamic scenes. Compare the new pattern p_{new} to all the codewords in the codebook, and get the nearest one cb_k and the distance d_k between them, the subscript k is the index number of the corresponding codeword. Compare the distance d_k with the match threshold parameter T_m and update the codebook in the following way.

If d_k is equal to 0. A same pattern is found and update the matched codeword cb_k by increasing its coefficient as Equ.7 shows, and decreasing the other codewords' coefficients by Equ.8 to make sure that the sum of all the codewords' coefficients in a codebook approximate 1.

$$c_{k,t} = (1 - \alpha)c_{k,t-1} + \alpha \ . \tag{7}$$

$$c_{i,t} = (1 - \alpha)c_{i,t-1}, i = 1...L, i! = k \ . \tag{8}$$

Where α is a learning rate. Then if d_k is between zero and T_m, the new pattern is supposed to be similar to the background. Create a new codeword $cb_{L+1} = \{p_{L+1} = p_{new}, c_{L+1} = \omega_1\}$ and add it to the codebook. Otherwise, add a new codeword $cb_{L+1} = \{p_{L+1} = p_{new}, c_{L+1} = \omega_0\}$ to the codebook. ω_0 is a low initial weight and ω_1 is a higher initial weight. Since if the new comer is similar to one of the codeword, we believe that it has a higher probability to be background, so initialize it with a higher weight ω_1.

4   Experimental Results

Four common videos for background subtraction algorithms from I2R [14] are employed to evaluate the performance of the proposed approach. These videos nearly cover all the challenges in background subtraction, such as dynamic scenes, illumination changes and moving cast shadows.

The proposed method is compared to four typical background subtraction algorithms, the classical background model based on the mixture of Gaussians [1],the recently developed $\text{PKDE}_{\text{siltp}}^{\omega=1}$ and $\text{PKDE}_{\text{mb-siltp}}^{\omega=1+2+3}$ in [7], the background model based on hybrid feature space called VKS-lab+siltp [10]. In the experiments, the scale factor τ is set to be 0.06 and choose 8 pixels from the 3*3 neighborhood region to get the MC-SILTPs. For every codebook $L_{max} = 24$, training number T = 50, initial weighting coefficient $w_0 = 1/T = 0.02$. The updating parameters are set as: $T_b = 0.7, T_s = 12, T_m = 6, w_1 = 0.05, \alpha = 0.02$. We utilize a filter to clear up the small noises less than 15 pixels in foreground mask just as the other methods have done in [7,10].

Fig.3 shows the foreground detection results of all these approaches in four representative videos, which are also used by Liao et al.[7] and Narayana et al.[10]. In the first scene AirPort, there are moving cast shadows along with pedestrians. All the methods using texture features can deal with this problem, except for MOG who is based on the gray value. Because the gray characteristics of the background completely change as soft shadow casts on the floor, and differ from the Gaussian distributions. Texture descriptors can keep invariant in such circumstances, since they are based on the gray differences between the center pixel and its neighborhoods. These differences do not vary, as the gray values of a soft shadow region change to a certain extent simultaneously.

As for the second scene Lobby, $\text{PKDE}_{\text{mb-siltp}}^{\omega=1+2+3}$ and MC-SILTP do a better job than the others. Although MOG can adapt to the illumination changes after a long time of learning, it results in a very bad performance during the learning time, and a large part of the background is misclassified as foreground. $\text{PKDE}_{\text{siltp}}^{\omega=1}$ and VKS-lab+siltp misclassify some background pixels as foreground, because SILTP can not work robustly in the flat and dark areas. The gray values of pixels in such areas are small and similar, so SILTP is very variable and even results in a bad description of the texture with zero patterns. $\text{PKDE}_{\text{mb-siltp}}^{\omega=1+2+3}$ can fix this problem by taking a fusion of three different scale SILTP operators to capture more structure information. MC-SILTP adapts to this situation in feature level without any complicated background model strategy. This operator extends from spatial space to feature space, and compares the differences between different color channels. So it can work robustly in flat and dark areas, where the differences in feature space are more reliable than that in spatial space.

In the third and forth columns of Fig.3, they are dynamic scenes indoor and outdoor. Except that MOG misclassifies the sky and part of white board, all of these methods do well with the dynamic scenes. This results demonstrate that the proposed framework for background subtraction is performing well, it can deal with dynamic scenes with a group of codewords.

According to the visual results in Fig.3, $\text{PKDE}_{\text{mb-siltp}}^{\omega=1+2+3}$ and MC-SILTP outperform the others and they are equally matched. A statistical comparison of them is done by F-measure [7], which can measure the accuracy of foreground segmentation, and it is defined as:

$$F = \frac{2*\text{TP}}{2*\text{TP} + \text{FN} + \text{FP}} . \tag{9}$$

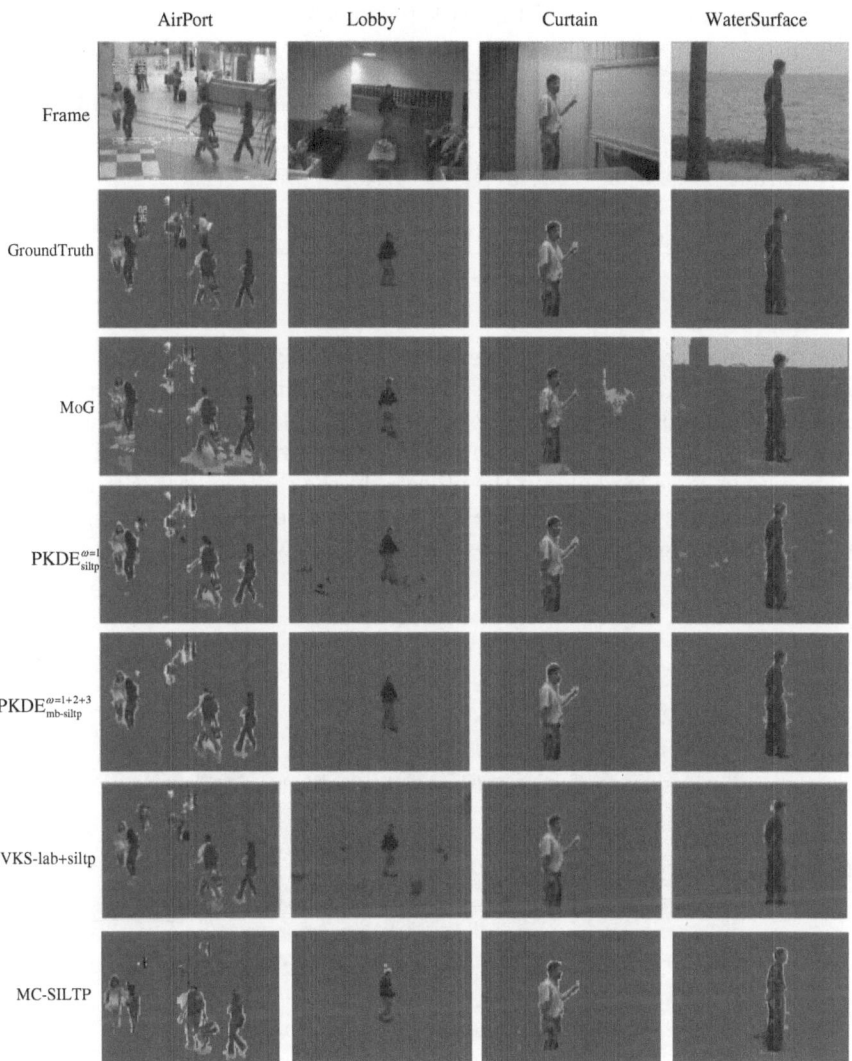

Fig. 3. Detection results on four videos

where TP is true positives in the video sequence, FN is false negatives, FP is false positives. The F-measure of these background subtraction algorithm is given by Table 1, it indicates that MC-SILTP can do better in some situations. Fig.4 gives a more detail illustration of $\mathrm{PKDE}_{\mathrm{mb-siltp}}^{\omega=1+2+3}$ and MC-SILTP.

Fig.4 shows the comparison of $\mathrm{PKDE}_{\mathrm{mb-siltp}}^{\omega=1+2+3}$ and MC-SILTP, in which $\mathrm{PKDE}_{\mathrm{mb-siltp}}^{\omega=1+2+3}$ represents the highest level of the background subtraction

Table 1. F-measure scores of five background subtraction algorithms on the four videos

Video	MoG	PKDE$^{\omega=1}_{siltp}$	PKDE$^{\omega=1+2+3}_{mb-siltp}$	VKS-lab+siltp	MC-SILTP
AirportHall	57.86	68.14	68.02	71.28	70.68
Curtain	50.53	91.16	92.40	94.07	94.43
Lobby	68.42	78.80	79.21	60.82	83.54
WaterSurface	63.52	74.30	83.15	92.16	75.72

algorithms recently. In the experiments, video "Lobby" and "Curtain" are chosen to do the research, and there are a lot of objects with textureless parts in these scenes. We can conclude from the results that MC-SILTP provides a much better performance than PKDE$^{\omega=1+2+3}_{mb-siltp}$ .PKDE$^{\omega=1+2+3}_{mb-siltp}$ often loses parts of the people and results in some big holes in the objects. In the contrast, MC-SILTP changes from spatial space to feature space. So it can not be bound by the spatial non-saliency, and still keep sensitive in textureless areas. MC-SILTP detects the entire objects.

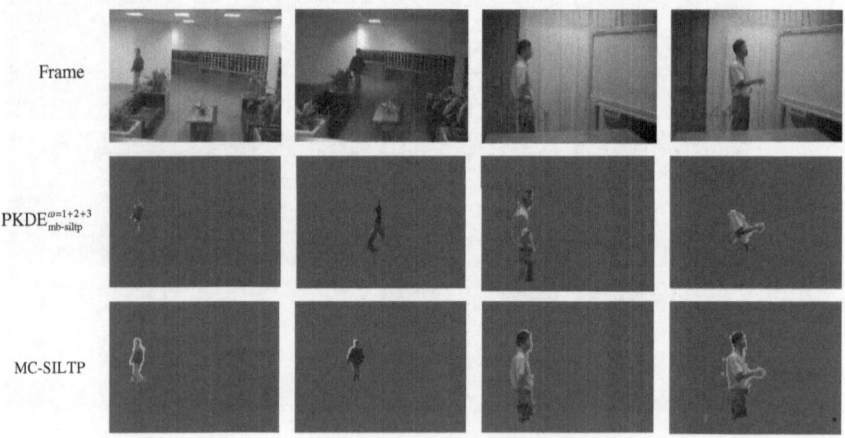

Fig. 4. Detection results of PKDE$^{\omega=1+2+3}_{mb-siltp}$ and MC-SILTP

5 Conclusion

In this paper, we propose a pixelwise background subtraction algorithm based on a novel texture feature descriptor. The feature is extracted with an operator called MC-SILTP, which is developed from a powerful local image descriptor demonstrated in [7] recently. MC-SILTP is proved to be better than SILTP by experiments, and it can work robustly on both the areas with abundant texture information and textureless areas. Then we introduce MC-SILTP to background subtraction, and propose to model the background with a group of real observed MC-SILTP samples. Qualitative and quantitative comparisons have been done between the proposed method with four other popular methods, results demonstrate that our approach outperform the state-of-the-art. Besides, this approach is very fast, because it mainly computes the distance of binary pattern and does not include much floating point operation, except for updating weighting coefficients.

In the future, we hope to do some research on the background subtraction framework, and introduce fuse multiscale spatial information to get a better performance. Furthermore, we believe that the proposed operator can be used in face recognition.

References

1. Stauffer, C., Grimson, W.: Adaptive background mixture models for real-time tracking. In: IEEE Conference on Computer Vision and Pattern Recognition, vol. 2, pp. 246–252 (1999)
2. Elgammal, A., Duraiswami, R.: Background and foreground modeling using non-parametric kernel density estimation for visual surveillance. Proceeding of the IEEE 90, 1151–1163 (2002)
3. Sheikh, Y., Shah, M.: Bayesian modeling of dynamic scenes for object detection. IEEE Transaction on Pattern Analysis and Machine Intelligence 27, 1778–1792 (2005)
4. Heikkilä, M., Pietikäinen, M.: A texture-based method for modeling the background and detecting moving objects. IEEE Transaction on Pattern Analysis and Machine Intelligence 28, 657–662 (2006)
5. Hu, W., Li, X., Zhang, X., Shi, X., Maybank, S., Zhang, Z.: Incremental tensor subspace learning and its applications to foreground segmentation and tracking. International Journal of Computer Vision 91, 303–327 (2011)
6. Wang, L., Wu, H., Pan, C.: Adaptive εLBP for Background Subtraction. In: Kimmel, R., Klette, R., Sugimoto, A. (eds.) ACCV 2010, Part III. LNCS, vol. 6494, pp. 560–571. Springer, Heidelberg (2011)
7. Liao, S., Zhao, G., Kellokumpu, V., Pietikainen, M., Li, S.: Modeling pixel process with scale invariant local patterns for background subtraction in complex scenes. In: IEEE Conference on Computer Vision and Pattern Recognition, pp. 1301–1306 (2010)
8. Liu, L., Sang, N., Huang, R.: Background subtraction using shape and colour information. Electronics Letters 46, 41–43 (2010)
9. Yao, J., Odobez, J.: Multi-layer background subtraction based on color and texture. In: IEEE Workshop on Computer Vision and Pattern Recognition, pp. 1–8 (2007)
10. Narayana, M., Hanson, A., Learned-Miller, E.: Background modeling using adaptive pixelwise kernel variances in a hubrid feature space. In: IEEE Conference on Computer Vision and Pattern Recognition (2012)
11. Kim, K., Chalidabhongse, T.: Background modeling and subtraction by codebook construction. In: International Conference on Image Processing, vol. 5, pp. 3061–3064 (2004)

12. Prati, A., Mikic, I., Trivedi, M.: Detecting moving shadows:algorithms and evaluation. IEEE Transaction on Pattern Analysis and Machine Intelligence 25, 918–923 (2003)
13. Jabri, S., Duric, Z., Wechsler, H., Rosenfeld, A.: Detection and location of people in video images using adaptive fusion of color and edge information. In: Proceeding of the IEEE International Conference on Pattern Recognition, vol. 4, pp. 627–630 (2000)
14. Li, L., Huang, W., Gu, I., Tian, Q.: Foreground object detection from videos containing complex background. In: Proceedings of the Eleventh ACM International Conference on Multimedia, pp. 2–10 (2011)
15. Barnich, O., Van Droogenbroeck, M.: A universal background subtraction algorithm for video sequences. IEEE Transactions on Image Processing 20, 1709–1724 (2011)

Elliptical Local Binary Patterns for Face Recognition

Huu-Tuan Nguyen and Alice Caplier

GIPSA Lab, Grenoble Institute of Technology, France

Abstract. In this paper, we propose a novel variant of Local Binary Patterns (LBP) so-called Elliptical Local Binary Patterns (ELBP) for face recognition. In ELBP, we use horizontal and vertical ellipse patterns to capture micro facial feature of face images in both horizontal and vertical directions. ELBP is applied in face recognition with dimension reduction step by Whitened Principal Component Analysis (WPCA). Our experiment results upon AR, FERET and Surveillance Cameras Face (SCface) databases prove the advantages of ELBP over LBP for face recognition under different conditions and with ELBP WPCA we can get very remarkable results.

1 Introduction

Face recognition is an interesting research study with many researchers from computer vision and biometrics fields. In a face recognition system, feature extraction and dimension reduction are the most important phases. LBP [1] and Gabor wavelets based are the most widely used methods for feature extraction. LBP label of each pixel of face image is derived by comparing its gray value with neighboring pixels that lie on a circle whose center is the pixel itself. The LBP image obtained by LBP operator is then divided into WxH ($3 \leq W, H \leq 9$) non-overlapped rectangular subregions to calculate theirs histograms. The LBP feature vector of face image is built by concatenating those histogram sequences. In order to reduce the LBP feature vectors' size, uniform patterns [1] are used. It is surprised that the first purpose of LBP was not for face recognition but texture analysis [2]. Ahonen et al. [1] used LBP to extract micro feature of facial images and then used template matching for classification and got very promising results. Other applications of LBP (related to face recognition) includes face detection [1], facial expression recognition [3], age estimation [4], gender classification [5], etc. But after all, LBP method was most successfully applied to face recognition. In comparison with Gabor wavelets based methods, LBP has some advantages: simple, low computation cost, robust to illumination variations.

From the original LBP [1], numerous variants were proposed for face recognition in recent years. In [6], Multi-scale Block LBP (MB-LBP) is formed by using block regions instead of single pixel from input images. LBP can be considered as a special case of MB-LBP when block region is one pixel. MB-LBP encodes both microstructures and macrostructures of face image and provides a better presentation for face images. Improved LBP (ILBP) is proposed in [7], the authors thresholding surrounding pixels of each pixels with theirs mean gray values. ILBP is proved more effective than LBP in face detection. In [8], Heikkil et al. compared center symmetric pairs of pixels to form Center Symmetric LBP (CS-LBP). CS-LBP captures both micro feature and gradient

J.-I. Park and J. Kim (Eds.): ACCV 2012 Workshops, Part I, LNCS 7728, pp. 85–96, 2013.

feature of face images. CS-LBP feature vector'size is half of LBP feature vector'size using the same circular pattern.

Besides the appearance of LBP variants, many researchers combined LBP with other feature extraction methods for better performance. Zhang et al. [9] adopted the same technique of using LBP as in [1] was adopted by on 40 Gabor Magnitude Pictures (GMPs) to generate Local Gabor Binary Pattern Histogram Sequence (LGBPHS) feature vectors and performed template matching with histogram intersection distance function for classification. The idea of applying LBP directly on Gabor Magnitude Images (GMI) was also introduced by Nguyen et al. [10], but this time the authors exploited WPCA for dimension reduction with cosine distance for recognition phase and archived better performance. The LGBP feature vectors' length is 40 times longer than LBP but the recognition rates (in both [9] and [10]) were still very modest. Zhang et al. [11] applied LBP operator on 90 Gabor Phase images (10 global real and imagine images, 80 local real and imagine images) to form the Histogram of Gabor Phase Patterns (HGPP) feature vector. The HGPP vector's length is much longer than LGBP but the performance of HGPP is even worse than LGBP [10] (see table 2 for more details). Tan et al. [12] used feature fusion method to exploit both Gabor wavelets and LBP feature extraction methods. After building Gabor wavelets (at 5 scales and 8 orientations) and LBP feature vectors, PCA was applied to reduce vectors' length, the resulted vectors were normalized by z-core normalization procedure. Finally, classification step was proceeded with Kernel Discriminative Common Vectors (KDCV). This fusing method gained higher recognition rates than LGBP and HGPP, but it did not solve the essential drawback of Gabor wavelets based methods: heavy computation cost.

The purpose of feature extraction step in face recognition is to capture the most intrinsic and discriminative facial features of face images. The most important facial parts of the human face are these eyes and the mouth. The natural shapes of human eyes and mouth are ellipses. Furthermore, horizontal information play a very important role in face recognition and the recognition performance is improved when we combines horizontal with vertical information [13]. In this work, we propose a novel variant of LBP so-called Elliptical LBP (ELBP) which use horizontal and vertical ellipse patterns to form the ELBP feature presentation for face recognition. The concept of applying elliptical patterns in LBP was also used by S. Liao and A.C.S Chung [14] to build the Elongated LBP. The authors used weighted factors for six regions of the face image and four different elliptical patterns (in four directions) to encode the anisotropic information of the image. In our ELBP, we use only one horizontal ellipse and one vertical ellipse for capturing the micro facial features of face image. Weighted factors are not used in producing the histogram sequence of ELBP images. In applying ELBP for face recognition, we use the template matching method (the same as Ahonen et al. [1]) and advanced method that uses cosine distance function for classification and WPCA for dimension reduction. The recognition results on AR, FERET and SCface database show that the using of ELBP can get very good performance in different conditions. The rest of this paper is organized as follow. In section 2 we present the details of ELBP for face recognition. Experiment results, discussion on ELBP parameters and running time of ELBP WPCA are given in Section 3. Section 4 is the conclusion.

2   ELBP for Face Recognition

2.1   Feature Extraction by ELBP

Local Binary Pattern Overview: At each pixel (x_c, y_c) of input image with gray value g_c, its LBP label is calculated by comparing g_c with gray values of its P surrounding pixels at R distance (these pixels are located on a circle of radius R-see Fig. 1 (a-b) for more details) as:

$$LBP^{P,R}(x_c, y_c) = \sum_{i=1}^{P} s(g_i^{P,R} - g_c)2^{i-1} \tag{1}$$

where s(x) is defined as:

$$s(x) = \begin{cases} 1 \text{ if } x \geq 0; \\ 0 \text{ if } x < 0. \end{cases} \tag{2}$$

In Fig. 3 one can see an image and its $LBP^{8,1}$ version.

Elliptical Local Binary Pattern (ELBP): In ELBP, at each pixel (x_c, y_c), we consider its neighboring pixels that lie on an ellipse (see Fig. 1 (c-d)) with (x_c, y_c) is the center. ELBP of (x_c, y_c) with P surrounding pixels at (R1, R2) distances is computed as:

$$ELBP^{P,R1,R2}(x_c, y_c) = \sum_{i=1}^{P} s(g_i^{P,R1,R2} - g_c)2^{i-1} \tag{3}$$

S(x) function is defined as Eq. (2).

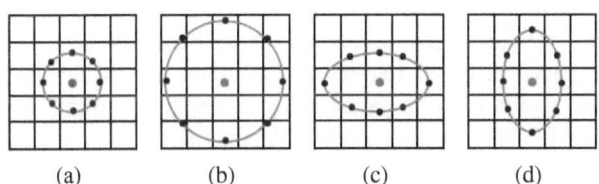

(a)   (b)   (c)   (d)

Fig. 1. $\mathbf{LBP^{8,1}}$, $\mathbf{LBP^{8,2}}$, $\mathbf{ELBP^{8,2,1}}$ and $\mathbf{ELBP^{8,1,2}}$ patterns

In details, coordinate of the ith neighboring pixel of (x_c, y_c) is calculated using the formulas:

$$angle\_step = 2 * \pi / P \tag{4}$$
$$x_i = x_c + R1 * cos((i - 1) * angle\_step)) \tag{5}$$
$$y_i = y_c - R2 * sin((i - 1) * angle\_step)) \tag{6}$$

Illustration of ELBP calculation for one pixel can be seen in Fig. 2.

In [15], authors indicated that eyes and mouth are the most important facial features in face recognition. The natural shapes of human eyes and mouth are ellipses. So horizontal ELBP is more suitable and more efficient than LBP in features extraction for

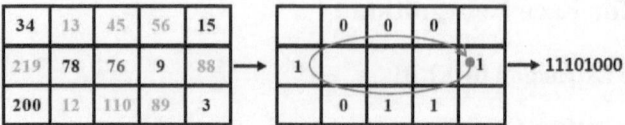

Fig. 2. Compute ELBP label of one pixel using $ELBP^{8,2,1}$

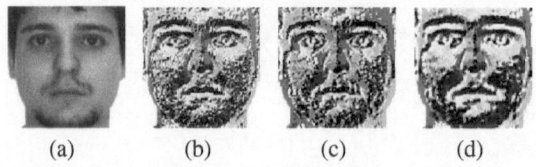

(a)   (b)   (c)   (d)

Fig. 3. An image (a) and its $LBP^{8,1}$ (b), $ELBP^{8,3,1}$ (c), $ELBP^{8,4,3}$ (d)

face recognition. When $R1 = R2$, ELBP is LBP, when $R1 < R2$ we have vertical ellipse and if $R1 > R2$ we have horizontal ellipse, which matches most for human eyes and mouth. In this work we use both horizontal and vertical ELBP to encode the micro facial feature in both directions because the combination of horizontal and vertical information of the face image gives the best recognition performance [13].

Building ELBP Feature Vector: For building the ELBP feature vector of input face images, we use ELBP operator to generate ELBP image (in Fig. 3 one can see an image and its ELBP images) and apply the similar methods as Ahonen et al. [1]. When only horizontal ELBP is used, we firstly generate the ELBP image for the input image, then the ELBP image is divided into sub non-overlapped rectangular regions. In the next step, histogram sequences of sub regions are calculated and then concatenated to form the ELBP feature vector, uniform patterns [1] are employed in this step to reduce the vector's length. In the case of using both horizontal and vertical ELBP, we apply two symmetric ELBP operators $ELBP^{P,R1,R2}$ and $ELBP^{P,R2,R1}$ to produce two ELBP images. Then each ELBP feature vector corresponding to ELBP image is computed. After that the two vectors are concatenated to form the complete horizontal and vertical ELBP feature vector for the given face image. All these steps are illustrated in Fig. 4.

The ELBP image is divided into WxH sub regions to build feature vector. So normally, with (8, R1, R2) neighborhood patterns the horizontal ELBP feature vector length is W*H*256 and the complete (both horizontal and vertical) ELBP feature vector length is 2*W*H*256. A LBP value is called uniform pattern [1] if its binary representation has no more than two bitwise transitions from 0 to 1 and vice versa. When uniform patterns are applied, the ELBP feature vector length is reduced about 4 times (from W*H*256 down to W*H*59 and from 2*W*H*256 down to 2*W*H*59). In this paper, we use uniform patterns to speed up the ELBP calculations and to save required memory for storing ELBP feature vectors.

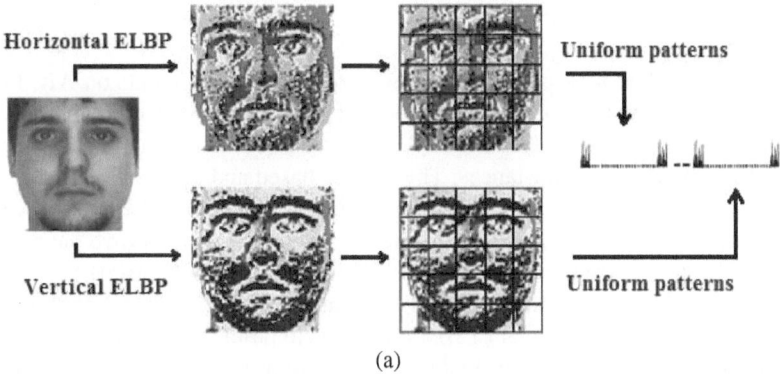

(a)

Fig. 4. Steps for ELBP feature vector computation

2.2 ELBP Template Matching for Face Recognition

Ahonen et al. [1] used template matching method with k-Nearest Neighbor (kNN) and Chi Square distance functions (non-weighted and weighted) for classification. Our tests in this work use non-weighted Chi Square distance. The Chi Square distance between two vectors $X = [x_1 x_2 ... x_M]$ and $Y = [y_1 y_2 ... y_M]$ is:

$$dist_{chi}(X, Y) = \sum_{i=0}^{M} \frac{(x_i - y_i)^2}{x_i + y_i} \tag{7}$$

2.3 ELBP WPCA Face Recognition

PCA is the most popular method used in dimension reduction step of single image per person face recognition systems. WPCA is PCA with an extra step: whitening the eigenvectors by eigenvalues. PCA has two shortcomings: the performance of PCA is degraded when using its leading eigenvalues and the weak discriminating in its eigenvectors. The whitening step is simple but very effective, it helps to rectify the shortcomings of PCA and thus makes the face recognition systems get better performance.

Whitened PCA: From N input face images, we produce N feature vectors X_i by applying ELBP operator. Then these vectors are mean-subtracted: $\Phi_i = X_i - \bar{X}$, $\bar{X} = \frac{1}{N} \sum_{i=1}^{N} X_i$. After that, we compute the covariance matrix C: $C = A^T A$, $A = [\Phi_1 \Phi_2 ... \Phi_N]$, $C_{ij} = \Phi_i^T \Phi_j$. N eigenvectors v_i corresponding to N principal eigenvalues λ_i are calculated by eigenvalue decomposition of matrix C and the projection matrix U_{PCA} is generated: $U_{PCA} = [u_1 u_2 ... u_N], u_i = A v_i / \sqrt{\lambda_i}$. That is the standard PCA algorithm [16], in WPCA, the eigenvectors are normalized by whitening factor $1/\sqrt{\lambda_i}$ as: $U_{WPCA} = [u_1 u_2 ... u_N], u_i = A v_i / \lambda_i$.

In the classification step, we use cosine distance function to classify the face images. The cosine distance between two vectors X and Y is calculated as:

$$dist_{cos}(X, Y) = -\frac{X^T Y}{||X|| ||Y||} \tag{8}$$

3   Experiment Results

For evaluating ELBP in face recognition, we proceeded experiments on AR, FERET and SCface face databases and compared the recognition rates with other methods. The template matching method was used on AR database for testing the facial expression changes and occlusion circumstances. The WPCA based and template matching methods performed on FERET database to evaluate the performance of ELBP in a large scale face database. Experiments on SCface database was conducted to prove the effectiveness of ELBP in video surveillance context. We used notation "ELBP (h)" to indicate that experiment use only horizontal ELBP descriptor and "ELBP (h+v)" for the case of using both horizontal and vertical ELBP descriptors to build the feature vector. We also made some benchmark tests to evaluate running time of ELBP WPCA. All of our experiments in this paper were conducted on a Dell Precision T3400 desktop (Intel Core 2 Duo E8400 @ 3.00 Ghz, 3.2 Gb RAM) with Windows XP SP3, the programming environment is Matlab 2010b.

3.1   Results on AR Database

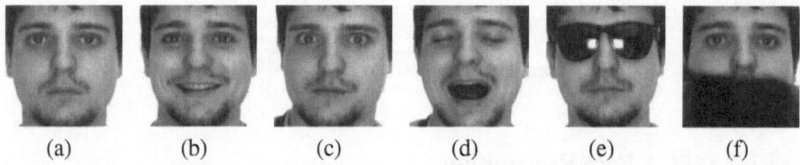

(a)   (b)   (c)   (d)   (e)   (f)

Fig. 5. Sample images from AR database

The AR face database [17] has about 4000 color face images of 134 people (75 men and 59 women). These images were taken under two sessions with similar conditions. We proceed three experiments: experiment 1 uses images from session 1, experiment 2 uses images from session 2 and experiment 3 uses images from both sessions. Experiment 1 and experiments 2 use the first images (neutral expression - see Fig. 5 (a)) of each session for gallery set and all others images from the same session with different facial expression variations and occlusions: smile, anger, scream, sun-glasses (Glasses), and scarf (see Fig. 5 (b-f)) are used for probe sets. Experiment 3 is formed by choosing first images of session 1 for the gallery and images of session 2 for probe sets. We cropped the images down to 128x128 resolution and then preprocessed them by retinal processing [18] for illumination normalization.

From table 1 we can conclude that horizontal ELBP is more powerful than LBP in encoding face micro features in facial expressions (especially in Scream expression in session 1 test) and occlusions (Glasses and Scarf). Our results with horizontal ELBP are even more better than other state-of-the-art methods: S-LNMF [19] and method in [20]. Another observation from table 1 is that the fusion of horizontal and vertical ELBP in feature extraction gives better performance than using single horizontal pattern. The

Table 1. Rank-1 RRs (%) comparison with other modern systems on AR database using the same evaluation method

Method	Smile	Anger	Scream	Glasses	Scarf
Experiment 1					
S-LNMF [19]	96.0	N/A[1]	49.0	84.0	87.0
LGBPHS [9]	N/A[1]			80.0	98.0
LBP	100		74.4	76.9	98.5
ELBP (h)			79.7	85.1	98.5
ELBP (h+v)			81.2	91.0	99.3
Experiment 2					
S-LNMF [19]	96.0	N/A[1]	54.0	66.0	89.0
LGBPHS [9]	N/A[1]			62.0	96.0
LBP	100		74.8	81.5	97.5
ELBP (h)			75.6	84.9	98.3
ELBP (h+v)			79.0	87.4	98.3
Experiment 3					
S-LNMF [19]	62.0	N/A[1]	27.0	49.0	55.0
Results of [20]	N/A[1]		52.3	54.2	81.3
LBP	95.0	96.6	56.3	57.2	95.0
ELBP (h)	95.0	98.3	57.2	57.2	95.0
ELBP (h+v)	96.6	98.3	62.2	65.6	95.8

[1] N/A: Not available result

recognition rates with scream image sets are lowest because the shapes of human eyes and mouth change most when they scream. The recognition rates on scarf probe sets are higher than on sun-glasses probe sets because with glasses the most important facial feature for face recognition, the eyes, are hidden. The combination of high and stable results on scarf sets (the minimum is 95%) and corresponding rates upon glasses probe sets points out that the upper part (above the mouth) of the face is much more important than the lower part in face recognition. The results of experiment 3 show that aging condition can degrade face recognition performance dramatically.

3.2 Results on FERET Database

The FERET face database is one of the most widely databases used for evaluating the performance of face recognition systems. In this paper, we used the same protocol as

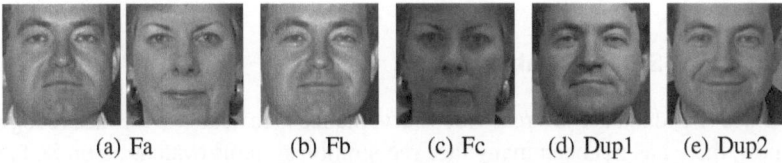

(a) Fa   (b) Fb   (c) Fc   (d) Dup1   (e) Dup2

Fig. 6. Sample images from FERET database

in [21]. FERET contains five frontal image sets named Fa (1196 images of 1196 sub-jects), Fb (1195 images of 1195 subjects with different facial expressions), Fc (194 images of 194 subjects under varying illumination conditions), Dup 1 (722 images of 722 subjects) and Dup 2 (234 images of 234 subjects) (see Fig. 6 (a-e)). The images in Dup 1 and Dup 2 sets are taken within one year and two years time span after those in Fa set. Fa set is used for gallery set. Fb, Fc, Dup 1 and Dup 2 sets are used for probe sets in Fb, Fc, Dup 1, Dup 2 tests respectively. All the images were cropped to 128x128 pixels resolution and then preprocessed by the retinal model [18] for reducing the effect of the illumination variations.

Table 2. Rank-1 RRs (%) comparison with other state-of-the-art results on FERET database using the standard evaluation protocol [21]

Method	Fb	Fc	Dup 1	Dup 2	Average
LBP	96.2	92.3	70.4	68.4	85.2
ELBP(h)	96.7	94.9	71.3	70.1	86.1
ELBP(h+v)	97.0	95.4	72.0	71.0	86.6
LGBPHS [9]	98.0	97.0	74.0	71.0	87.8
HGPP [11]	97.6	89.9	77.3	76.1	88.7
LGBP [10]	98.1	98.9	83.8	81.6	92.1
LBP WPCA	98.7	99.0	83.9	78.2	92.1
ELBP(h) WPCA	99.3	99.0	87.7	83.8	94.2
FGLBP [12]	98.0	98.0	90.0	85.0	94.2
ELBP(h+v) WPCA	99.4	100	89.1	86.8	95.0

The result of [14] was not included in this table be-cause: the authors only provided the average RR (93.2%) and they did not follow the standard pro-tocol [21] (They used a small subset of FERET database).

The comparison results in table 2 confirm that horizontal ELBP is more robust than LBP in micro facial features extraction (in both template matching and WPCA meth-ods), especially in Dup 2 case. It is obvious that the using of horizontal and vertical ELBP gives very impressive improvement recognition rates in comparison with origi-nal LBP and single horizontal ELBP (the most significant improvement cases are in the aging condition: Dup 1 and Dup 2 experiments). The perfect recognition rate (100%) upon Fc probe set of ELBP(h+v) WPCA illustrates the effectiveness of ELBP under illumination variations.

3.3   Results on SCface Database

The SCface [22] database is a real video surveillance face database for face recognition. Until now, there has not been many face recognition systems evaluating on SCface due to its challenging real-world conditions. The database contains images of 130 subjects taken from five cameras in daylight time, two cameras in night mode and one camera in

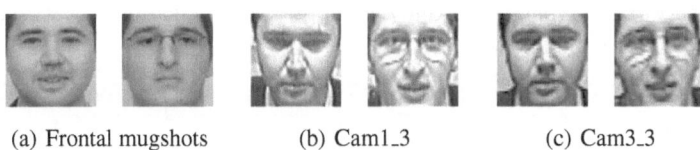

(a) Frontal mugshots   (b) Cam1_3   (c) Cam3_3

Fig. 7. Sample images from SCface database

IR mode at three different distances. In this work, we follow the same proposed test protocols as in [22] to proceed two experiments: DayTime and NightTime. Frontal facial mugshot images (Figs. 7 (a) show examples) were used for gallery set and other images (see Figs. 7 (b-c) for some examples) were used for probe sets. In DayTime experiment we have 15 probe sets containing images from 5 cameras at 3 distances under day time condition. NightTime test has 6 probe sets that include images from 2 cameras working in nigh time at 3 distances. In total, we use 130 images for training (also for the Gallery set), 1950 probe images in DayTime test and 900 probe images in NightTime protocol. All experiment images were cropped to 48x48 pixels resolution (using the eyes' coordinates from the database) and then preprocessed by standard histogram equalization for illumination normalization.

Table 3. Rank-1 RRs (%) comparison with other state-of-the-art results on SCface database using the DayTime protocol [22]

Camera/Distance	PCA[22]	DSR[23]	LBP WPCA	ELBP(h) WPCA	ELBP(h+v) WPCA
cam1_1	2.3		43.1	43.1	43.1
cam1_2	7.7		50.0	51.5	56.2
cam1_3	5.4		41.5	41.5	45.4
cam2_1	3.1		31.5	36.2	36.9
cam2_2	7.7		44.6	48.5	50.8
cam2_3	3.9		34.6	35.4	42.3
cam3_1	1.5		20.8	25.4	34.6
cam3_2	3.9		38.5	37.7	46.9
cam3_3	7.7		49.2	49.2	51.5
cam4_1	0.7		30.0	32.3	32.3
cam4_2	3.9		50.0	50.0	50.0
cam4_3	8.5	N/A	44.6	46.2	50.8
cam5_1	1.5		28.5	31.5	36.2
cam5_2	7.7		26.9	30.8	32.3
cam5_3	5.4		23.9	29.2	31.5
Average	**4.7**	**20.2**	**37.2**	**39.2**	**42.7**

[1] N/A: Not available result.

The results from table 3 and table 4 show that our ELBP WPCA method outperforms other state of the art systems, especially when compared to the baseline PCA [22] (our average result in DayTime experiment is about nine times higher than in [22]). These

Table 4. Rank-1 RRs (%) comparison with other state-of-the-art results on SCface database using the NightTime protocol [22]

Camera/Distance	PCA[22]	LBP WPCA	ELBP(h) WPCA	ELBP(h+v) WPCA
cam6_1	1.5	6.9	9.2	9.2
cam6_2	3.1	13.9	14.6	15.4
cam6_3	3.9	19.2	19.2	25.4
cam7_1	0.7	10.0	10.8	13.1
cam7_2	5.4	11.5	10.8	13.1
cam7_3	4.6	9.2	13.9	13.9
Average	**3.2**	**11.8**	**13.1**	**15.0**

results (table 3 and table 4) also prove that horizontal ELBP descriptor is more robust than LBP in micro facial features extraction (under both day time and night time conditions at three distances) and again (as evaluations on AR and FERRET databases), the combination of horizontal and vertical ELBP brings the best performance. To the best of our knowledge, our results on SCface database are the first complete and highest results reported in the literature so far.

It is clear that the results on SCface database are much lower than the recognition rates on AR database (table 1) and on FERET database (table 2). The very low resolution (small in size and very poor quality) of probe images in SCface database is the cause of those results.

3.4 ELBP Parameters

The original LBP [1] for face recognition used $LBP^{8,1}$ and $LBP^{8,2}$ operators on 7x7 sub regions of input images (128x128 resolution) to get the best performance. Our best results on AR database use $LBP^{8,5}$ (9x9 sub regions), $ELBP^{8,5,3}$ and $ELBP^{8,3,5}$ (9x9 sub regions). The $LBP^{8,5}$ (9x9 sub regions), $ELBP^{8,5,3}$ and $ELBP^{8,3,5}$ (9x9 sub regions) are used with FERET database. On SCface database, the $LBP^{8,3}$ (6x6 sub regions), $ELBP^{8,3,5}$ and $ELBP^{8,5,3}$ (6x6 sub regions) give the highest recognition rates. All this information about ELBP's parameters indicates that the best ratio between horizontal radius and vertical radius of ELBP is $1.67(5/3)$.

3.5 ELBP WPCA Running Time

One of the most important aspect of applying a face recognition system in real life is the computation cost, particularly under real-time applications like camera surveillance. For estimating computational cost of ELBP WPCA, we compare its running time (the total time for feature extraction, dimension reduction and classification phases) on FERET and SCface databases with required time for generating Gabor wavelets representations (at 5 scales and 8 orientations) of 1196 Fa set's images from FERET database. We use 80x88 resolution images for doing Gabor wavelets like in [9] for a fair comparison. Each benchmark is run for 10 times and the average results are reported in table 5.

Table 5. Running time (s) of ELBP WPCA in comparison with Gabor wavelets calculation

Database	Train/test images	Sub regions	Image size	ELBP(h) WPCA	ELBP(h+v) WPCA
FERET	1196-2345	9x9	128x128	50.4	94.3
SCface	130-2730	6x6	48x48	16.4	32.1
Gabor wavelets on Fa set of FERET			80x88	108.1	

It can be seen from table 5 that ELBP WPCA is very fast, horizontal ELBP WPCA even finishes recognition all four probe sets of FERET within a minute (just 50.4 secs). In addition, ELBP (h+v) WPCA is faster than generating Gabor wavelets output for 1196 images of Fa set (although the image size in Gabor wavelets calculation is much smaller then in ELBP). We also notice that ELBP feature extraction process only uses one CPU core while Gabor wavelets uses both 2 CPU cores on our machine. On SCface database, ELBP WPCA is super fast because the image size is quite small.

4   Conclusion

This work introduces a novel variant of LBP operator so-called ELBP for face recognition. We use a horizontal and a vertical ellipse patterns to form the ELBP face descriptor for feature extraction. Then ELBP images are divided into sub rectangular regions to build their ELBP histogram sequences. The ELBP feature vector is generated by concatenating sub regions' histogram sequences. In dimension reduction phase, we use WPCA for better recognition performance. The experimental evaluations on AR, and FERET databases show that, ELBP is more efficient than LBP in encoding micro facial features and ELBP can work well under various conditions such as partial occlusion, expression variations and aging. Additionally, the recognition performance on SCface database proves the effectiveness of ELBP for the problem of face recognition in video surveillance context. The original LBP is popular for its robustness to rotation because it uses circular patterns. While our results in this paper demonstrate advantages of ELBP over LBP for face recognition, we do not suggest that ELBP is robust against rotation. We strongly believe that ELBP can archive better results in the research fields related to face recognition, where LBP was applied.

References

1. Ahonen, T., Hadid, A., Pietikainen, M.: Face description with local binary patterns: Application to face recognition. Trans. PAMI 28, 2037–2041 (2006)
2. Ojala, T., Pietikainen, M., Maenpaa, T.: Multiresolution gray-scale and rotation invariant texture classification with local binary patterns. Trans. PAMI 24, 971–987 (2002)
3. Feng, X., Pietikäinen, M., Hadid, A.: Facial expression recognition based on local binary patterns. Pattern Recognition and Image Analysis 17, 592–598 (2007)
4. Karthigayani, P., Sridhar, S.: A novel approach for face recognition and age estimation using local binary pattern, discriminative approach using two layered back propagation network. In: Proc. Trendz in Information Sciences and Computing (TISC), pp. 11–16 (2011)

5. Lian, H.-C., Lu, B.-L.: Multi-view Gender Classification Using Local Binary Patterns and Support Vector Machines. In: Wang, J., Yi, Z., Żurada, J.M., Lu, B.-L., Yin, H. (eds.) ISNN 2006. LNCS, vol. 3972, pp. 202–209. Springer, Heidelberg (2006)
6. Liao, S., Zhu, X., Lei, Z., Zhang, L., Li, S.Z.: Learning Multi-scale Block Local Binary Patterns for Face Recognition. In: Lee, S.-W., Li, S.Z. (eds.) ICB 2007. LNCS, vol. 4642, pp. 828–837. Springer, Heidelberg (2007)
7. Jin, H., Liu, Q., Lu, H., Tong, X.: Face detection using improved LBP under bayesian framework. In: Proc. Image and Graphics, pp. 306–309 (2004)
8. Heikkilä, M., Pietikäinen, M., Schmid, C.: Description of Interest Regions with Center-Symmetric Local Binary Patterns. In: Kalra, P.K., Peleg, S. (eds.) ICVGIP 2006. LNCS, vol. 4338, pp. 58–69. Springer, Heidelberg (2006)
9. Zhang, W., Shan, S., Gao, W., Chen, X., Zhang, H.: Local gabor binary pattern histogram sequence (lgbphs): A novel non-statistical model for face representation and recognition. In: Proc. IEEE ICCV, vol. 1, pp. 786–791 (2005)
10. Nguyen, H.V., Bai, L., Shen, L.: Local Gabor Binary Pattern Whitened PCA: A Novel Approach for Face Recognition from Single Image Per Person. In: Tistarelli, M., Nixon, M.S. (eds.) ICB 2009. LNCS, vol. 5558, pp. 269–278. Springer, Heidelberg (2009)
11. Zhang, B., Shan, S., Chen, X., Gao, W.: Histogram of gabor phase patterns (HGPP): a novel object representation approach for face recognition. Trans. IEEE 16, 57–68 (2007)
12. Tan, X., Triggs, B.: Fusing Gabor and LBP Feature Sets for Kernel-Based Face Recognition. In: Zhou, S.K., Zhao, W., Tang, X., Gong, S. (eds.) AMFG 2007. LNCS, vol. 4778, pp. 235–249. Springer, Heidelberg (2007)
13. Goffaux, V., Dakin, S.: Horizontal information drives the behavioral signatures of face processing. Frontiers in Psychology 1 (2010)
14. Liao, S., Chung, A.C.S.: Face Recognition by Using Elongated Local Binary Patterns with Average Maximum Distance Gradient Magnitude. In: Yagi, Y., Kang, S.B., Kweon, I.S., Zha, H. (eds.) ACCV 2007, Part II. LNCS, vol. 4844, pp. 672–679. Springer, Heidelberg (2007)
15. Zhao, W., Chellappa, R., Phillips, P., Rosenfeld, A.: Face recognition: A literature survey. Acm Computing Surveys (CSUR) 35, 399–458 (2003)
16. Turk, M., Pentland, A.: Face recognition using eigenfaces. In: Proc. IEEE CVPR, pp. 586–591 (1991)
17. Martinez, A.M., Benavente., R.: The AR face database. CVC Technical Report 24 (1998)
18. Vu, N., Caplier, A.: Illumination-robust face recognition using retina modeling. In: IEEE ICIP 2009, pp. 3289–3292 (2009)
19. Oh, H., Lee, K., Lee, S.: Occlusion invariant face recognition using selective local non-negative matrix factorization basis images. Image and Vision Computing 26, 1515–1523 (2008)
20. Min, R., Hadid, A., Dugelay, J.: Improving the recognition of faces occluded by facial accessories. In: 9th IEEE Conf. on AFGR, Santa Barbara, United States (2011)
21. Phillips, P., Moon, H., Rizvi, S., Rauss, P.: The FERET evaluation methodology for face-recognition algorithms. Trans. PAMI 22, 1090–1104 (2000)
22. Grgic, M., Delac, K., Grgic, S.: SCface surveillance cameras face database. Multimedia Tools and Applications 51, 863–879 (2009)
23. Zou, W., Yuen, P.: Very low resolution face recognition problem. In: Proc. BTAS, pp. 1–6 (2010)

Block LBP Displacement Based Local Matching Approach for Human Face Recognition

Liang Chen and Ling Yan

University of Northern British Columbia, Prince George, BC, Canada V2N 4Z9
{chen.liang.97,yanttian}@gmail.com

Abstract. A local matching approach, known as Electoral College, where each block contributes one single vote to the final decision, which is generated by a simply majority voting from all local binary decisions, has been proved to be stable for political elections as well as general pattern recognition. Given the registration difficulties caused by the non-rigidity of human face images, block LBP displacement is introduced so that an Electoral College, where a local decision is made on LBP statistics for each block, can be applied to face recognition problems. Extensive experiments are carried out and have demonstrated the outstanding performances of the block LBP displacement based Electoral College in comparison with the original LBP approach. It is expected and shown by experiments that the approach also applies to descriptor approaches other than LBP.

1 Introduction

As known as the two most famous voting schemes in political election, national voting, also known as direct popular voting, gets its name by directly appointing the final winner the one who gets most votes, where each voter of the nation contributes one vote to the final decision in a single winner election case; while in regional voting, or Electoral College, a direct popular voting is firstly performed within each region to generate the vote of this region on a winner-take-all basis, and then a second stage of popular voting by the regions takes place over the nation to finally select the winner. It has been proved that Electoral College is more stable than the national voting in political elections as well as in pattern recognition [1,2]. If we can take each pixel of an image as one voter of a nation, the above mentioned two voting schemes can be applied to patten recognition problems where the answer to "do the two images belong to the same category" counts on the voting from the pixels of the two images. Due to the fact that direct pixel to pixel comparison does not make sense in pattern recognition, the decision making process in popular voting and the local decision making process in each region in Electoral College usually perform in the form of feature or descriptor matching, where features or descriptors are extracted from an entire image or each region of an image.

Face recognition is a special pattern recognition. Subspace approaches constitute a traditional category of approaches for face recognition. It has been shown [3,4] that the performance of a subspace approach can be improved when it

J.-I. Park and J. Kim (Eds.): ACCV 2012 Workshops, Part I, LNCS 7728, pp. 97–108, 2013.
© Springer-Verlag Berlin Heidelberg 2013

is embedded into Electoral College. Descriptor based approaches are now very popular in face recognition research. A key step of a "sucessful" descriptor based approach always is finding a "better" descriptor for the face images, on which the similarity calculations are based. Though finding a "better" descriptor is an open issue [1], in recent years, descriptor approaches gained great interests, with LBP (local binary patterns) [5] as one of the most known well performing texture descriptors. In LBP approach, pixels over a face image is codified into an LBP map using the LBP operators, the LBP map is then divided into windows, where histograms of pixel LBP codes are obtained, and the concatenation of window level histograms are used as the descriptor of the face image. We can see that, LBP approach presents three levels of locality [5]: pixel level LBP code for each pixel, regional level statistics for each window and a global level descriptor of an entire face image.

Considering Electoral College and LBP descriptor, the former aiming at a more stable final decision given a certain description of the images and the latter offering an effective descriptor for the face image represented by pixel values, this paper aims to adopt LBP in Electoral College scheme in face recognition. That is, we first divide a face image into blocks, apply LBP approach to every block to generate a local decision for this block, and then use Electoral College to make the final decision from all decisions of the blocks. A similar approach has been introduced by Ji *etal.* [6], where Borda count is used instead of Electoral College. However, it is easy to see that the performance of such an integration works are seriously depends on whether the corresponding blocks of face images are corresponding to same face areas. That is to say, alignment difficulties should be taken into consideration. As we know that human faces are never rigid: people have face expressions showing their feelings, they gain or lose weight on the faces and they grow older[2]. When it is reflected on the pictures, more factors come into affection: angle of the picture taken, accessories such as glasses and earrings, etc.. This all lead to the fact that there is no exactly precise alignment of face images manually or automatically, even if the images are from the same person. Though we are aiming at an impossible ideal alignment, there are always approaches that lead us close. Considering the biological feather of human faces, we should admit the fact that faces are "regionally variable, globally stable", that is, when we try to align two images from the same person, there would always be a kind of "upper" limit of the offset.

Based on this fact, we adopt the idea of displacement into our algorithm. Specifically, when we calculate the similarity of two corresponding blocks of images, aiming at approaching the better alignment, we try to shift one or both of the blocks within a broader area so as to cover all possible positions that a precise alignment would be, given a certain upper bound of the offset.

Based on discussion above, we apply LBP approach as descriptor of two face images to be compared with in the Electoral College Framework, which we name as LBP Displacement Local Matching Approach (LBP-DLMA). In

---

[1] Indeed there is no definition of "better".

[2] Or, age backwards as Benjamin Button did.

this approach, to simulate a relatively precise alignment, the idea of shifting is adopted.

Since the LBP-DLMA retains its advantage of both LBP descriptor and Electoral College, a better performance over original LBP approach can be expected. We can also expect that this approach will still work if LBP is replaced with other descriptors.

The remaining of the paper is organized as follows: The concepts and the approach are given in Section 2; the experimental results are presented in Section 3; we further demonstrate the extensibility of our approach to other descriptors in Section 4; the discussions and conclusion remarks shall be discussed in Section 5.

2   LBP Displacement and LBP-DLMA

2.1   LBP Displacement Concepts

Assume that we have an $(m+2s) \times (n+2s)$ LBP map of a face image. By removing i, j, k, l pixels from top, bottom, leftmost and rightmost margins ($i, j, k, l \geq 0$, $i + j = 2s$, $k + l = 2s$), we obtain $(2s + 1)^2$ slightly smaller LBP maps of size $m \times n$. For each of these $m \times n$ LBP maps, we partition it into $K \times L$ blocks (K blocks per column, and L blocks per row), The set of corresponding blocks of all these $m \times n$ maps are called a pile of LBP displacement blocks, or a LBP displacement pile.

Note that, there is a concept *window* in LBP approach, where the statistic of LBP labels is obtained. The concept *"window"* still exists in LBP displacement: in each block, there are one or more windows where the statistics of LBP labels are obtained. The concatenation of the statistics of all windows in a block is called a block level LBP description.

The set of all the piles of LBP displacement blocks for a face image is called the LBP displacement description of the face image.

As an example, Figure 1(a) is a 17×22 LBP map of a face image, we let $s = 1$. By removing 2 pixels from top, one pixel each from leftmost and rightmost margins, a smaller 15×20 sized LBP map can be obtained as shown in Figure 1(b). We partition this map into 12 blocks, each of size 5×5. There are a total of 9 such 15×20 sized LBP maps corresponding to different values of i, j, k and l, each of which can be partitioned into 12 blocks. A pile of LBP displacement blocks, consisting the blocks of all these partitioned 15×20 sized LBP maps corresponding to the green colored block of Figure 1(b), are shown in Figures 2(a) - 2(i). Note that the 8th block, Figure 2(h), in the pile is the green colored block of the 15×20 sized LBP map of Figure 1(b). We shall have 12 such LBP displacement piles – the set of all these 12 LBP displacement piles is the LBP displacement description of the face image.

2.2   LBP Displacement Based Local Matching Approach

A descriptor approach for pattern recognition is always associated with a similarity metric. A typical LBP approach works as follows: first the LBP maps of

38	38	39	39	32	39	58	58	57	57	58	58	19	18	18	25	53
58	38	39	58	6	6	1	0	58	58	4	0	13	26	26	25	58
58	58	58	6	58	6	0	58	58	58	0	4	13	13	13	12	25
38	0	16	15	15	10	10	9	10	14	15	15	58	7	0	11	25
22	16	20	21	21	10	33	33	58	58	20	14	15	34	33	8	7
0	58	19	13	14	57	35	33	58	58	20	57	57	10	33	58	7
17	19	27	57	57	46	46	39	22	22	27	57	57	46	32	32	4
26	26	57	57	58	43	57	45	0	23	55	53	58	50	57	46	4
54	55	54	42	48	48	49	44	22	58	51	52	48	42	45	57	58
51	51	47	48	48	58	44	38	4	23	51	51	48	48	43	44	58
52	52	0	16	17	58	58	31	4	0	58	58	30	58	48	23	16
58	17	0	11	11	35	58	3	4	0	58	58	30	29	0	17	17
58	58	58	13	28	15	3	6	5	8	58	15	34	1	0	0	19
58	20	14	41	27	58	57	57	9	40	57	58	46	6	4	8	58
19	20	57	58	58	56	57	56	58	57	57	58	46	57	6	8	58
19	57	28	0	0	11	11	58	58	1	0	0	58	57	57	33	33
19	26	58	28	1	0	0	58	58	0	58	11	11	58	41	58	33
26	18	58	56	56	57	57	58	58	57	57	46	46	58	58	32	39
26	18	58	58	58	58	58	58	58	56	57	57	46	58	58	32	41
26	19	58	58	0	0	58	11	0	58	11	0	58	58	58	40	38
25	28	35	46	58	0	21	15	58	58	0	0	58	35	34	40	38
18	54	57	56	58	56	56	57	57	46	46	58	58	57	57	38	38

(a) Original LBP Map of Size 17 × 22

58	58	6	58	6	0	58	58	58	0	4	13	13	13	12
0	16	15	15	10	10	9	10	14	15	15	58	7	0	11
16	20	21	21	10	33	33	58	58	20	14	15	34	33	8
58	19	13	14	57	35	33	58	58	20	57	57	10	33	58
19	27	57	57	46	46	39	22	22	27	57	57	46	32	32
26	57	57	58	43	57	45	0	23	55	53	58	50	57	46
55	54	42	48	48	49	44	22	58	51	52	48	42	45	57
51	47	48	48	58	44	38	4	23	51	51	48	48	43	44
52	0	16	17	58	58	31	4	0	58	58	30	58	48	23
17	0	11	11	35	58	3	4	0	58	58	30	29	0	17
58	58	13	28	15	3	6	5	8	58	15	34	1	0	0
20	14	41	27	58	57	57	9	40	57	58	46	6	4	8
20	57	58	58	56	57	56	58	57	57	58	46	57	6	8
57	28	0	0	11	11	58	58	1	0	0	58	57	57	33
26	58	28	1	0	0	58	58	0	58	11	11	58	41	58
18	58	56	56	57	57	58	58	57	57	46	46	58	58	32
18	58	58	58	58	58	58	58	56	57	57	46	58	58	32
19	58	58	0	0	58	11	0	58	11	0	58	58	58	40
28	35	46	58	0	21	15	58	58	0	0	58	35	34	40
54	57	56	58	56	56	57	57	46	46	58	58	57	57	38

(b) 15 × 20 Sized LBP Map From (a)

Fig. 1. LBP Map

39 58 58 57 57	58 58 57 57 58	58 57 57 58 58	6 1 0 58 58	1 0 58 58 4
6 1 0 58 58	1 0 58 58 4	0 58 58 4 0	6 0 58 58 58	0 58 58 58 0
6 0 58 58 58	0 58 58 58 0	58 58 58 0 4	10 10 9 10 14	10 9 10 14 15
10 10 9 10 14	10 9 10 14 15	9 10 14 15 15	10 33 33 58 58	33 33 58 58 20
10 33 33 58 58	33 33 58 58 20	33 58 58 20 14	57 35 33 58 58	35 33 58 58 20
(a) Block 1	(b) Block 2	(c) Block 3	(d) Block 4	(e) Block 5

0 58 58 4 0	6 0 58 58 58	0 58 58 58 0	58 58 58 0 4
58 58 58 0 4	10 10 9 10 14	10 9 10 14 15	9 10 14 15 15
9 10 14 15 15	10 33 33 58 58	33 33 58 58 20	33 58 58 20 14
33 58 58 20 14	57 35 33 58 58	35 33 58 58 20	33 58 58 20 57
33 58 58 20 57	46 46 39 22 22	46 39 22 22 27	39 22 22 27 57
(f) Block 6	(g) Block 7	(h) Block 8	(i) Block 9

Fig. 2. A Pile of LBP Displacement Blocks of the LBP Map in Figure 1(a)

images are generated and divided into windows where statistics of LBP labels are obtained; the window statistics of all windows in each LBP map are concatenated into a global LBP description; finally the similarities among these global LBP descriptions are calculated in order to make recognition decisions.

The typical metrics for calculation the similarities between two global LBP descriptions (of two images, of course) are Euclidean Distance, Histogram Intersection and Chi square statistic [5]. We adopt them to the calculation of the similarities between the block level LBP descriptions of two LBP blocks. Note that in each block, there are one or more windows; the block level LBP description is the concatenation of window level LBP statistics.

Assuming that the similarity between two LBP blocks b_1 and b_2 is defined by $\mathrm{Sim}(b_1, b_2)$, we define the similaity between two piles P_1 and P_2 to be

$$\mathrm{Sim}(P_1, P_2) = \min_{b_1 \in P_1, b_2 \in P_2} \mathrm{Sim}(b_1, b_2). \tag{1}$$

The purpose of the introducing block LBP displacement is to conquer the (local) registration difficulties. We can understand that, when comparing a probe with a gallery, since "shifting" the position of a block in the probe is "almost" equivalent to "shifting" the position of the corresponding block in the gallery. Therefore, to reduce the time complexity, we always let $s = 0$ for a probe image, that is, we let each LBP block pile in a probe consist of only one LBP block.

Assume that we want to match a probe $\mathcal{P}$ against a gallery set of T images, the LBP Displacement based Local Matching Approach is shown in Table 1.

3   Experiments

Our experiments are carried on FERET [7], "Faces in the Wild" (LFW) [17] and FRGC [8]. LBP descriptor involves a few parameters. In all our experiments, as suggested in [5], we choose the circle of radius to be 2 and the number of sampling points distributed a circle of radius 2 to be 8. In all our experiments, we divide images into 5×5 blocks, each with 7×7 windows. For the parameter s, we use $s = 3$. That is, in the first step of generating the smaller LBP map (see Item 2.1.1 of Section A in Table 1), we allow the removal of i and $6 - i$ pixels from the top and the bottom margins respectively, j and $6 - j$ pixels from the left and the right margins respectively, where $0 \leq i, j \leq 6$. To further reduce the number of LBP displacement blocks in each pile, we restrict the relative offset by restricting $|3 - i| + |3 - j| \leq 4$. It is understandable that, we may further improve the accuracies if we further adjust these parameters on a "trial and error" base. But we trust that such adjustment in an academic paper does not have much academic values.[3]

3.1   FERET

FERET database [7], the datebase we carry our experiments on, consists of 14051 gray-scale images from 1199 individuals. The images vary in lighting conditions, facial expressions, poses, etc. Following most of the protocols of the experiments on FERET (eg. [5]), five sets of FERET are used: Fa gallery set that contains images of 1196 subjects, one image for each subject; Fb probe set that contains 1195 face images of the subjects in Fa but with alternative facial expressions; Fc probe set that contains 194 face images taken under different illumination conditions on the same day as the corresponding image in Fa was taken; Dup1

---

[3] "If you torture the data long enough, it will confess."—Ronald Coase.

Table 1. LBP Displacement based Local Matching Approach

Perimeters Chosen: Number of Piles in Each Image $c \times l$ (c piles per column, l piles per row), Shifting Value s, Number of Windows per Block $w_c \times w_l$ (w_c piles per column, w_l piles per row).

A. Off-Line Gallary Image LBP Displacement Discription Construction:
Require: $\mathcal{G}$, a gallery of $m \times n$ sized face images; the size of the gallery is T.
For each image in $\mathcal{G}$

1. Obtain the pixel label map by calculating the LBP pattern of each pixel (Note: The label map is slight smaller than $m \times n$ since the pixels on the boundaries may not have a label.)
2. For $i = 0$ to $2s$
   2.1. For $j = 0$ to $2s$
      2.1.1. Remove i, $2s - i$, j and $2s - j$ pixels from the leftmost, rightmost, topmost and bottommost boundaries of the label map. (Note: as a total, there are $(2s + 1)^2$ label maps.)
      2.1.2. Partition the label map into $c \times l$ blocks; partition each block into $w_c \times w_l$ windows, where we obtain the LBP label statistics (histogram of pixel labels); then concatenate the LBP label statistics of all windows in each block into a block level LBP description.
3. obtain the LBP displacement description of the galary image by piling up the corresponding block level LBP descriptions into each pile.

B. On-Line Face Recognition:
Require: $\mathcal{P}$ is an $m \times n$ sized probe image.

1. Obtain the LBP displacement description for $\mathcal{P}$ as follows:
   1.1. Obtain the pixel label map by calculating the LBP pattern of each pixel.
   1.2. Remove s pixels from all four sides of the label map.
   1.3. Partition the label map into $c \times l$ blocks.
   1.4. Partition each block into $w_c \times w_l$ windows, where we obtain window level LBP statistics, then concatenate the window level LBP statistics into a block level LBP description; each block LBP description constructs a LBP displacement pile; the set of all LBP piles is the LBP displacement description.
2. Do classification as follows:
   2.1 Set vote counters $V_t = 0$ for all $t \in \{1, 2, \cdots, T\}$.
   2.2 For each pile P_i of the LBP displacement description of the probe image, do followings:
      2.2.1 For the corresponding LBP displacment pile G_{t_i} of each gallery image g_t (where $t \in \{1, 2, \cdots, T\}$)
         2.2.1.1 Calculate $\text{Sim}(P_i, G_{t_i})$, according to Equation (1).
      2.2.2 Find image index $I = \arg\max_{t \in \{1,2,\cdots,T\}} \text{Sim}(P_i, G_{t_i})$.
      2.2.3 Increase V_I by 1.
   2.3 Classify the image as the identity of image g_J in the gallery set, where $J = \arg\max_{t \in \{1,2,\cdots,T\}} V_t$.

---

probe set that contains 722 face images taken anywhere between one minute and 1031 days after the corresponding image in Fa was taken; Dup2 probe set being a subset of dup1 that contains 234 face images taken at least 18 months after the Fa image was taken.

All faces are first normalized into a standard size 150×130 pixels (150 pixels per column, 130 pixels per row), where the distance between the centers of the two eyes is 56 pixels and the line between two eyes lies on the 53rd pixel below the top boundary. The standard 150 by 130 elliptical mask from FERET data collection is used to exclude non-face areas from the LBP maps, a few pixels are removed from each side of the mask since the LBP map of an image is always smaller than the image.

Following [5], permutation test with 95% confidence level is also carried out using the image list, list640.srt, in the CSU face identification evaluation system package [9]. list640.srt contains 4 images each for 160 subjects. 10000 permutations are tested, with each containing one image per subject in the gallery set and another in the probe set.

It was explained in [10] that a preprocessing stage can significantly improve the performance of LBP approach. Therefore, we also do the experiments with the preprocessing as suggested in [10].

The results are shown in Table 2. The results of a few famous approaches are also shown in the Table in order to meet the requirements of certain readers' / reviewers' interests.

We can easily conclude that LBP-DLMA not only improves LBP approach, but also achieves the results at least comparable to the state of the art results.

Table 2. The recognition rates of original LBP and weighted LBP, the LBP-DLMA, and LBP-DLMA boosted by PreProcessing for the FERET probe sets, the mean recognition rate of the Fb+Fc+Dup1, and results of permutation test with a 95% confidence level

Method		Fb	Fc	Dup1	Dup2	Fb,Fc & Dup1	Permutation Test lower	mean	upper
LBP, no weight [11]		93%	51%	61%	50%	78.20%	71%	76%	81%
LBP, weighted [11]		97%	79%	66%	64%	84.74%	76%	81%	85%
LBP DLMA	Euclidean Distance	99.37%	93.60%	79.66%	75.56%	92.10%	84.92%	89.24%	93.31%
	Histogram intersection	99.39%	96.16%	82.52%	80.31%	93.32%	87.21%	91.22%	95.09%
	Chi square statistic	99.31%	96.20%	82.23%	80.53%	93.18%	87.34%	91.33%	95.18%
Preproceed LBP DLMA	Euclidean Distance	99.29%	98.97%	85.37%	82.29%	94.50%	88.86%	92.77%	96.41%
	Histogram intersection	99.37%	99.48%	88.40%	85.89%	95.63%	91.43%	94.92%	98.12%
	Chi square statistic	99.37%	99.25%	88.71%	86.89%	95.71%	91.51%	94.95%	98.12%
LGBPHS[12]		98%	97%	74 %	71%	89.70%	/	/	/
HGPP[13]		97.6%	98.9%	77.7 %	76.1%	90.91%	/	/	/
SIS [14]		91%	90%	68 %	68%	83.04%	/	/	/
Schwartz [15]		95.7%	99.0%	80.3 %	80.3%	90.74%	/	/	/

Table 3. Recognition rates of LBP-DLMA approaches on FRGC Experiment 104

LBP[16]	LBP DLMA		
	Euclidean	Histogram intersection	Chi square statistics
28.1%	34.38%	32.17%	33.23%
LBP with Preprocessing[16]	LBP DLMA with Preprocessing		
	Euclidean	Histogram intersection	Chi square statistics
58.1%	58.31%	67.47%	67.20%

3.2 FRGC 104

FRGC experiment 104 [8] is generally considered the most challenging in this FRGC V1 dataset. It is required to recognize 608 uncontrolled faces from 152 controlled gallery faces. We normalize the face images into size 150×130 as we did for FERET experiments. The results are shown in Figure 3. We also include the resutls of LBP-DLMA with a "preprocessing" stage, as suggested by Tan et al [10]. We can see that LBP DLMA with and without preprocessing improve LBP with and without preprocessing significantly.

We should emphasize here that, our intension is to improve LBP approaches by using local displacement based local matching scheme. It is not our intension to show that our approach is better than all possible approaches in all datasets. We understand some other approaches, such as [15], get better results for this experiment; we should add that those approaches actually use the settings more flexible than ours – they uses a training approach while we do not use.

3.3 LFW

We have also carried experiments on "Labeled Faces in the Wild" (LFW)[4] [17]. We test the performance of our approach on the 10 folds of view 2. All the face images were taken in unconstrained environments, exhibiting " 'natural' variability in pose, lighting, focus, resolution, facial expression, age, gender, race, accessories, make-up, occlusions, background, and photographic quality" [17]. In this task, given two face images, the goal is to decide whether two images are of the same person. This is a binary classification problem, with two possible outcomes: "same" or "different". LFW view 2 provides 10 folds of face sets where the sets of people in different folds are disjoint; when testing on one fold, the other nine folds can be used for training. Results of various approaches have been reported at LFW official site. [5]

We use LFW-a version of images (the images aligned using a commercial face alignment software) [18]. The images are of size 250×250. We first crop them into images of size 90×78 (by removing 88 pixel margins from top, 72 from

---

[4] The set is available via LFW official site
http://vis-www.cs.umass.edu/lfw/results.html

[5] Note that most of the approaches reported were developed only for the specific binary classification task; our approach was not intended to be applicable only to this kind of tasks.

bottom, and 86 pixel margins from both left and right sides). Note that, there were errors in the alignment of many images; we just keep them as they were (so some of the final cropped faces indeed are not correctly aligned).

Since a "voting" is required in each pile, we need a few "reference faces" to find relative values. Here, our "reference faces" use a dummy set: for the experiments in the i-th fold, we use the first images (named "***..-0001.jpg") of the first 10 individuals in the $(i - 1)$th fold (when $i - 1 = 0$, we use the 10th fold) as the dummy set.

For a pair of images x and y, for each pile, we first obtain the similarity array between x and the set consists of y and dummy set, then obtain the similarity array between y and the set consists of x and the dummy set; the average to these two arrays are taken so as to make local decision according the this array.

Our results are shown in Figure 3 and Table 4.

Due to the nature that our LBP DLMA does not have a training process, Our approach should be compared to other no-training approaches as suggested in LFW site; we also include the ROCs of all these no-training approaches SD-MATCHES (L & R system with SIFT descriptors and MATCHES flavour), H-XS-40 (Histogram of LBP features with Chi Square similarity measure and 40 windows), GJD-BC-100 (Gabor Jets Descriptors with Borda Count measure and 100 reference images) and LARK representation without supervision [19], which are available in both LFW site and [20], in Figure 3 and Table 4. We can see that the LBP DLMA, regardless the similarity metrics that it uses, is significantly better than all other approaches.

Table 4. The accuracies of LBP DLMA and a few no-training approaches for LFW

Approach		Accuracy
SD-MATCHES		0.6410 ± 0.0062
H-XS-40		0.6945 ± 0.0048
GJD-BC-100		0.6847 ± 0.0065
LARK unsupervised		0.7223 ± 0.0049
LBP DLMA	Euclidean	0.7517 ± 0.0122
	Histogram intersection	0.7648 ± 0.0186
	Chi square statistic	0.7622 ± 0.0206

4 Extensionablity

We expect that our approach can be applied for other descriptor approaches. – Just replacing LBP in Table 1 by any descriptor approach $\mathcal{A}$, we should be able to have $\mathcal{A}$ DLMA. As example, we have done experiments of TPLBP DLMA and FPLBP DLMA on FERET datasets. Three-Patch LBP (TPLBP) and Four-Patch LBP (FPLBP) of [21] have a few parameters, including patch size, ring radii, the number of the additional patches distributed in the ring, etc. We use the default values of [21].

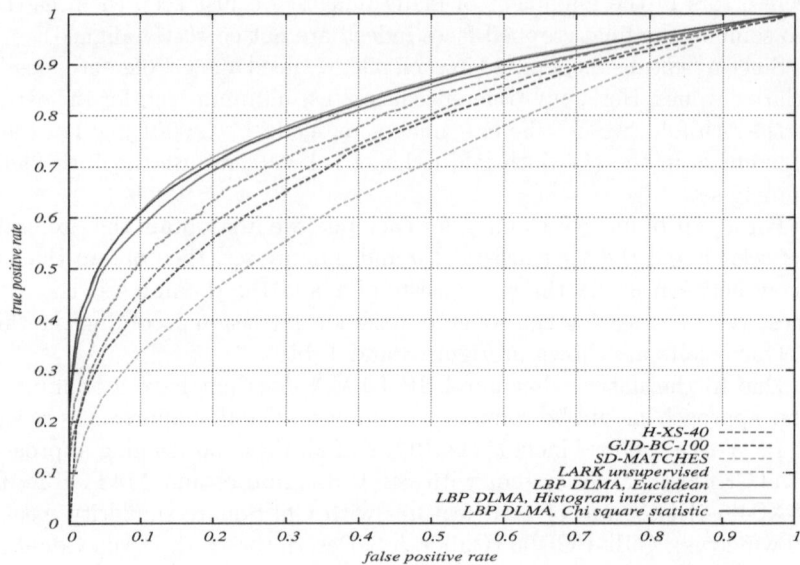

Fig. 3. ROC Curves of the Results over View 2 of LFW

Table 5. The recognition rates of original TPLBP, FPLBP, and TPLBP DLMA and FPLBP DLMA without / with Preprocessing [10] for the FERET probe sets, the mean recognition rate of the Fb+Fc+Dup1, and results of permutation test with a 95% confidence level

	Method	Fb	Fc	Dup1	Dup2	Fb,Fc & Dup1	Permutation Test		
							lower	mean	upper
TPLBP	Euclidean Distance	94.64%	74.23%	62.33%	55.98%	81.71%	68.13%	74.12%	80.00%
	Histogram intersection	96.44%	86.08%	74.65%	69.23%	88.04%	80.00%	85.06%	90.00%
	Chi square statistic	95.98%	86.08%	74.79%	69.66%	87.83%	79.38%	84.50%	89.38%
TPLBP DLMA	Euclidean Distance	99.26%	91.90%	75.97%	71.80%	90.62%	83.05%	87.51%	91.77%
	Histogram intersection	99.48%	95.15%	79.79%	75.7%	92.35%	85.68%	89.83%	93.91%
	Chi square statistic	99.38%	93.27%	78.83%	74.30%	91.79%	85.75%	89.90%	93.96%
Preprocessed TLBP DLMA	Euclidean Distance	98.88%	98.39%	77.56%	73.54%	91.54%	84.92%	89.27%	93.48%
	Histogram intersection	99.14%	98.23%	83.17%	81.98%	93.60%	87.87%	91.88%	95.68%
	Chi square statistic	99.15%	98.99%	82.31%	81.46%	93.38%	87.85%	91.85%	95.68%
FPLBP	Euclidean Distance	95.73%	69.59%	64.13%	54.70%	82.52%	72.50%	78.07%	83.13%
	Histogram intersection	96.65%	74.23%	67.45%	56.84%	84.60%	75.94%	81.19%	86.25%
	Chi square statistic	96.65%	74.23%	67.73%	56.41%	84.70%	75.63%	81.16%	86.25%
FPLBP DLMA	Euclidean Distance	98.89%	76.16%	68.68%	57.11%	86.47%	79.64%	84.32%	88.91%
	Histogram intersection	98.82%	81.09%	69.62%	60.98%	87.21%	80.84%	85.51%	90.09%
	Chi square statistic	99.04%	84.38%	70.56%	60.50%	87.95%	81.12%	85.78%	90.31%
Preprocessed FPLBP DLMA	Euclidean Distance	98.74%	98.24%	75.10%	69.65%	90.61%	84.01%	88.27%	92.45%
	Histogram intersection	99.00%	98.23%	76.96%	73.49%	91.39%	84.79%	89.07%	93.25%
	Chi square statistic	98.94%	98.22%	77.19%	73.08%	91.44%	85.05%	89.33%	93.49%

The results are shown in Figure 5.

We can easily see that the performances of TPLBP DLMA and FPLBP DLMA are significantly better than TPLBP and FPLBP respectively.

5 Discussions and Conclusion

We introduce a LBP displacement concept so that LBP can be embedded into an Electoral College framework. The integration of LBP and Electoral College, LBP DLMA, improves significantly the performances of the original LBP. Experiments also show that our approach can also be applied to descriptor approaches other than LBP.

The LBP DLMA adopts Electoral College, where winner-take-all is applied to select one "winner" when a pile of a probe is matched with a pile of a gallery image. An immediate question is whether we can replace the Electoral College by "soft-combination", where the similarities of corresponding LBP displacement piles are added up to form the similarity between the LBP displacement descriptions of a pair of faces. Indeed we can prove that the answer to this question is positive [22]. It may be interesting to investigate the adoption of more complex strategies, such as the randomized decision trees [23] for constructing / representing LBP displacement pile, and the learning of a similarity metrics [24] for exploiting the similarity values or assessments of all LBP displacement piles of a pair of LBP displacement descriptions.

Acknowledgement. This work is partly supported by NSERC Discovery Grant (Grant No. 261403-2011 RGPIN).

References

1. Chen, L., Tokuda, N.: Regional voting versus national voting –stability of regional voting (extended abstract). In: Int. ICSC Symposium on Advances in Intelligent Data Analysis, Rochester, New York, USA (1999)
2. Chen, L., Tokuda, N.: A general stability analysis on regional and national voting schemes against noise – why is an electoral college more stable than a direct popular election? Artificial Intelligence 163, 47–66 (2005)
3. Chen, L.: Electoral college and direct popular vote for multi-candidate election. In: Proceedings of the British Machine Vision Conference, pp. 100.1–100.11. BMVA Press (2010), doi:10.5244/C.24
4. Chen, L., Tokuda, N.: A unified framework for improving the accuracy of all holistic face identification algorithms –electoral college for human face identification by computing machinery. Artificial Intelligence Review 33 (2010)
5. Ahonen, T., Hadid, A., Pietikäinen, M.: Face Recognition with Local Binary Patterns. In: Pajdla, T., Matas, J(G.) (eds.) ECCV 2004, Part I. LNCS, vol. 3021, pp. 469–481. Springer, Heidelberg (2004)
6. Zou, J., Ji, Q., Nagy, G.: A comparative study of local matching approach for face recognition. IEEE Transactions on Image Processing 16, 2617–2628 (2007)
7. Phillips, P., Moon, H., Rizvi, S.A., Rauss, P.: The FERET evaluation methodology for face-recognition algorithms. IEEE Trans. Pattern Analysis & Machine Ingelligence 22, 1090–1104 (2000)

8. Phillips, P.J., Flynn, P.J., Scruggs, T., Bowyer, K.W., Chang, J., Hoffman, K., Marques, J., Min, J., Worek, W.: Overview of the face recognition grand challenge. In: Proc. of Computer Vision and Pattern Recognition, San Diego, vol. I, pp. 947–954 (2005)
9. Beveridge, J.R., She, K., Draper, B., Givens, G.H.: A nonparametric statistical comparison of principal component and linear discriminant subspaces for face recognition. In: Proceedings of the IEEE Conference on Computer Vision and Pattern Recognition, pp. 535–542 (2001)
10. Tan, X., Triggs, B.: Enhanced Local Texture Feature Sets for Face Recognition Under Difficult Lighting Conditions. In: Zhou, S.K., Zhao, W., Tang, X., Gong, S. (eds.) AMFG 2007. LNCS, vol. 4778, pp. 168–182. Springer, Heidelberg (2007)
11. Ahonen, T., Hadid, A., Pietikäinen, M.: Face description with local binary patterns: Application to face recognition. IEEE Transactions on Pattern Analysis and Machine Intelligence 28, 2037–2041 (2006)
12. Zhang, W., Shan, S., Gao, W., Chen, X.: Local gabor binary pattern histogram sequence (LGBPH): A novel non-statistical model for face representation and recognition. In: Proc. of Int. Conf. on Computer Vision. (2005) 786–791
13. Zhang, B., Shan, S., Chen, X., Gao, W.: Histogram of gabor phase patterns (hgpp): A novel object representation approach for face recognition. IEEE Transactions on Image Processing 16, 57–68 (2007)
14. Liu, J., Chen, S., Zhou, Z.-H., Tan, X.: Single Image Subspace for Face Recognition. In: Zhou, S.K., Zhao, W., Tang, X., Gong, S. (eds.) AMFG 2007. LNCS, vol. 4778, pp. 205–219. Springer, Heidelberg (2007)
15. Schwartz, W.R., Guo, H., Davis, L.S.: A Robust and Scalable Approach to Face Identification. In: Daniilidis, K., Maragos, P., Paragios, N. (eds.) ECCV 2010, Part VI. LNCS, vol. 6316, pp. 476–489. Springer, Heidelberg (2010)
16. Holappa, J., Ahonen, T., Pietikainen, M.: An optimized illumination normalization method for face recognition. In: 2nd IEEE International Conference on Biometrics: Theory, Applications and Systems, BTAS 2008, pp. 1–6 (2008)
17. Huang, G.B., Ramesh, M., Berg, T., Learned-Miller, E.: Labeled faces in the wild: A database for studying face recognition in unconstrained environments. Technical Report 07-49, University of Massachusetts, Amherst (2007)
18. Taigman, Y., Wolf, L., Hassner, T.: Multiple one-shots for utilizing class label information. In: The British Machine Vision Conference (BMVC), London (2009)
19. Seo, H., Milanfar, P.: Face verification using the lark representation. IEEE Transactions on Information Forensics and Security 6, 1275–1286 (2011)
20. Ruiz-del Solar, J., Verschae, R., Correa, M.: Recognition of faces in unconstrained environments: A comparative study. EURASIP Journal on Advances in Signal Processing, Article ID 184617, 19 pages (2009)
21. Wolf, L., Hassner, T., Taigman, Y.: Descriptor based methods in the wild. In: Real-Life Images Workshop at the European Conference on Computer Vision, ECCV (2008)
22. Chen, L., Yan, L., Liu, Y., Gao, L., Zhang, X.: Displacement Template with Divide-&-Conquer Algorithm for Significantly Improving Descriptor Based Face Recognition Approaches. In: Fitzgibbon, A., Lazebnik, S., Perona, P., Sato, Y., Schmid, C. (eds.) ECCV 2012, Part V. LNCS, vol. 7576, pp. 214–227. Springer, Heidelberg (2012)
23. Nowak, E., Jurie, F.: Learning visual similarity measures for comparing never seen objects. In: IEEE Conference on Computer Vision and Pattern Recognition, pp. 1–8 (2007)
24. Huang, C., Zhu, S., Yu, K.: Large scale strongly supervised ensemble metric learning, with applications to face verification and retrieval. Technical Report TR115, NEC (2011)

Face Recognition with Learned Local Curvelet Patterns and 2-Directional L1-Norm Based 2DPCA

Wei Zhou and Sei-ichiro Kamata

Waseda University, Japan

Abstract. In this paper, we propose Learned Local Curvelet Patterns (LLCP) for presenting the local features of facial images. The proposed method is based on curvelet transform which can overcome the weakness of traditional Gabor wavelets in higher dimension, and better capture the curve singularities and hyperplane singularities of facial images. Different from wavelet transform, curvelet transform can effectively and efficiently approximate the curved edges with very few coefficients as well as taking space-frequency information into consideration. First, LLCP designs several learned codebooks from Curvelet filtered facial images. Then each facial image can be encoded into multiple pattern maps and finally block-based histograms of these patterns are concatenated into an histogram sequence to be used as a face descriptor. In order to reduce the face feature descriptor, 2-Directional L1-Norm Based 2DPCA $((2D)^2\text{PCA-L1})$ is proposed which is simultaneously considering the row and column directions for efficient face representation and recognition. Performance assessment in several face recognition problem shows that the proposed approach is superior to traditional ones.

1 Introduction

During the last several decades, face recognition has become a popular area of research in computer vision. Compared with other biometrics [1], such as fingerprint and iris, face recognition has great advantage in high-universality, high-collectability, high-acceptability, and low-circumvention. Hence, face recognition is widely used in a variety of fields such as image analysis, classification, forensic identification and access control. However, due to the fact that the facial appearances are easily affected by the variations of expression, illumination, pose and other factors, it is still an active and challenging research topic.

Recently, local matching approaches are presented in face recognition with invariant to illumination and expression issues. The most famous method is called Local Binary Patterns (LBP) [2]. Since Gabor wavelet has good characteristics in space frequency, space position and direction selectivity, local patterns based on Gabor feature have also been proposed for face representation, such as Local Gabor Binary Patterns (LGBP) [3] and Learned Local Gabor Patterns (LLGP) [4]. Different from LBP or LGBP, in LLGP, first, the patterns are learned by

J.-I. Park and J. Kim (Eds.): ACCV 2012 Workshops, Part I, LNCS 7728, pp. 109–120, 2013.
© Springer-Verlag Berlin Heidelberg 2013

applying the clustering approach to the set of patches to generate several code-books. Second, the facial image is encoded into some pattern maps based on the learned codebooks. However, the common issue of these methods is that the feature dimension is very large due to Gabor decomposition and Gabor trans-form cannot well represent curve singularity of human face images since Gabor wavelets are very effective in representing objects with isolated point singulari-ties, but failed to represent line or curve singularities.

To overcome the weakness of Gabor wavelets, and to better capture the curve singularities and hyperplane singularities of high dimensional signals, Candes and Donoho [5] proposed curvelet transform, which directly takes edges as the basic representation elements and is strongly anisotropic. It is optimal in repre-senting curved singularities in images or higher dimensional signals. The detail and fine coefficients of curvelet are strongly orientation-sensitive, which is a use-ful property for detecting curves in images. In [6], comparison of wavelet, Ga-bor wavelet and curvelet transform for face recognition under illumination and expression changes is discussed and concluded that curvelet is a better choice compared with wavelet and Gabor wavelet, since the curvelet transform has a more sparse representation of the image than wavelet, thus offering a descrip-tion with higher time frequency resolution and high degree of directionality and anisotropy, which is particularly appropriate for many images rich with edges and curves. In a word, Curvelet transform not only captures time-frequency lo-calization property of wavelets but also shows a very high degree of directionality and anisotropy as well as its singularities can be well approximated with very few coefficients.

In traditional curvelet based face recognition problems, researches [7] [8] [9] are only applied some feature reduction methods into curvelet coefficients and do not consider the special patterns in our face images [10]. In this paper, unlike the common methods which used Gabor wavelet to transform facial images into frequency space and overcome the problems of traditional Wavelet and Curvelet feature based face recognition [7], Local Curvelet Patterns are studied. First, according to curvelet filtered facial images, Learned Local Curvelet Patterns (LLCP) is proposed. Then, based on the learned patterns, the facial image can be encoded into multiple pattern maps. At last, the input facial image can be described as a histogram sequence by combining the block-based histograms of the learned patterns together. The proposed patterns have several advantages: first, face-special patterns are designed to encode the face images. Second, it can better obtain curve or line singularities in our face images. Third, it consid-ered both scale and orientation information of our face. Fourth, multi-mapping is used to obtain the final more robust histogram sequence. Fifth, less feature dimension will be generated compared to Gabor wavelet based descriptor. In order to further reduce the feature dimension, the 2-Directional L1-Norm Based 2DPCA($(2D)^2$PCA-L1) for efficient face representation and recognition is devel-oped by simultaneously considering the row and column directions in our face image. Experimental results based on two famous and challenging databases-AR [11] and FERET [12] show the effectiveness of the proposed methods.

The remainder of this paper is organized as follows: LLCP will be described in section 2. In section 3, $(2D)^2$PCA-L1 will be introduced and experiments are presented in section 4. Finally, conclusions are discussed in section 5.

2   Learned Local Curvelet Patterns

Curvelet aims to deal with interesting phenomena occurring along curved edges in a 2D image. As illustrated in [13], curvelet needs fewer coefficients for representation and the edge produced from curvelet is smoother than wavelet edge. Curvelet has several advantages compared to other transforms as follows:

1.) Optimal sparse representation of objects with edges
2.) Optimal image reconstruction in severely ill-posed problems

Curvelet transform is a special member of the multi-scale geometric transforms. It is a transform with multi-scale pyramid with many directions at each length scale. In our study, the facial image is decomposed into coarse, detail and fine coefficients and some reconstructed images including coarse layer, two detailed layer and one fine layer are illustrated in Fig.1. Further, reconstructed images by four orientations of detailed 2 layer are shown in Fig.2. Here, CurveLab 2.1.2 which is available at [14] is used.

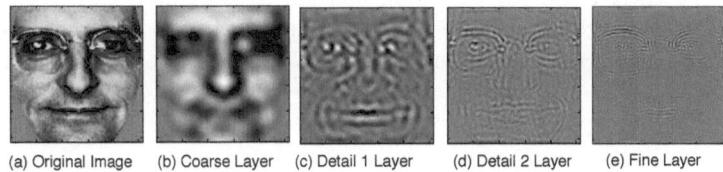

(a) Original Image   (b) Coarse Layer   (c) Detail 1 Layer   (d) Detail 2 Layer   (e) Fine Layer

Fig. 1. Reconstructed Images from different curvelet coefficients

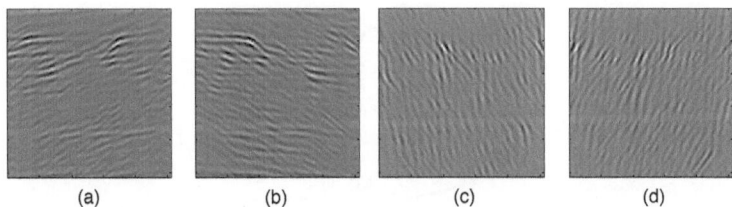

(a)   (b)   (c)   (d)

Fig. 2. Reconstructed Images from detailed 2 layer by four orientations

Generally speaking, face encoding by LLCP can be divided into learning phase and representing phase. In the learning phase, several particular codebooks are constructed while facial images are encoded in the representing phase. In advance, several particular codebook patterns should be learned (shown in Fig.3).

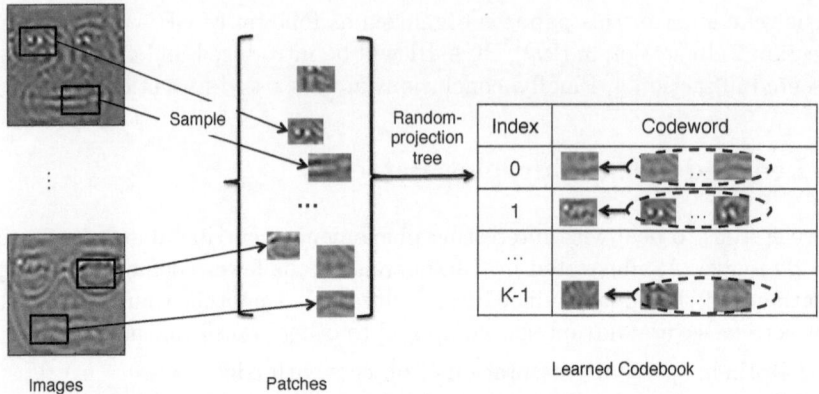

Images   Patches   Learned Codebook

Fig. 3. The process of learning a particular codebook

In this step, images are first sampled into many patches and then all the patches are clustered into K patterns by random-projection tree [15].

In the learning phase (shown in Fig.5), each image in the training set is reconstructed to Curvelet Patterns (CPs) with different curvelet coefficients. Then based on all CPs in the same layer, one patch set can be constructed by sampling patches. Here, the sampling method is illustrated as Fig.4). At last, by using random-projection tree clustering approach to each patch set, LLCP learned codebooks can be constituted. Thus, C LLCP codebooks can be obtained with C layers.

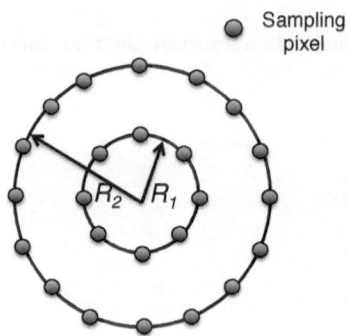

Fig. 4. Sampling method used in our implementation, $R1 = 1$, $R2 = 2$

In the representing phase (shown in Fig.6), first, each facial image is reconstructed to several CPs with different curvelet coefficients. Then each CP is encoded into T pattern maps by mapping its patches to the corresponding LLCP codebook pixel by pixel (Note: here the intensity of mapped image is

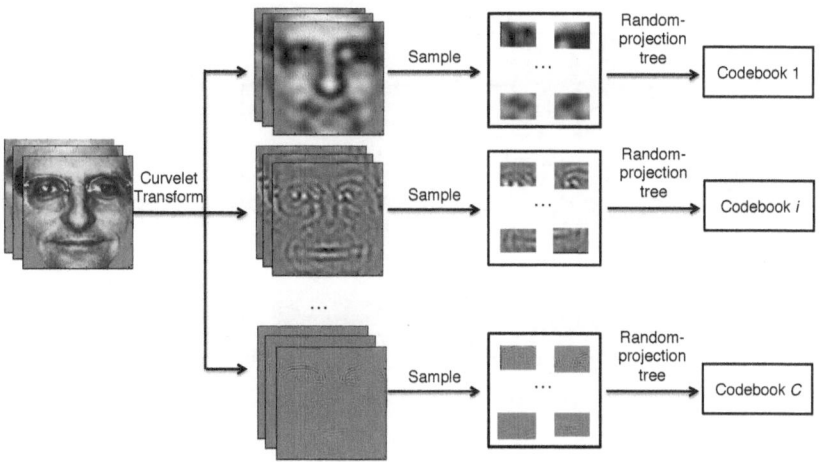

Fig. 5. The learning phase for face encoding by LLCP

corresponding to the type of learned patterns, and also the same intensity in different LLCP maps stands for different type of patterns since the codebook is different.) One point should be noted during representing phase is that the input patch can match top T-th smallest distance codewords in the codebook since the input patch is very similar with the top T-th smallest distance codewords in some cases. Thus, one CP can be encoded by T LLCP maps. And totally $C \times T$ LLCP maps can be obtained. Finally, these pattern maps are spatially divided into many blocks and the histograms of all the blocks are concatenated together to form one enhanced histogram sequence which is considered to represent the input facial image.

3   2-Directional L1-Norm-Based 2DPCA

Recently, a new technique called L1-Norm-Based 2DPCA (2DPCA-L1) was proposed for feature reduction and image representation [16]. As discussed in [16], 2DPCA-L1 avoids computation of the eigenvalue decomposition process and its iteration step is also easy to be performed. Compared to traditional L2-Norm-Based 2DPCA, it not only makes good use of structural information of image but is also robust to outliers. However in [16], 2DPCA-L1 is just works in the row direction of images, in this section, we extend it into both row and column directions.

First, an optimal matrix U can be learned by 2DPCA-L1 from a set of training images to reflect the corresponding row information of images, and then we can project a h by w image X onto U, yielding an h by s matrix $Y^U = XU$. Similarly, the alternate 2DPCA-L1 can learn an optimal matrix V to reflect the corresponding column information between images, and then we can also project X onto V, yielding a t by w matrix $Y^V = V^T X$. In this study, we are about to present an approach to use the projection matrices U and V simultaneously.

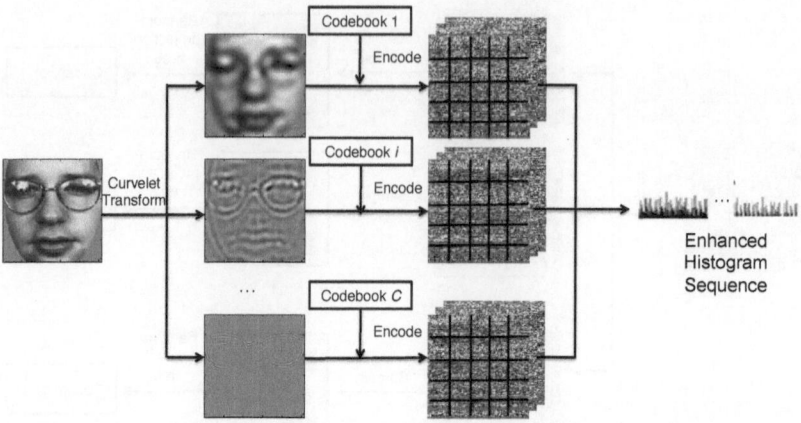

Fig. 6. The representing phase for face encoding by LLCP

Suppose now we have obtained the projection matrices U and V, and then we can project the h by w image X onto U and V simultaneously, yielding a t by s matrix Y as

$$Y = V^T X U \tag{1}$$

The matrix Y can also be named the coefficient matrix in the image representation task, which could be used to reconstruct the original image X by the following equation:

$$\bar{X} = V Y U^T \tag{2}$$

4   Experiments

In this section, two large and widely used database - AR [11] and FERET [12] datasets are considered to evaluate the proposed methods. For AR dataset, the performance is focused on the proposed feature reducing method while LLCP is estimated in FERET dataset detailedly. And in general, 1-NN is used for simple classification.

4.1   Evaluation on AR Database

The AR [11] dataset consists of over 3,200 color images of the frontal images of faces of 126 subjects. Each subject has 26 different images, including frontal views of with different facial expressions, lighting conditions and occlusions. For each subject, these images were recorded in two different sessions which are separated by two weeks, each session consisting of 13 images. For the experiments

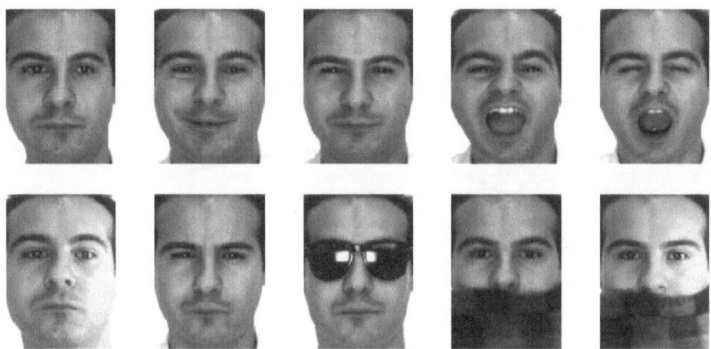

Fig. 7. Some samples from AR dataset

reported in this section, 60 different individuals were randomly selected from this database. Then there are totally 1560 images in our experiments. All the images are manually cropped and resized to 88×80 pixels and divided into 8 by 8 blocks in our study. Some example images of one person are shown in Fig.7.

In the first evaluation, the recognition performance of the subspace algorithms on AR database are compared to judge whether the proposed methods are powerful or not. Four samples of each individual are randomly selected as gallery (training images), and the remaining ones are used for probe (testing images). In our study, we perform 5 times to randomly choose the training set and calculate the average recognition rates.

In our implementation, the codebooks are learned from gallery and K is fixed to 128 according to the following experiments, so totally the feature dimension is $128 \times 64 \times 4 = 32768$ for LLCP based encoding and $256 \times 64 = 16384$ for LBP based encoding. Some results are list in Fig. 8 and Fig. 9. Here $h = 256$, $w = 64$ and $h = 128$, $w = 256$ for LBP and LLCP, respectively.

From these two figures, we can see $(2D)^2$PCA-L1 based feature reduction is better than 2DPCA-L1 based feature reduction with fewer coefficients. Additionally, $(2D)^2$PCA-L1 based approach can reduce feature dimension significantly where the feature dimension is just $50 \times 60 = 3000$ by cutting 90.8% compared to the original feature vector for LLCP encoding.

In the next experiment, some $d(d = 1, 2, 3, 4)$ images of each person are randomly chosen for training, while the remaining images for testing. To compare our method with LBP, LGBP and LLGP, five tests are performed with a varying number of training samples and mean rate is recorded. Table 1 shows the accuracy. The optimal recognition rate by the feature reduction methods is recorded. It can be seen that the proposed methods achieve better performance. Thus, feature reduction approaches can obtain discriminant feature space and $(2D)^2$PCA-L1 is the outstanding one compared to LDA [17], 2DPCA [18] and 2DPCA-L1.

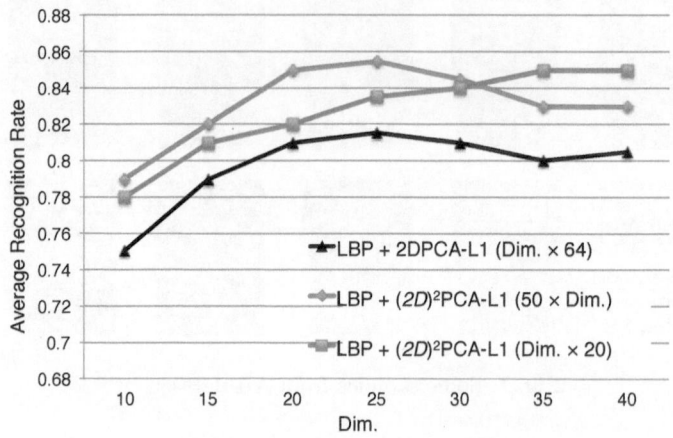

Fig. 8. Performance on AR dataset by LBP+2DPCA-L1 and LBP+$(2D)^2$PCA-L1

Table 1. Accuracy on AR- d (1,2,3,4) for gallery and others for probe

	1	2	3	4
LBP	0.72	0.75	0.77	0.80
LBP+DCT	0.72	0.77	0.78	0.82
LBP+2DPCA	0.73	0.75	0.78	0.81
LBP+2DPCA-L1	0.73	0.76	0.79	0.82
LBP+$(2D)^2$PCA-L1	0.78	0.79	0.82	0.85
LGBP	0.78	0.81	0.83	0.85
LLGP	0.82	0.84	0.85	0.88
LLCP	0.85	0.86	0.89	0.91
LLCP+DCT	0.85	0.88	0.89	0.92
LLCP+2DPCA	0.85	0.87	0.90	0.92
LLCP+2DPCA-L1	0.86	0.88	0.90	0.92
LLCP+$(2D)^2$PCA-L1	0.88	0.91	0.92	0.95

4.2   FERET Database

The FERET database consists of a total of 14,051 gray-scale images representing more than 1,100 individuals. These images contain variations in lighting, facial expressions, pose angle and so on. In this study, only frontal faces are selected. These facial images can be divided into five sets as follows: 1.) Fa set, which is generally used as a gallery set, containing frontal images of 1,196 people. 2.) Fb set which has 1,195 images and the subjects were asked for an alternative facial expression than that in set Fa. 3.) Fc set which were taken under different lighting conditions and contains 194 images. 4.) Dup I set (722 images). The images were taken later compare with Fa or Fb set. 5.) Dup II set (234 images).

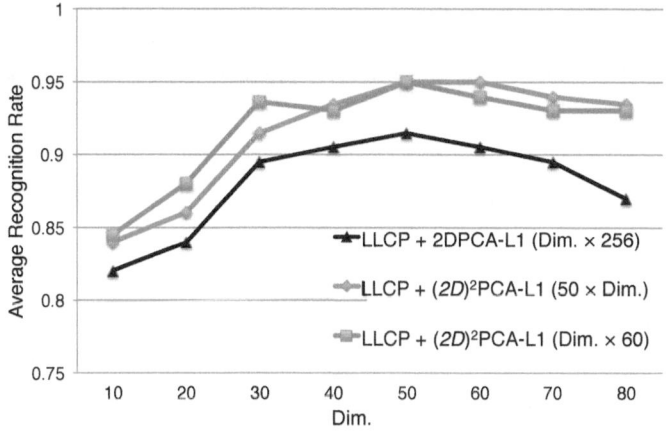

Fig. 9. Performance on AR dataset by LLCP+2DPCA-L1 and LLCP+$(2D)^2$PCA-L1

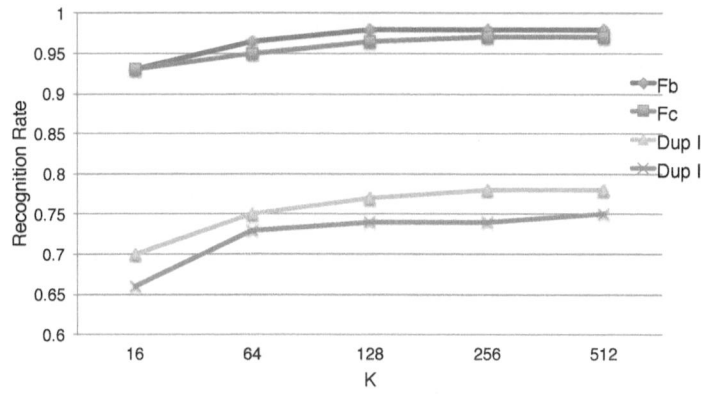

Fig. 10. Performance of LLCP on FERET dataset with different size of codebook

This is a subset of the dup I set containing those images which were taken at least a year after the corresponding gallery image. And in our study all facial images are normalized to 88×80 pixels according to the eye positions provided in the database.

In order to determine how the parameters K affect the final recognition rate, one experiment is evaluated on FERET dataset based on LLCP method. Fig. 10 shows the recognition rates change with the size of codebook K (Here, the block size is fixed to 8×8 and histogram intersection is used for classification). From this figure, we can find that the performance becomes better with the increase of K and slightly change or equal when K is larger than 128.

Next experiment is based on the change of block size while the K is fixed to 128. The results are shown in Fig.11. From this figure, we can seen that small or large block size can decrease the performance, especially with larger one.

When the block size is about 8×8, the best accuracy can be obtained. The possible reason may be that larger block size can not preserve enough spatial information in the facial images while the patterns in smaller block can not discriminate effectively.

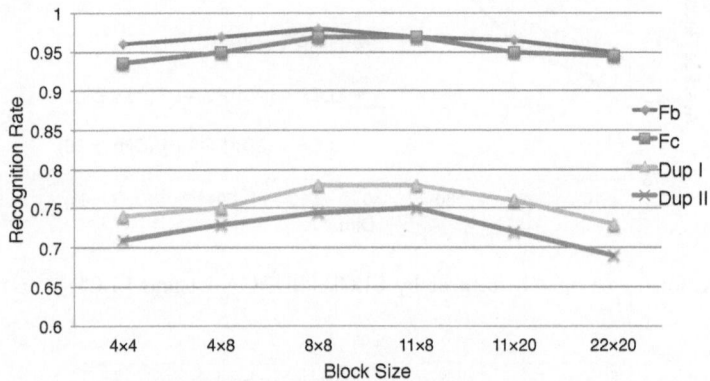

Fig. 11. Performance of LLCP on FERET dataset with different block size

The third experiment is designed to judge whether our multi-mapping is useful or not. The evaluation is list in Fig.12. We can see that a little larger T can improve our performance which is same as our thinking. And if T is so large, that means the input patch is also encoded by some dissimilar patterns which can confuse the distribution of final histogram and decrease our final result.

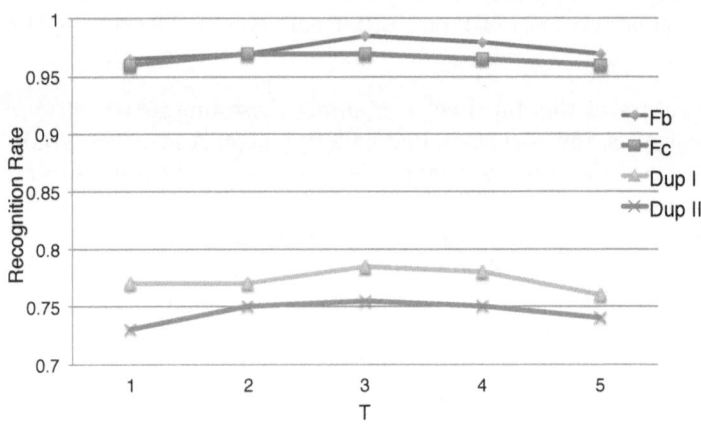

Fig. 12. Performance of LLCP onFERET dataset with different parameter T

Table 2. Precision in FERET [12] database

	Fb	Fc	Dup I	Dup II
LBP	0.91	0.43	0.58	0.42
LGBP [3]	0.94	0.97	0.68	0.53
LLGP [4]	0.97	0.97	0.75	0.71
POEM-HS [19]	0.98	0.96	0.78	0.76
Retina filter + POEM-HS [19]	0.98	0.99	0.80	0.79
LLCP	0.98	0.97	0.78	0.75
LLCP+DCT	0.98	0.96	0.79	0.77
LLCP+2DPCA	0.98	0.97	0.80	0.77
LLCP+2DPCA-L1	0.99	0.97	0.81	0.79
LLCP+$(2D)^2$PCA-L1	0.99	0.98	0.83	0.81

At last, the total performance on FERET dataset is list in Table 2. From Table 2, we can get the effectiveness of Curvelet transform and the proposed feature dimension reduction method.

5   Conclusions

In this paper, first, Learned Local Curvelet Patterns (LLCP) for presenting the local patterns is proposed. The represented facial images can better capture the curve singularities and hyperplane singularities than some traditional methods, such as LGBP, LLGP. Some codebooks are learned from sampled patches which is regraded as face-specific. Second, one feature reduction method called $(2D)^2$PCA-L1 is proposed, which is simultaneously considering the row and column directions for efficient face representation and recognition. Experiments in face recognition show the effectiveness of our proposed local descriptors and illustrated the powerful of our feature reduction method.

References

1. Jain, A.K.: Biometric recognition: how do i know who you are? In: Proceedings of 12th IEEE Signal Processing and Communications Applications Conference, pp. 3–5 (2004)
2. Ahonen, T., Hadid, A., Pietikainen, M.: Face description with local binary patterns: Application to face recognition. IEEE Trans. Pattern Anal. Mach. Intell. 28, 2037–2041 (2006)
3. Zhang, W., Shan, S., Gao, W., Chen, X., Zhang, H.: Local gabor binary pattern histogram sequence (lgbphs): a novel non-statistical model for face representation and recognition. In: ICCV, pp. 786–791 (2005)
4. Xie, S., Shan, S., Chen, X., Meng, X., Gao, W.: Learned local gabor patterns for face representation and recognition. Signal Processing, 2333–2344 (2009)
5. Candes, E., Demanet, L., Donoho, D., Ying, L.: Fast discrete curvelet transforms. Multiscale Modeling and Simulation 5, 861–899 (2006)

6. Zhang, J., Wang, Y., Zhang, Z., Xia, C.: Comparison of wavelet, gabor and curvelet transform for face recognition. Optica Applicata XLI, 183–193 (2011)
7. Mandal, T., Wu, Q.J.: Face recognition using curvelet based pca. In: ICPR, pp. 1–4 (2008)
8. Huo, H., Song, E.: Face recognition using curvelet and selective pca. In: ICICIP, pp. 348–351 (2010)
9. Zhang, J., Wang, Y.: A comparative study of wavelet and curvelet transform for face recognition. In: CISP, pp. 1718–1722 (2010)
10. Saha, A., Wu, Q.: Facial expression recognition using curvelet based local binary patterns. In: 2010 IEEE International Conference on Acoustics Speech and Signal Processing (ICASSP), pp. 2470–2473 (2010)
11. Martinez, A., Benavente, R.: The ar-face database. CVC Technical Report 24 (1998)
12. Phillips, P.J., Moon, H., Rauss, P.J., Rizvi, S.: The feret evaluation methodology for face recognition algorithms. IEEE Trans. Pattern Anal. Mach. Intell. 22 (2000)
13. Boubchir, L., Fadili, J.: Multivariate statistical modelling of images with the curvelet transform. Image Processing Group, 747–750 (2005)
14. http://www.curvelet.org
15. http://cseweb.ucsd.edu/naverma/RPTrees/index.html
16. Li, X., Pang, Y., Yuan, Y.: L1-norm-based 2dpca. IEEE Transactions on Systems, Man, and Cybernetics, Part B: Cybernetics 40, 1170–1175 (2009)
17. Zhao, W., Chellappa, R., Krishnaswamy, A.: Discriminant analysis of principal components for face recognition. In: 3rd International Conference on Automatic Face and Gesture Recognition (1998)
18. Yang, J., Zhang, D., Frangi, A.F., Yang, J.Y.: Two-dimensional pca: a new approach to appearance-based face representation and recognition. IEEE Trans. Pattern Anal. Mach. Intell. 26, 131–137 (2004)
19. Vu, N.S., Dee, H.M., Caplier, A.: Face recognition using the POEM descriptor. Pattern Recognition 45, 2478–2488 (2012)

LBP − TOP Based Countermeasure against Face Spoofing Attacks

Tiago de Freitas Pereira[1,2], André Anjos[3],
José Mario De Martino[1], and Sébastien Marcel[3]

[1] School of Electrical and Computer Engineering - University of Campinas
(UNICAMP)
[2] CPqD Telecom & IT Solutions
[3] IDIAP Research Institute
tiagofrepereira@gmail.com, {andre.anjos,marcel}@idiap.ch,
martino@fee.unicamp.br

Abstract. User authentication is an important step to protect information and in this field face biometrics is advantageous. Face biometrics
is natural, easy to use and less human-invasive. Unfortunately, recent
work has revealed that face biometrics is vulnerable to spoofing attacks
using low-tech cheap equipments. This article presents a countermeasure
against such attacks based on the $LBP − TOP$ operator combining both
space and time information into a single multiresolution texture descriptor. Experiments carried out with the REPLAY ATTACK database show
a Half Total Error Rate ($HTER$) improvement from 15.16% to 7.60%.

1 Introduction

Despite the progress in the last years, automatic face recognition is still an active
research area. Many tasks, such as recognition under occlusion or recognition in
a crowd and with complex illumination conditions still represent unsolved challenges. Advances in the area were extensively reported in [8] and [16]. However,
the issue of verifying if the face presented to a camera is indeed a face from a
real person and not an attempt to deceive (spoof) the system has received less
attention.

A spoofing attack consists in the use of forged biometric traits to gain illegitimate access to secured resources protected by a biometric authentication
system. The lack of resistance to attacks is not exclusive to face biometrics. [23],
[14] and [18] indicate that fingerprint authentication systems suffer from similar
weakness. [11], [12] and [19] diagnose the same shortcoming on iris recognition
systems. Finally, [5] and [7] address spoofing attacks to speaker biometrics. The
literature review for spoofing in face recognition systems will be presented in
Section 2.

In authentication systems based on face biometrics, spoofing attacks are usually perpetrated using photographs, videos or forged masks. Moreover, with the
increasing popularity of social networks websites (facebook, flicker, youtube, in-
stagram and others) a great deal of multimedia content is available on the web

J.-I. Park and J. Kim (Eds.): ACCV 2012 Workshops, Part I, LNCS 7728, pp. 121–132, 2013.
© Springer-Verlag Berlin Heidelberg 2013

that can be used to spoof a face authentication system. In order to mitigate the vulnerability of face authentication systems, effective countermeasures against face spoofing have must be deployed.

In this context, we proposed a novel countermeasure against face spoofing. Our approach uses an operator called Local Binary Patterns from Three Orthogonal Planes (LBP-TOP) that combines space and time information into a single descriptor with a multiresolution strategy. Experiments conducted using the REPLAY ATTACK database [6] indicate that our approach has a better performance in detecting face spoofing attacks using photographs and videos than state-of-the-art techniques.

The remainder of the paper is organized as follows: Section 2 briefly review the relevant literature. Section 3 discusses the application of Local Binary Patterns (LBP) in space and time domains. Section 4 presents our approach against facial spoofing attacks. Our experimental set-up and results are discussed in Section 5. Finally, in Section 6 we summarize this work highlighting its main contributions.

2   Prior Work

Considering the type of countermeasures that do not require user collaboration, Chakka et al. in [4] made a classification considering the following cues in spoofing attacks:

- Presence of vitality (liveness);
- Differences in motion patterns;
- Differences in image quality assessment.

Presence of vitality or **liveness** detection consists in the search of features that only live faces can possess. For example, Pan et al. in [20] develop a countermeasure based on eye-blink.

The countermeasures based on differences in **motion** patterns rely on the fact that real faces displays different motion behavior compared to a spoof attempt. Kollreider et al. [13] present a motion based countermeasure that estimates the correlation between different regions of the face using optical flow. In that countermeasure, the input is considered a spoof if the optical flow field on the center of the face and on the center of the ears present the same direction. The performance was evaluated using the subset "Head Rotation Shot" of the XM2VTS database whose real access was the videos of this subset and the attacks were generated with hard copies of those data. With this database, that was not made publicly available, an Equal Error Rate (EER) of 0.5% was achieved. Anjos et al. [3] present a motion based countermeasure measuring the correlation between the face and the background through simple frame differences. With the PRINT ATTACK database, that approach presented a good discrimination power ($HTER$ equals to 9%).

Countermeasures based on differences in **image quality assessment** rely on the presence of artifacts intrinsically present at the attack media. Such remarkable properties can be originated from media quality issues or differences in reflectance properties. Li et al. [15] hypothesize that fraudulent photographs have less high frequency components than real ones. To test the hypothesis a small database was built with 4 identities containing both real access and printed photo attacks. With this **private** database, an accuracy of 100% was achieved. Because of differences in reflectance properties, real faces very likely present different texture patterns compared with fake faces. Following that hypothesis, Maatta et al. [17] and Chingovska et al. [6] explored the power of Local Binary Patterns (LBP) as a countermeasure. Maatta et al. combined 3 different LBP configurations ($LBP_{8,2}^{u2}$, $LBP_{16,2}^{u2}$ and $LBP_{8,1}^{u2}$) in a normalized face image and trained a SVM classifier to discriminate real and fake faces. Evaluations carried out with NUAA Photograph Impostor Database [21] showed a good discrimination power (2.9% in EER). Chingovska et al. analyzed the effectiveness of $LBP_{8,1}^{u2}$ and set of extended LBPs [22] in still images to discriminate real and fake faces. Evaluations carried out with three different databases, the NUAA Photograph Impostor Database, REPLAY ATTACK database and CASIA - Face Anti-spoofing Database [24] showed a good discrimination power with $HTER$ equals to 15.16%, 19.03% and 18.17% respectively. Assuming that real access images concentrate more information in a specific frequency band, Zhang et al. [24] used, as countermeasure, a set of DoG filters to select a specific frequency band to discriminate attacks and non attacks. Evaluations carried out with the CASIA - Face Anti-spoofing Database showed an Equal Error Rate of 17.00%.

3 LBP in Space and Time Domain

Maatta et al. [17] and Chingovska et al. [6] propose LBP based countermeasures to spoofing attacks based on the hypothesis that real faces present different texture patterns in comparison with fake ones. However, the proposed techniques analyze each frame in isolation, not considering the behavior over time. As pointed out in Section 2, motion is a cue widely used and in combination with texture can generate a powerful countermeasure.

The first attempt to extend LBP to image sequences, exploring the space and time information, was introduced with the concept of Volume Local Binary Patterns ($VLBP$) [25]. To capture interframe patterns in textures, $VLBP$ considers the frame sequence as a parallel sequence. Considering a 3×3 kernel and thresholding the surroundings of each pixel with the central pixel of the frame sequence, the result is considered a binary value and its decimal representation is:

$$VLBP_{L,P,R} = \sum_{q=0}^{3P+1} f(i_c - i_q)2^q, \tag{1}$$

where L corresponds to the number of predecessors and successors frames, P is the number of neighbors of i_c that corresponds to the gray intensity of the evaluated pixel, i_q corresponds to the gray intensity of a specific neighbor of i_c, R is the radius of considered neighborhood and $f(x)$ is defined as follows:

$$f(x) = \begin{cases} 0 \ if \ x < 0 \\ 1 \ if \ x \geq 0 \end{cases}. \tag{2}$$

An histogram of this descriptor, contains 2^{3P+1} elements. Considering $P = 8$ (the most common configuration [6] [17] [1]) the number of bins in such histogram will be $33,554,432$ which is not computationally tractable.

To address this issue, [25] presented a simplification of the $VLBP$ operator; the so called LBP from Three Orthogonal Planes ($LBP - TOP$). Instead of considering the frame sequence as a three parallel planes, the $LBP - TOP$ consider three orthogonal planes intersecting the center of a pixel in the XY direction (normal LBP [1]), XT direction and YT direction, where T is the time axis (the frame sequence). Considering three orthogonal planes intersecting each pixel in a frame sequence, three different histograms are generated and then concatenated, as it can be seen in Fig. 1. With this approach, the size of the histogram decreases to $3 * 2^P$.

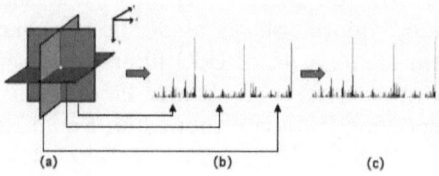

(a)   (b)   (c)

Fig. 1. (a) Three planes intersecting one pixel (b) LBP histogram of each plane (c) Concatenating the histograms (courtesy of [25])

In the $LBP - TOP$ representation, the radii in each direction (R_X, R_Y and R_T) and the number of sampling points in each plane (P_{XY}, P_{XT} and P_{YT}) can be different as well as the type of LBP operator in each plane. They can follow the normal, the uniform pattern ($u2$) or rotation invariant uniform pattern ($riu2$) approaches [10], for example. The representation of the $LBP-TOP$ descriptor is denoted as $LBP - TOP^{operator}_{P_{XY},P_{XT},P_{YT},R_X,R_Y,R_T}$. In addition to the computational simplification, compared with $VLBP$, $LBP - TOP$ has the advantage to generate independent histograms for each of intersecting planes, in space and time, which can be treated in combination or individually.

Because of the aforementioned complexity issues on the implementation of a $VLBP$ based processor, the developed countermeasure uses $LBP - TOP$ to extract spatio-temporal information from video sequences.

4   The Proposed Countermeasure

Fig. 2 shows a block diagram of the proposed countermeasure. First, each frame of the *original frame sequence* was gray-scaled and passed through a face detector using MCT features [9]. Only *detected faces* with more than 50 pixels of width

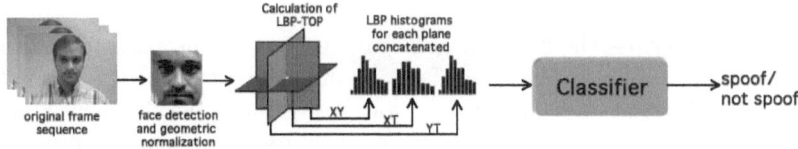

Fig. 2. Block diagram of the proposed countermeasure

and height were considered. The detected faces were geometric normalized to 64×64 pixels. In order to reduce the face detector noise, for each set of frames used in the $LBP - TOP$ calculation, the same face bounding box was used. As can be seen in the Fig. 3, the middle frame was chosen. Unfortunately, the face detector is not error free and in case of error in the middle frame face detection, the nearest detection was chosen otherwise the observation was discarded.

After face detection step, the LBP *operators* were calculated for each plane (XY, XT and YT) and the *histograms* were computed and then concatenated.

To generate a multiresolution description, the histograms in time domain (XT and YT) are concatenated for different values of R_t. The notation chosen to represent these settings is using brackets for the multiresolution data. For example, $R_t = [1, 3]$ means that the $LBP - TOP$ operator will be calculated for $R_t = 1$, $R_t = 2$ and $R_t = 3$ and all resultant histograms will be concatenated. After the feature extraction step, this data is ready for binary *classification* to discriminate spoofing attacks from real accesses.

In order to be comparable with [6], each observation in the original frame sequence will generate a score independent of the rest of the frame sequence.

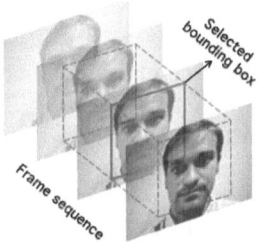

Fig. 3. Face detection strategy for $R_t = 1$

The proposed countermeasure was implemented using the free signal-processing and machine learning toolbox Bob [2] and the source code of the algorithm is available as an add-on package to this framework[1].

---

[1] http://pypi.python.org/pypi/antispoofing.lbptop

5 Experiments

This section describes the performance evaluation of the proposed countermeasure on the REPLAY-ATTACK database [6] and using its defined protocol. Such protocol defines 3 non-overlapped partitions for training, development and testing countermeasures. The training set should be used to train the countermeasure, the development set is used to tune the parameters of the countermeasure and to estimate a threshold value to be used in the test set. The protocol defines the Equal Error Rate (EER) as a decision threshold. Finally, the test set must be used only to report results. As performance measurement, the protocol suggests to report the Half Total Error Rate ($HTER$) on the test data.

5.1 Evaluation Methodology

In order to measure the effectiveness of this countermeasure, each parameter was tuned solely (fixing other elements) using the development set. For this, 5 experiments were carried out evaluating the effectiveness of:

1. Each $LBP - TOP$ plane;
2. Different classifiers;
3. Different LBP operators;
4. Different numbers of sampling points in the $LBP - TOP$ operator
5. Multiresolution approach.

Inspired on [6], the $LBP - TOP$ operator chosen to start the evaluation was $LBP - TOP^{u2}_{8,8,8,1,1,R_T}$.

5.2 Effectiveness of Each $LBP - TOP$ Plane

Fig. 4 shows the evolution of the test set $HTER$ considering individual and combined histograms of $LBP - TOP$ planes. First, it was analyzed the effectiveness of each individual plane and then combinations when the multiresolution area (R_t) is increased. We used, as binary classifier, a linear projection derived from Linear Discriminant Analysis LDA as is [6].

It can be seen that, by combining the time components (XT and YT planes) the results were improved. This suggests that the time information is an important cue. The combination of the three planes generated the best results which suggests that both spatial and time information are important to classify real and fake faces. For that reason, next results will be presented always with a combination of the three $LBP - TOP$ planes (XY, XT and YT).

5.3 Effectiveness of Different Classifiers

Fig. 5 shows the performance of this countermeasure, in $HTER$ terms, with different classifiers when the multiresolution area (R_t) is increased. The first classifier applied was the χ^2 distance, since the feature vectors are histograms.

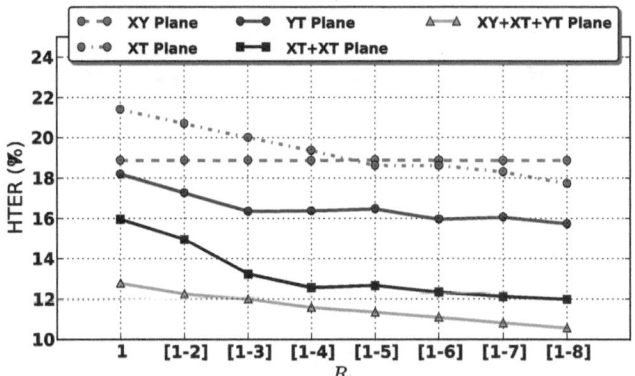

Fig. 4. (Color online) Evaluation of HTER(%) in each plane when the multiresolution area (R_t) is increased with $LBP - TOP^{u2}_{8,8,8,1,1,R_t}$ and LDA classifier - test-set

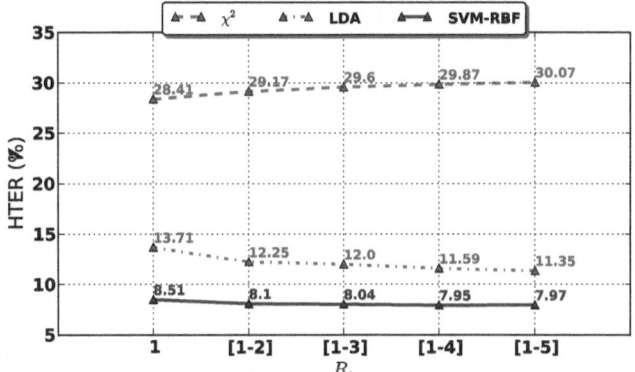

Fig. 5. (Color online) Evaluation of $HTER(\%)$ with $LBP - TOP^{u2}_{8,8,8,1,1,R_t}$ using different classifiers

For that, the same strategy adopted in [6] was carried out. A reference histogram only with real accesses was created averaging the histograms in the training set. Experiments using more complex classifiers were carried out as well. For that, Linear Discriminant Analysis (LDA) and Support Vector Machines (SVM) with a radial basis function kernel (RBF) were chosen.

It can be seen that best results were obtained with the non linear SVM using RBF kernel. It is important to remark that results presented with SVM, should be analyzed carefully for overtraining. The final machine uses ~ 25000 support vectors to achieve 7.97%. This number represents $\sim 33\%$ of the training set size. A simple comparison with the same $LBP - TOP$ configuration with LDA classifier resulted in an $HTER$ equal to 11.35%. This is not a huge gap and the classifier is far simpler.

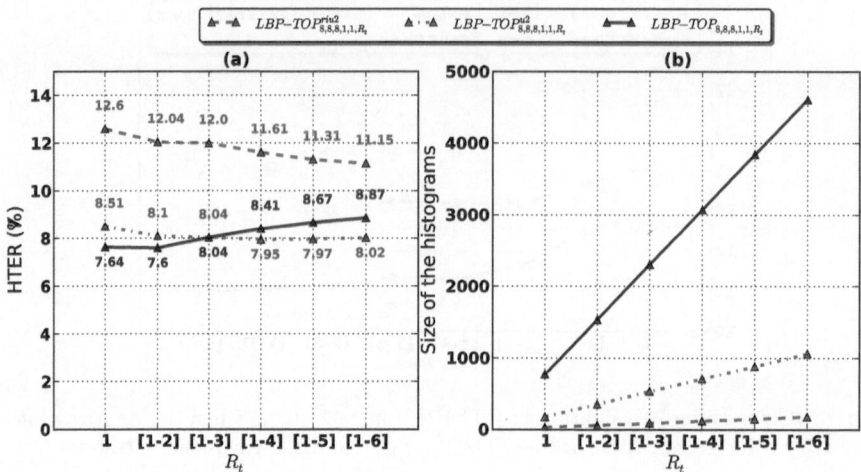

Fig. 6. (Color online) (a) Evaluation of $HTER(\%)$ with $LBP - TOP_{8,8,8,1,1,R_t}$ using different LBP configurations in the planes with SVM classifier (b) Evaluation of the histogram size when (R_t) is increased

5.4   Effectiveness of Different LBP Operators

The size of the histogram in a multiresolution analysis, in time domain, increases linearly with R_t. The choice of an appropriate LBP representation in the planes is an important issue since this choice impacts the size of the histograms. Using uniform patterns or rotation invariant extensions, in one or multiple planes, may bring a significative advantage in computational complexity. Fig. 6 (a) shows the performance, in $HTER$ terms, configuring each plane as normal LBP (with 256 bins for $P = 8$), LBP^{u2} and LBP^{riu2} when the multiresolution area (R_t) is increased. Results must be interpreted with the support with the Fig. 6 (b), which shows the number of bins on the histograms used for classifications in each configuration.

It can be seen that, when R_t is increased, the $HTER$ saturates in $\sim 11\%$ and $\sim 8\%$ for LBP^{riu2} and LBP^{u2} respectivelly. The normal LBP operator presents a minimum in 7.60% with $R_t = [1, 2]$ (the best result achieved in this paper). Results with LBP and LBP^{u2} presented similar performance and even the LBP presented the best result, using LBP^{u2} seems a reasonable tradeoff between computational complexity and performance (in $HTER$ terms). Hence we will still proceed with LBP^{u2}.

5.5   Effectiveness of Different Numbers of Sampling Points in the $LBP - TOP$ Operator

Another parameter that impacts in the size of the histograms is the number of sampling points (P) in each plane. Fig. 7(a) and (b) show the performance, in

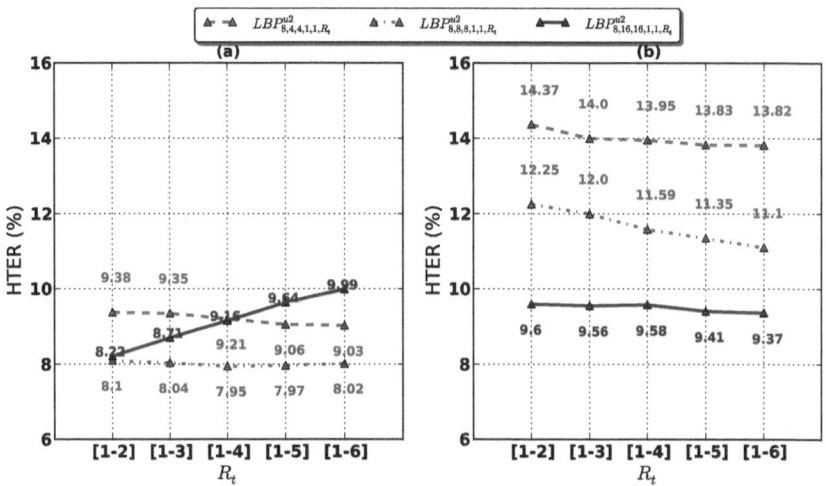

Fig. 7. (Color online) Evaluation of $HTER(\%)$ with $LBP - TOP^{u2}$ using different values for P_{XT} and P_{YT} in the time planes using **(a)** SVM classifier **(b)** LDA classifier

$HTER$, varying the values of P_{XT} and P_{YT} to 4, 8 and 16 when the multiresolution area (R_t) is increased with SVM and LDA classifiers respectively.

It can be seen that results with $LBP - TOP^{u2}_{8,8,8,1,1,R_t}$ achieved the best performance (saturating around 8%), using an SVM classifier (see Fig. 7(a)). However, it was expected good performance using P_{XT} and P_{YT} set to 16 when the multiresolution analysis (R_t) is increased, since more points were extracted over the time. Observing the Fig. 7 (b), with LDA as a classifier, the best performance was achieved with P_{XT} and P_{YT} equal to 16. These results suggests that, when the multiresolution area is increased with P_{XT} and P_{YT} equals to 16, the SVM classifier loses generalization power. In order to track that hypothesis, a simple observation in the number of support vectors can be done. Not surprisingly, the number of support vectors increases from ~ 30000 to ~ 35000 for R_t equals to $[1,2]$ and $[1,6]$ respectively. That increase, in the final SVM, represents $\sim 32\%$ and $\sim 39\%$ of the training set size respectively, re-assign the overtraining hypothesis. Hence we will still proceed with $LBP^{u2}_{8,8,8,1,1,R_t}$ for the next experiment.

5.6   Effectiveness of Multiresolution Approach

Fig. 8 shows the performance of this countermeaure considering a multiresolution approach compared with a single resolution approach. The single resolution approach consists in using only fixed values for R_t, without concatenating histograms for each R_t. With this approach the size of the histograms will be constant along R_t increase, what decreases the computational complexity.

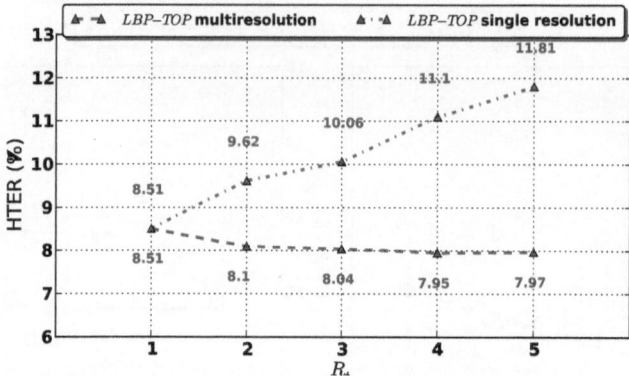

Fig. 8. (Color online) Evalutation of $HTER\%$ using $LBP-TOP^{u2}_{8,8,8,1,1,R_t}$ with and without histogram concatenation using SVM classifier

It can be seen that, when the single resolution approach is considered, the HTER increases with R_t whereas the multiresolution approach helps to keep the HTER low with the increasing value of R_t. It is possible to suggest that, for the $LBP-TOP$ descriptor, motion patterns between closest frames carry more information for spoofing detection than distant ones. Nevertheless, information from distant frames are important as well and thats help to explain why the best results were achieved with the multiresolution approach.

5.7   Summary

Table 5.7 summarizes all results obtained compared with the state of art results. The two first rows are results presented in [6] and the third row was a counter-measure based on [3] whose source code is freely available for comparison. It can be seen that the proposed countermeasure presented the best results, overtaking the state of art results in the REPLAY ATTACK database.

Table 1. HTER(%) of classification with different countermeasures

	$HTER(\%)$	
	dev	**test**
$LBP^{u2}_{8,1} + SVM$ [6]	14.84	15.16
$(LBP^{u2}_{8,2}+LBP^{u2}_{16,2}+ LBP^{u2}_{8,1}) + $ SVM [6]	13.90	13.87
Motion coefficient based [3]	11.78	11.79
$LBP-TOP^{riu2}_{8,8,8,1,1,[1-6]} + SVM$	9.78	11.15
$LBP-TOP^{u2}_{8,4,4,1,1,[1-6]} + SVM$	8.49	9.03
$LBP-TOP^{u2}_{8,8,8,1,1,[1-4]} + SVM$	8.49	7.95
$LBP-TOP_{8,8,8,1,1,[1-2]} + SVM$	7.88	**7.60**
$LBP-TOP_{8,16,16,1,1,[1-2]} + SVM$	9.16	8.22

6   Conclusion

This article presented a countermeasure against face spoofing attacks using the *LBP − TOP* descriptor combining both space and time information into a single descriptor. Experiments carried out with the REPLAY ATTACK database showed that an analysis in time domain improved the results comparing to the still frame analysis presented in [6] and [17]. A multiresolution analysis in time domain shows even better results, achieving 7.60% when combined with an SVM classifier (the best result achieved). It is important to remark that results with SVM classifier should be taken with care because with the increase of the multiresolution area, the SVM classifier tends to overtrain on the data. However, experiments with simpler classifiers, such as LDA, showed that the *LBP − TOP* multiresolution approach still demonstrated a great potential against face spoofing in different kind of attacks scenarios, beating the state of art results.

Acknowledgement. The authors would like to thank the Swiss Innovation Agency (CTI Project Replay), the FP7 European TABULA RASA Project (257289), FUNTTEL (Brazilian Telecommunication Technological Development Fund) and CPqD Telecom & IT Solutions for their financial support.

References

1. Ahonen, T., Hadid, A., Pietikainen, M.: Face Description with Local Binary Patterns: Application to Face Recognition. IEEE Transactions on Pattern Analysis and Machine Inteligence 28, 2037–2041 (2006)
2. Anjos, A., El Shafey, L., Wallace, R., Günther, M., McCool, C., Marcel, S.: Bob: a free signal processing and machine learning toolbox for researchers. In: 20th ACM Conference on Multimedia Systems (ACMMM), Nara, Japan (2012)
3. Anjos, A., Marcel, S.: Counter-measures to photo attacks in face recognition: a public database and a baseline. In: IAPR IEEE International Joint Conference on Biometrics (IJCB), Washington DC, USA (2011)
4. Chakka, M., Anjos, A., Marcel, S., Tronci, R., Muntoni, D., Fadda, G., Pili, M., Sirena, N., Murgia, G., Ristori, M., Roli, F., Yan, J., Yi, D., Lei, Z., Zhang, Z.: Competition on counter measures to 2-d facial spoofing attacks. In: IAPR IEEE International Joint Conference on Biometrics (IJCB), Washington DC, USA (2011)
5. Chetty, G., Wagner, M.: Liveness verification in audio-video speaker authentication. In: Proceeding of International Conference on Spoken Language Processing, ICSLP, vol. 4, pp. 2509–2512 (2004)
6. Chingovska, I., Anjos, A., Marcel, S.: On the effectiveness of local binary patterns in face anti-spoofing. In: IEEE BIOSIG 2012 (2012)
7. Eveno, N., Besacier, L.: A speaker independent "liveness" test for audio-visual biometrics. In: Ninth European Conference on Speech Communication and Technology (2005)
8. Flynn, P., Jain, A., Ross, A.: Handbook of biometrics. Springer (2008)
9. Froba, B., Ernst, A.: Face detection with the modified census transform. In: 2004 Proceedings of the Sixth IEEE International Conference on Automatic Face and Gesture Recognition, pp. 91–96. IEEE (2004)

10. Inen, M., Pietikäinen, M., Hadid, A., Zhao, G., Ahonen, T.: Computer Vision Using Local Binary Patterns, vol. 40. Springer (2011)
11. Johnson, P., Tan, B., Schuckers, S.: Multimodal fusion vulnerability to non-zero effort (spoof) imposters. In: 2010 IEEE International Workshop on Information Forensics and Security (WIFS), pp. 1–5. IEEE (2010)
12. Kanematsu, M., Takano, H., Nakamura, K.: Highly reliable liveness detection method for iris recognition. In: 2007 Annual Conference, SICE, pp. 361–364. IEEE (2007)
13. Kollreider, K., Fronthaler, H., Bigun, J.: Non-intrusive liveness detection by face images. Image and Vision Computing 27, 233–244 (2009)
14. Leyden, J.: Gummi bears defeat fingerprint sensors. The Register - 16 (2002)
15. Li, J., Wang, Y., Tan, T., Jain, A.: Live face detection based on the analysis of fourier spectra. Biometric Technology for Human Identification 5404, 296–303 (2004)
16. Li, S., Jain, A.: Handbook of face recognition. Springer (2011)
17. Maatta, J., Hadid, A., Pietikäinen, M.: Face spoofing detection from single images using texture and local shape analysis. IET Biometrics 1, 3–10 (2012)
18. Matsumoto, T., Matsumoto, H., Yamada, K., Hoshino, S.: Impact of artificial gummy fingers on fingerprint systems. In: Proceedings of SPIE, vol. 4677, pp. 275–289 (2002)
19. Pacut, A., Czajka, A.: Aliveness detection for iris biometrics. In: Proceedings of the 2006 40th Annual IEEE International Carnahan Conferences Security Technology, pp. 122–129. IEEE (2006)
20. Pan, G., Sun, L., Wu, Z., Lao, S.: Eyeblink-based anti-spoofing in face recognition from a generic webcamera. In: IEEE 11th International Conference on Computer Vision, ICCV 2007, pp. 1–8. IEEE (2007)
21. Tan, X., Li, Y., Liu, J., Jiang, L.: Face Liveness Detection from a Single Image with Sparse Low Rank Bilinear Discriminative Model. In: Daniilidis, K., Maragos, P., Paragios, N. (eds.) ECCV 2010, Part VI. LNCS, vol. 6316, pp. 504–517. Springer, Heidelberg (2010)
22. Trefný, J., Matas, J.: Extended set of local binary patterns for rapid object detection. In: Proceedings of the Computer Vision Winter Workshop, vol. 2010 (2010)
23. Uludag, U., Jain, A.: Attacks on biometric systems: a case study in fingerprints. In: Proc. SPIE-EI, pp. 622–633 (2004)
24. Zhang, Z., Yan, J., Liu, S., Lei, Z., Yi, D., Li, S.: A face antispoofing database with diverse attacks. In: 2012 5th IAPR International Conference on Biometrics (ICB), pp. 26–31. IEEE (2012)
25. Zhao, G., Pietikainen, M.: Dynamic texture recognition using local binary patterns with an application to facial expressions. IEEE Transactions on Pattern Analysis and Machine Intelligence 29, 915–928 (2007)

An Efficient LBP-Based Descriptor for Facial Depth Images Applied to Gender Recognition Using RGB-D Face Data

Tri Huynh, Rui Min, and Jean-Luc Dugelay

Department of Multimedia Communications, EURECOM, Sophia Antipolis, France
{huynhq,min,jld}@eurecom.fr

Abstract. RGB-D is a powerful source of data providing the aligned depth information which has great potentials in improving the performance of various problems in image understanding, while Local Binary Patterns (LBP) have shown excellent results in representing faces. In this paper, we propose a novel efficient LBP-based descriptor, namely Gradient-LBP (G-LBP), specialized to encode the facial depth information inspired by 3DLBP, yet resolves its inherent drawbacks. The proposed descriptor is applied to gender recognition task and shows its superiority to 3DLBP in all the experimental setups on both Kinect and range scanner databases. Furthermore, a weighted combination scheme of the proposed descriptor for depth images and the state-of-the-art LBP^{U2} for grayscale images applied in gender recognition is proposed and evaluated. The result reinforces the effectiveness of the proposed descriptor in complementing the source of information from the luminous intensity. All the experiments are carried out on both the high quality 3D range scanner database - Texas 3DFR and images of lower quality obtained from Kinect - EURECOM Kinect Face Dataset to show the consistency of the performance on different sources of RGB-D data.

1 Introduction

Originally proposed by Ojala et al. [1] for texture analysis, Local Binary Patterns (LBP) has now shown its leading performance in a wide range of applications, especially in facial image processing. A large number of works demonstrating excellent results in applying LBP variants to various tasks ranging from face recognition [2], facial expression analysis [3] to age estimation [4], gender and ethnicity classification [5][6], etc. could be widely found in literature recently.

Due to its simplicity yet very powerful discriminative capability, many LBP variants have been developed since its first introduction. Most of them focus solely on luminous intensity [7][8]. Some other methods also extend the LBP approach to 3D data [9] or spatio-temporal signals [10]. However, there are very few efforts in customizing the technique for depth images and RGB-D data. Meanwhile, the explosive development of 3D content and devices recently has made the depth information widely available and successfully exploited in many applications. The access to the RGB-D source of information of home customers

J.-I. Park and J. Kim (Eds.): ACCV 2012 Workshops, Part I, LNCS 7728, pp. 133–145, 2013.

has never been as easy with the introduction of Kinect-like devices. Depth data is a special source of information that could characterize the object shape while being fully invariant to textures and lighting condition, which has been proved to be consistently improving the performance of various tasks in computer vision [11][12]. In [13], Huang et al. put a pioneering effort in developing an LBP-based descriptor, namely 3DLBP, specialized for facial depth images utilizing a special characteristics of the smoothness of facial depth images comparing to grayscale images. This work can be seen as the current state-of-the-art LBP-based feature specially developed for human facial depth images. However, the method suffers from some shortcomings as the feature length is much larger while the performance gain compared to LBP^{U2} is not significant. The encoding method is unstable when separating and encoding each digit of the binary form of the depth differences individually. Furthermore, the encoding scheme only uses the absolute value of the depth difference and ignores its true signed measure. These shortcomings should potentially reduce the performance of the approach.

With the above analysis, in this paper we introduce a novel efficient LBP-based descriptor, namely Gradient-LBP, for facial depth images which is proven to be superior to 3DLBP and resolves its inherent drawbacks. The proposed descriptor is applied to the gender recognition task and demonstrate its efficiency in outperforming 3DLBP in all the experimental setups on both the Kinect and range scanner images. Furthermore, we propose and evaluate a weighted combination scheme of the proposed descriptor for depth images and LBP^{U2} for grayscale images in gender recognition using different RGB-D sources of information. Experimental results reinforce the effectiveness of the proposed descriptor in complementing the result on grayscale images and confirm the efficiency of the combination of LBP-based approaches on RGB-D data for facial analysis.

In short, the contributions of the paper are as follow:

- Proposition of an efficient descriptor for facial depth images: the descriptor is much more compact and outperforms 3DLBP in all the experimental setups on both Kinect and range scanner images in gender recognition task.
- Proposition and analysis of a weighted combination scheme of the proposed descriptor for facial depth images and the state-of-the-art LBP^{U2} feature for grayscale images in gender recognition using different sources of RGB-D data: the experimentation is carried out on both high quality 3D range scanner database and images of lower quality from Kinect device. The experimental results reinforce the effectiveness of the proposed descriptor in complementing the information from grayscale images and confirm the efficiency of the combination of LBP-based approaches on RGB-D data.

The rest of the paper is organized as follows. Section 2 briefly reviews the related works in literature. The definition of LBP and 3DLBP is recapulated in section 3. Section 4 presents our proposed descriptor for human facial depth images. Section 5 introduces the proposed weighted combination scheme on RGB-D data. The experimental setups and results are given in section 6. Finally, the conclusion and future works are presented in section 7.

2 Related Work

LBP is originally proposed as a simple yet efficient operator that encodes the sign of the differences between the central pixel and its eight surrounding neighbors. Since then, the method has been continuously improved and now there are many variants applied in a vast domain of applications. In [14], Jin et al. proposed an improved LBP (ILBP) which compares all the pixels with the mean intensity of the patch to enhance its discriminative capability. Since LBP only encode the signs of the gray-value differences (GDs), Guo et al. proposed a complete LBP (CLBP) [7] to partly encode all the information from the sign, the GDs and also the gray values of the central pixels. Also to compensate the information of the gray values of neighboring pixels in the patch, Ylioinas et al. introduced the combination of LBP and the variance of the gray values of surrounding pixels in LBP/VAR [15] and showed consistent improvement. A complete survey of these methods could be found in [16].

Most of the variants are solely introduced for grayscale images. Some other efforts tried to extend the approach to 3D and spatio-temporal data. In [9], Fehr exploited the spherical harmonic transform to compute LBP for 3D volume data in frequency domain. Whereas Zhao and Pietikäinen successfully extended the approach to spatio-temporal data with the introduction of volumn LBP (VLBP) [10], in which it combines motion and appearance information in image sequences. However, very few variants are found in the domain of depth images and RGB-D data. In [13], Huang et al. made a pioneering attempt to extend the LBP approach to facial depth images with the introduction of 3DLBP. The method utilizes the special characteristics of the smoothness of the facial depth images comparing to grayscale images, where most of the depth differences (DDs) of neighboring pixels are very small. Therefore, 3DLBP uses a limited number of bits to represent the DDs and encodes them in an LBP-like way. This approach shows its efficiency when encoding most of the depth difference information into the feature, besides the original LBP. However, it suffers from some drawbacks as the feature length is large, the encoding scheme is unstable when transforming the DDs into the binary form and encoding each digit separately, breaking the integrity of the values, causing a very little change of the DDs would create a big difference in the coded values. Furthermore, the method only uses the absolute values of the DDs, ignoring their signed nature with positive and negative entities. These shortcomings should potentially affect the performance of the approach.

3 Face Representation Using LBP and 3DLBP

In this section, LBP and 3DLBP are reviewed as the background for the comprehension of our proposed approach in section 4.

3.1   LBP

LBP operator performs by thresholding the differences of the center value and the neighborhood in the 3x3 grid surrounding one pixel. The resulting values are then considered as an 8-bit binary number represented for that pixel (Fig. 1). The histogram of these binary numbers in the whole image can be used as a descriptor for the image.

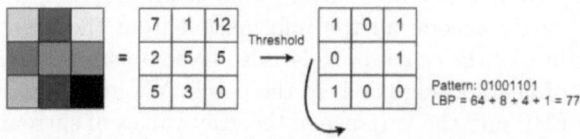

Fig. 1. An example of the original LBP operator [15]

The operator was then extended and generalized for any radius and number of points in the neighborhood. The notation (P, R) is used to indicate the use of P sample points in the neighborhood on the circle of radius R. The value of the LBP code at the pixel (x_c, y_c) is given by:

$$LBP_{P,R} = \sum_{p=0}^{P-1} s(g_p - g_c)2^p \qquad (1)$$

where g_c is the gray value of the center pixel (x_c, y_c), g_p are the gray values of P pixels at the radius R, s defines the thresholding function as follow:

$$s(x) = \begin{cases} 1 & \text{if x} \geq 0 \\ 0 & \text{otherwise} \end{cases} \qquad (2)$$

Another remarkable improvement of LBP is the so called uniform pattern [1]. LBP codes are not uniformly distributed, some codes appear much more frequently than the others. These frequent codes have at most two transitions from 0 to 1 or vice versa when the pattern is traversed circularly, and are called uniform patterns. When computing the histogram, every uniform pattern is labeled with one distinguished value while all the non-uniform patterns are group into one category. The uniform LBP is denoted as $LBP_{P,R}^{U2}$. The $LBP_{8,1}^{U2}$ has 59 bins and was proven as much more efficient than the original LBP.

3.2   3DLBP

LBP is a powerful approach to analyze and discriminate textures. However, it just considers the sign of differences and ignores the difference values, which can be an important source of information. By just keeping the sign of the differences, two different textures could be misclassified as the same by LBP.

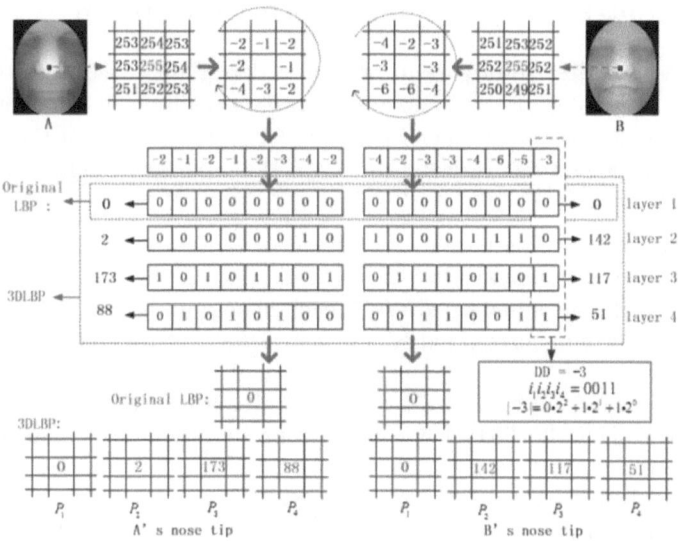

Fig. 2. An example of 3DLBP and its comparison to LBP [13]

In [13], Huang et al. extended the LBP approach to encode the extra information of the values of depth differences (DD) specialized for facial depth images.
From a statistical point of view, the authors observe that more than 93% of the
DD between points in R = 2 are smaller than 7. This is due to the smoothness
in depth transitions of human faces, which is not true for grayscale images in
general, where the neighboring points could be arbitrarily different depending
on the texture and environmental conditions. Hence, the authors then use just
three bits to represent the DD. Three binary units can characterize the absolute
value of DD from 0 to 7. All the $|DD| > 7$ are assigned to 7. The DD is then
binarized. Therefore, combining with one bit representing the sign of the DD,
for each pixel surrounding the center point, there are four bits representing that
position $\{i_1 i_2 i_3 i_4\}$, where $i_2 i_3 i_4$ represents the absolute value of the DD and i_1
represents the sign (encoded as the original LBP). Formally speaking, we have:

$$i_1 = \begin{cases} 1 & \text{if DD} \geq 0 \\ 0 & \text{otherwise} \end{cases} \tag{3}$$

$$|DD| = i_2 * 2^2 + i_3 * 2^1 + i_4 * 2^0 \tag{4}$$

The four bits are then separated into four layers. Then, for each layer, the corresponding bits of all the DD from the surrounding pixels are concatenated and
generate one LBP code. In total, there are four LBP codes {P1, P2, P3, P4},
where the first LBP code is the same as the original LBP. They are called 3D
Local Binary Patterns (3DLBP) (see Fig. 2). For matching, the histogram of
each LBP code is computed, then the four histograms are concatenated to form
a unique descriptor for the image.

4 Gradient-LBP for Facial Depth Description

3DLBP is a good descriptor that incorporates depth differences into the feature besides the original LBP. This feature works especially well for depth images thanks to the smoothness of facial depth images, where most of the depth differences are smaller than 7 levels. However, this approach suffers from several limitations:

- The feature length is large. At each pixel, there are four LBP codes. For the creation of the descriptor, each LBP code will then contribute to a histogram. With the standard $LBP_{8,1}$, each histogram is of size 256 bins. Four histograms would correspond to a feature length of 256 x 4 = 1024.
- The encoding scheme is unstable. A very small change of the depth difference (DD) in a position could lead to a big difference in the coded values. For example, when the DD of 3 (binary representation 011) increases into 4 (binary representation 100), the whole three last LBP codes will change. This problem is caused by the unconventional way of transforming the DD into binary form and forcefully encoding each binary digit separately in an LBP-like way.
- The DD are encoded on the basic of their absolute values, losing the information of the full range DD including negative and positive entities. Although this is compensated by the inclusion of the LBP from the signs of DD in the first LBP code, the separate encoding of this information into an LBP code and transforming into the histogram loosens the intrinsic connection of the sign and the absolute value parts of the DD.

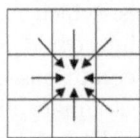

Fig. 3. The eight orientations in computing the standard $LBP_{8,1}$

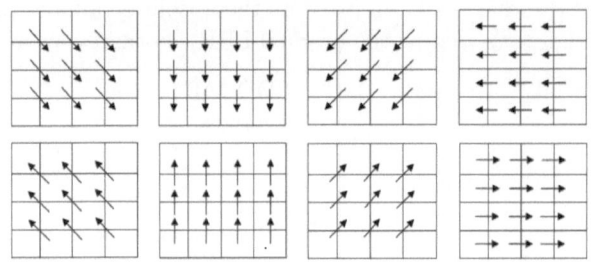

Fig. 4. The eight LBP8,1-based orientations of the depth differences. The example demonstrates the separated depth difference images in each orientation of the sample image patch of size 4x4.

With the above observations, we propose a novel efficient approach to incorporate the DD into the original LBP that overcomes all the mentioned shortcomings of 3DLBP. The proposed method has been proved to be superior to the 3DLBP approach in all the experimental setups carried out.

The proposed approach is based on a different orientation-based view towards the LBP operator. For the standard $LBP_{8,1}$, we can view the LBP operator under a different perspective where eight surrounding pixels are compared to the center value following eight orientations as shown in Fig. 3. The values of DD are indeed calculated through these eight orientations. As illustrated in Fig. 4, we consider each orientation in the computation of LBP separately for the whole image, which leads to the creation of eight depth difference images corresponding to eight orientations used by $LBP_{8,1}$ (Fig. 3). This notion of orientations and the oriented depth difference images can be generalized with the use of $LBP_{P,R}$ of any number of sample points and radius. P sample points used correspond to P orientations and would generate P oriented depth difference images, when we consider each orientation separately in the computing of DD of neighboring pixels. Each oriented depth difference image contains the DD values of neighboring pixels in one corresponding orientation. At each position (x_c, y_c) where the $LBP_{P,R}$ code is computed as in equation (1), the P oriented depth differences corresponding to that central pixel are provided as follow:

$$ODD_{p=0...P-1}^{P,R,p} = max(min(g_p - g_c, 7), -8) \tag{5}$$

where $ODD^{P,R,p}$ is the Oriented Depth Difference at pixel (x_c, y_c) in the depth difference image corresponding to the orientation p (the orientation formed by the point (x_p, y_p) and (x_c, y_c)), g_p is the depth value at position (x_p, y_p) on the circle of radius R surrounding the center pixel, g_c is the depth value at the center pixel, min(x,y) and max(x,y) are two functions that take the min and max value between two variables (x,y) respectively. This means that we clip the DD to be in the range -8 to 7, anything greater than 7 is set to be 7 and anything less than -8 is set to -8. The DD thus has sixteen possible values. This threshold is based on the statistical observation of 3DLBP that most of the DD of neighboring pixels are no more than 7. Notice that we use the true values of DD, not taking their absolute part.

After having P oriented depth difference images obtained as stated above, we can build the histogram of each depth difference image in each orientation, resulting in P histograms. Each histogram has 16 bins from -8 to 7. The information from P histograms is then combined by concatenation to form a unique oriented histogram of depth differences. For the creation of the image descriptor, the histogram of LBP^{U2} is also extracted from the original depth image. The descriptor is then the concatenation of the histogram of LBP^{U2} and the oriented histogram of depth differences. It should be noticed that, the P depth difference images corresponding to P orientations in computing LBP are pairwise symmetric (see Fig. 4), they are pairwise minus sign of the other. Thus, using all P depth difference images would be redundant. We propose to use only half of the P depth difference images in the computation of the final descriptor and call this Gradient-LBP (G-LBP). For the standard LBP of 8 sample points, the proposed descriptor has the length of $59 + 16*4 = 123$ (the histogram of LBP^{U2} with 59 bins concatenated with four histograms of four oriented depth difference images, each has 16 bins) which is much more compact compared to 3DLBP.

5   Weighted Combination of LBP and Gradient-LBP on RGB-D Face Data for Gender Recognition

Although the problem of Gender Recognition has been explored extensively in the scope of grayscale face images [6][5], there are very few works on Depth or RGB-D source of information. In [11], Lu et al. experimented the combination of range and luminous intensity data in gender and ethnicity classification and showed the improvement by this approach. However, the authors just use the simple pixel-based feature and the basic averaging fusion of the depth and luminous intensity information and demonstrate moderate results. Furthermore, the experiments are carried out only on range scanner data. The analysis on lower quality RGB-D data obtained from other widely used devices such as Kinect has not been examined.

Here, we apply a weighted combination scheme based on our proposed descriptor for depth images and the state-of-the-art LBP^{U2} feature for grayscale images, since the contribution of each part is unbalanced, usually grayscale images are more discriminative than depth images. The method is then evaluated on both professional 3D range scanner images and a Kinect database to evaluate the behavior of the approach on different sources of RGB-D data. Support Vector Machines is chosen to perform the classification task due to its superior efficiency in demographic classification as has been proven in [17]. More specifically, the classification is first performed separately for grayscale and depth images using SVM, which returns the probabilities of belonging to classes of male or female for each subject. The combination scheme of the results on two sources of information is formulated as follow:

$$p(male|s) = \frac{w_g * p(male|s_{gray}) + w_d * p(male|s_{depth})}{w_g + w_d} \tag{6}$$

$$p(female|s) = \frac{w_g * p(female|s_{gray}) + w_d * p(female|s_{depth})}{w_g + w_d} \tag{7}$$

where s is the subject to be classified, $p(male|s)$ and $p(female|s)$ are the final probabilities that the subject belongs to male or female class respectively, $p(male|s_{gray})$ and $p(female|s_{gray})$ are the resulting posterior probabilities returned by SVM for grayscale images while $p(male|s_{depth})$ and $p(female|s_{depth})$ are the results from depth images, w_g and w_d are the weighting factor for the grayscale and depth information respectively, they are the free parameters and could be adjusted according to the contribution of each part to the final decision.

In our experimentation, we propose to use these parameters as the resulting accuracy returned by SVM when using each source of information (grayscale or depth) separately for training and validating.

6   Experimental Analysis

6.1   Experimental Data

The EURECOM Kinect Face Dataset [19] and Texas 3DFR Dataset [18] are used for experimentation, both having color and depth images where the former is obtained using Kinect device while the latter is captured by range scanner.

The EURECOM Kinect Face dataset contains face images of 52 people (14 females, 38 males) taken in two sessions. In each session, the people are captured with nine states (neutral, smile, open mouth, left profile, right profile, occlusion eyes, occlusion mouth, occlusion paper, light on), besides the depth image, the raw depth level sensed by Kinect is also provided in a .txt file for better precision. The dataset also includes 6 manually located landmark points on the face (left eye, right eye, tip of the nose, left and right side of the mouth, the chin).

The Texas 3DFR dataset provides 1149 images (366 females, 783 males). The data includes both the raw images and the preprocessed data where the images underwent Gaussian smoothing, median filtering and hole filling steps. The 25 manually located anthropometric facial fiducial points are also provided.

6.2   Preprocessing

Based on the manual landmark points on the face, the images are first cropped into a square centered by the nose with the width and height two times the distance between the left and right eye centers.

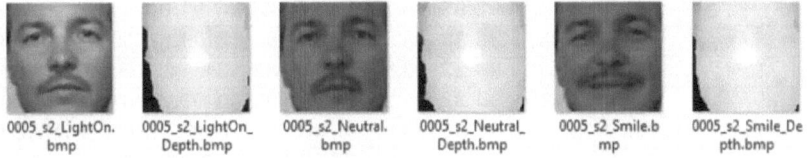

| 0005_s2_LightOn. | 0005_s2_LightOn_ | 0005_s2_Neutral. | 0005_s2_Neutral_ | 0005_s2_Smile.b | 0005_s2_Smile_De |
| bmp | Depth.bmp | bmp | Depth.bmp | mp | pth.bmp |

Fig. 5. The sample preprocessed images from EURECOM Kinect Face Dataset

For the depth information of EURECOM Kinect Face Dataset, we use raw depth levels in .txt files to have better representation. To fill holes, the closing operation is further applied to depth images. An illustration of the preprocessed images is shown in Fig. 5. For the Texas 3DFR dataset, we use the preprocessed images provided by the database. The cropped images in the EURECOM Kinect Face Dataset are then scaled to 96x96 pixels and the ones in Texas 3DFR dataset are scaled to 256x256 due to their higher resolution.

6.3   Settings

The images are divided into 8x8 blocks. The LBP^{U2}, 3DLBP and Gradient-LBP are extracted for each block and then concatenated to form a spatially enhanced feature for evaluation. Different configuration of (P,R) = (8,1) and (P,R) = (8,2) for all the three descriptors are experimented to obtain the in-depth evaluations.

For the classification task, we use SVM with non-linear RBF kernel as it has been proven to be a prominent technique in gender classification. We use 3 states in the EURECOM Kinect Face Dataset (Neutral, Smile and Light On) which cover different expressions and lighting conditions. For all the investigated methods, we carry out three experimental setups. In the first experiment (Kinect 1), we use the first session of EURECOM Kinect Face Dataset as the training set, the second session is the testing set. The second experiment (Kinect 2) is carried out by using first half number of males and females in both sessions of EURECOM Kinect Face Dataset as training set, the remaining are for testing. The third experiment (Range Scanner) is executed on the Texas 3DFR Dataset, where first half number of males and females are used as training and the remaining are used for testing, as in Kinect 2 setup.

6.4   Results and Analysis

To evaluate the performance of the proposed Gradient-LBP for depth images, the comparison between Gradient-LBP, 3DLBP and LBP^{U2} on depth images in three experimental setups as stated in section (6.3) are carried out. The detail

Table 1. The detailed comparison of the accuracy (in %) of the three investigated descriptors on depth images

	Kinect 1			Kinect 2			Range Scanner		
	Male	Female	Overall	Male	Female	Overal	Male	Female	Overal
$LBP^{U2}_{8,1}$	96.49	83.33	**92.95**	80.70	73.81	**78.85**	95.15	59.02	**83.65**
$LBP^{U2}_{8,2}$	98.25	78.57	**92.95**	83.33	71.43	**80.13**	96.17	59.19	**83.13**
$3DLBP_{8,1}$	95.61	90.48	**94.23**	83.33	83.33	**83.33**	95.66	60.11	**84.35**
$3DLBP_{8,2}$	96.49	88.10	**94.23**	84.21	90.48	**85.90**	97.96	63.39	**86.96**
Gradient-LBP$_{8,1}$	96.49	92.86	**95.51**	86.84	88.10	**87.18**	99.74	62.30	**87.83**
Gradient-LBP$_{8,2}$	96.49	92.86	**95.51**	85.09	88.10	**85.90**	100	68.31	**89.91**

Table 2. The accuracy (in %) of the combination scheme compared to using LBP-based descriptors on depth and grayscale images for the configuration of (P,R) = (8,1)

	Kinect 1			Kinect 2			Range Scanner		
	Male	Female	Overall	Male	Female	Overal	Male	Female	Overal
G-LBP$_{8,1}$ (Depth)	96.49	92.86	**95.51**	86.84	88.10	**87.18**	99.74	62.30	**87.83**
$LBP^{U2}_{8,1}$ (Gray)	98.25	97.62	**98.08**	94.74	69.05	**87.82**	95.15	92.90	**94.43**
Combination	99.12	100	**99.36**	95.61	76.19	**90.38**	98.98	91.80	**96.70**

results are shown in Table 1. From the experiments, we can draw two conclusions regarding the performance of the features specialized for depth images:

- LBP_{U2} alone is not a good descriptor for depth images, the extra depth difference information included in 3DLBP does improve the recognition performance for depth images.
- For the depth images, the Gradient-LBP outperforms the 3DLBP and LBP^{U2} approaches in all the experimental setups on both Kinect data and range scanner images, and for both radius of 1 and 2, this proves the consistent superiority of the proposed descriptor on facial depth data.

Table 3. The accuracy (in %) of the combination scheme compared to using LBP-based descriptors on depth and grayscale images for the configuration of (P,R) = (8,2)

	Kinect 1			Kinect 2			Range Scanner		
	Male	Female	Overall	Male	Female	Overal	Male	Female	Overal
G-LBP$_{8,2}$ (Depth)	96.49	92.86	**95.51**	85.09	88.10	**85.90**	100	68.31	**89.91**
LBP$_{8,2}^{U2}$ (Gray)	95.61	97.62	**96.15**	92.98	71.43	**87.18**	93.62	90.16	**92.52**
Combination	98.25	97.62	**98.08**	93.86	80.95	**90.38**	97.96	89.62	**95.30**

The evaluated effectiveness of the combination scheme of Gradient-LBP for depth images and LBP^{U2} for grayscale images comparing to the use of these methods alone on both Kinect and range scanner data are given in Table 2 and Table 3. The results demonstrate that, although LBP^{U2} is very robust and efficient in representing grayscale images in gender classification, the addition of Gradient-LBP source of information from depth images always improve the final performance. The results are very consistent in both the range scanner data and images of lower quality from home devices like Kinect. This result reinforces the effectiveness of the proposed feature for depth images in complementing the luminous intensity information and the efficiency of the combination of LBP-based approaches on RGB-D data.

It can also be noticed that, all the experimental results follow the same trend in which the experimental setup with the first session of the EURECOM Kinect Face Dataset used as training produces the highest accuracy, followed by the setup where half the images in Texas 3DFR are trained, the lowest result corresponds to using half the images in both sessions of the EURECOM Kinect Face Dataset as training. This can be explained since in Kinect 1 setup, all the people presented in the testing set also appear in the training set, which helps the classifier easily recognize the features. The result of the experimental setup Kinect 2 in overall is worse than Range Scanner because the resolution and quality of images in the Texas 3DFR database are better than Kinect data.

7   Conclusion and Future Works

In this paper, a novel feature descriptor specialized for facial depth images inspired by 3DLBP is introduced. The proposed descriptor is much more compact yet consistently outperforms 3DLBP in all the experimental setups carried out on both sets of images from Kinect device and range scanner. We further propose a weighted combination scheme and reinforce the effectiveness of the proposed descriptor by its efficient combination with the result from LBP^{U2} on grayscale images. Although LBP^{U2} is already an excellent descriptor for grayscale images, the combined scheme consistently shows the improvement across different experimental setups and RGB-D sources.

In the scope of this work, experimentations have been performed on a simple two-class problem, that is to say gender recognition, in order to validate the efficiency of the proposed approach. Next step would consist in extending our tests on multiple-class problems, e.g. age, ethnicity, identity classification.

References

1. Ojala, T., Pietkäinen, M., Mäenpää, T.: Multiresolution Gray-Scale and Rotation Invariant Texture Classification with Local Binary Patterns. IEEE Trans. Pattern Analysis and Machine Intelligence 24(7), 971–987 (2002)
2. Ahonen, T., Hadid, A., Pietikäinen, M.: Face description with local binary patterns: Application to face recognition. PAMI 28(12), 2037–2041 (2006)
3. Shan, C., Gong, S., McOwan, P.: Facial expression recognition based on Local Binary Patterns: A comprehensive study. Image and Vision Computing 27(6), 803–816 (2009)
4. Gunay, A., Nabiyev, V.V.: Automatic age classification with LBP. In: International Symposium on Computer and Information Sciences - ISCIS (2008)
5. Lian, H.-C., Lu, B.-L.: Multi-view Gender Classification Using Local Binary Patterns and Support Vector Machines. In: Wang, J., Yi, Z., Żurada, J.M., Lu, B.-L., Yin, H. (eds.) ISNN 2006. LNCS, vol. 3972, pp. 202–209. Springer, Heidelberg (2006)
6. Farinella, G., Dugelay, J.-L.: Demographic classification: Do gender and ethnicity affect each other? In: IAPR International Conference on Informatics, Electronics & Vision, ICIEV 2012, Dhaka, Bangladesh (2012)
7. Guo, Z., Zhang, L., Zhang, D.: A completed modeling of local binary pattern operator for texture classification. IEEE Trans. Image Process. 19(6), 1657–1663 (2010)
8. Yang, H., Wang, Y.: A LBP-based face recognition method with Hamming distance constraint. In: Proc. International Conference on Image and Graphics, pp. 645–649 (2007)
9. Fehr, J.: Rotational Invariant Uniform Local Binary Patterns For Full 3D Volume Texture Analysis. FINSIG (2007)
10. Zhao, G., Pietikäinen, M.: Dynamic texture recognition using volume local binary patterns. In: ECCV, Workshop on Dynamical Vision, pp. 12–23 (2006)
11. Lu, X., Chen, H., Jain, A.: Multimodal Facial Gender and Ethnicity Identification. In: Zhang, D., Jain, A.K. (eds.) ICB 2005. LNCS, vol. 3832, pp. 554–561. Springer, Heidelberg (2005)

12. Zhu, Y., Dariush, B., Fujimura, K.: Controlled human pose estimation from depth image streams. In: CVPR time-of-flight Workshop (2008)
13. Huang, Y., Wang, Y., Tan, T.: Combining statistics of geometrical and correlative features for 3D face recognition. In: Proceedings of the British Machine Vision Conference, pp. 879–888 (2006)
14. Jin, H., Liu, Q., Lu, H., Tong, X.: Face detection using improved lbp under Bayesian framework. In: Proceedings of the Third International Conference on Image and Graphics, pp. 306–309 (2004)
15. Ylioinas, J., Hadid, A., Pietikäinen, M.: Combining Contrast Information and Local Binary Patterns for Gender Classification. In: Heyden, A., Kahl, F. (eds.) SCIA 2011. LNCS, vol. 6688, pp. 676–686. Springer, Heidelberg (2011)
16. Huang, D., Shan, C., Ardabilian, M., Wang, Y., Chen, L.: Local Binary Patterns and Its Application to Facial Image Analysis: A Survey. IEEE Transactions on Systems, Man, and Cybernetics, Part C 41(6), 765–781 (2011)
17. Mäkinen, E., Raisamo, R.: Evaluation of Gender Classification Methods with Automatically Detected and Aligned Faces. IEEE Trans. Pattern Anal. Mach. Intell. 30(3), 541–547 (2008)
18. Gupta, S., Castleman, K.R., Markey, M.K., Bovik, A.C.: Texas 3D Face Recognition Database. In: IEEE Southwest Symposium on Image Analysis and Interpretation, Austin, TX, pp. 97–100 (2010)
19. EURECOM Kinect Face Dataset, http://RGB-D.eurecom.fr

Face Spoofing Detection Using Dynamic Texture

Jukka Komulainen, Abdenour Hadid, and Matti Pietikäinen

Center for Machine Vision Research,
Department of Computer Science and Engineering,
P. O. Box 4500 FI-90014 University of Oulu, Finland
{jukmaatt,hadid,mkp}@ee.oulu.fi

Abstract. While there is a significant number of works addressing e.g. pose and illumination variation problems in face recognition, the vulnerabilities to spoofing attacks were mostly unexplored until very recently when an increasing attention is started to be paid to this threat. A spoofing attack occurs when a person tries to masquerade as someone else e.g. by wearing a mask to gain illegitimate access and advantages. This work provides the first investigation in research literature on the use of dynamic texture for face spoofing detection. Unlike masks and 3D head models, real faces are indeed non-rigid objects with contractions of facial muscles which result in temporally deformed facial features such as eye lids and lips. Our key idea is to learn the structure and the dynamics of the facial micro-textures that characterise only real faces but not fake ones. Hence, we introduce a novel and appealing approach to face spoofing detection using the spatiotemporal (dynamic texture) extensions of the highly popular local binary pattern approach. We experiment with two publicly available databases consisting of several fake face attacks of different natures under varying conditions and imaging qualities. The experiments show excellent results beyond the state-of-the-art.

1 Introduction

Because of its natural and non-intrusive interaction, identity verification and recognition using facial information is among the most active and challenging areas in computer vision research. Despite the significant progress in the face recognition technology in the recent decades, wide range of viewpoints, aging of subjects and complex outdoor lighting are still research challenges. While there is a significant number of works addressing these issues, research on face biometric systems under spoofing attacks has mostly been overlooked although face recognition systems are known, since long time ago, to respond weakly to attacks. A spoofing attack occurs when a person tries to masquerade as someone else by falsifying data and thereby gaining illegitimate access. Very recently, an increasing attention is started to be paid to the problem of spoofing attacks against face biometric systems. This can be attested by the recently organized IJCB 2011 competition on counter measures to 2D facial spoofing attacks [1] which can be seen as a kick-off for studying best practices for non-intrusive spoofing detection.

J.-I. Park and J. Kim (Eds.): ACCV 2012 Workshops, Part I, LNCS 7728, pp. 146–157, 2013.
© Springer-Verlag Berlin Heidelberg 2013

One can spoof a face recognition system by presenting a photograph, a video or a 3D model of a targeted person to the camera. While one can also use make-up or plastic surgery as other means of spoofing, photographs are probably the most common sources of spoofing attacks because one can easily download and capture facial images. Typical countermeasure against spoofing is liveness detection that aims at detecting physiological signs of life such as eye blinking, facial expression changes, mouth movements, etc. For instance, Pan et al. [2] exploited the observation that humans blink once every 2-4 seconds and proposed an eye blink-based anti-spoofing method. Another commonly used countermeasure is motion analysis since it can be assumed that the movement of planar objects (e.g. displays and photographs) differs significantly from that of real human faces which are complex non-rigid 3D objects [3,4]. Obviously, such countermeasures can only be considered with photographs while nowadays videos are ubiquitous and hence can easily be used for spoofing attacks. Another category of anti-spoofing methods are based on the analysis of skin properties such as skin texture and skin reflectance. An intuitive approach is to explore the high frequency information in the facial region, since mobile phone displays and smaller photographs probably contain fewer high frequency components compared to real faces [5,6]. Such an approach is likely to fail with higher quality photographs and videos, as shown for example in [7]. Recently, also micro-texture analysis has been applied to measure facial texture quality with impressive results [8,9]. However, the evaluations were made using data sets with little variations and the used high frequency information depends strongly on the input image and fake face quality. Other countermeasures against face spoofing attacks include multi-modal analysis and multi-spectral methods. A system combining face recognition with other biometric modalities such as gait and speech is indeed intrinsically more difficult to spoof than uni-modal systems. Multi-spectral imaging can also be used for analyzing the reflectance of object surfaces and thus discriminating live faces from fake ones [10].

It appears that most of the existing methods for spoofing detection are either very complex (and hence not very practical for real-world face biometric systems requiring fast processing) or using non-conventional imaging systems (e.g. multi spectral imaging) and devices (e.g. thermal cameras). We therefore propose in this work a novel computationally fast approach based on highly discriminative dynamic micro-texture features, using conventional images and requiring no user-cooperation.

This work provides **the first investigation in research literature** on the use of dynamic texture for face spoofing detection. Unlike masks and 3D head models, real faces are indeed non-rigid objects with contractions of facial muscles which result in temporally deformed facial features such as eye lids and lips. Our key idea is to learn the structure **and** especially the **dynamics** of the facial micro-textures that characterize only real faces but not fake ones. Hence, we introduce the first and appealing spatio-temporal approach to face spoofing detection using the spatiotemporal (dynamic texture) extensions of the highly popular local binary pattern (LBP) approach [11]. Spatiotemporal LBP has shown very promising performance in various problems, including dynamic texture recognition, face and facial expression recognition, lip-reading, and activity and gait recognition [11].

Dynamic textures provide a new and very effective tool for motion analysis. The past research on motion analysis has been usually based on assumption that the scene is Lambertian, rigid and static. For example, the Lambertian assumption has been crucial when developing methods for tracking, determining optical flow or finding correspondences. This kind of constraints greatly limits the applicability of motion analysis. Recently, approaches based on dynamic textures have been proposed as a new and potentially very effective tool for motion analysis [11]. These developments have inspired us to approach face spoofing detection from dynamic texture point of view. We introduce below our novel approach and provide extensive experimental analysis on two publicly available databases (CASIA Face Anti-Spoofing Database [7] and Print-Attack Database [12]) consisting of several fake face attacks of different natures and under varying conditions and imaging qualities, showing excellent results beyond the state-of-the-art.

2   Spatiotemporal Face Liveness Description

For describing the face liveness for spoofing detection, we considered an elegant approach to face analysis from videos which is based on a spatiotemporal representation for combining facial appearance and dynamics. We adopted the LBP based spatiotemporal representation because of its recent excellent performance in modeling moving faces for face and facial expression recognition and also for dynamic texture recognition. More specifically, we considered local binary patterns from three orthogonal planes (LBP-TOP) which have shown to be very effective in describing the horizontal and vertical motion patterns in addition to appearance [13].

The LBP texture analysis operator, introduced by Ojala *et al.* [14,15], is defined as a gray-scale invariant texture measure, derived from a general definition of texture in a local neighborhood. It is a powerful texture descriptor and among its properties in real-world applications are its discriminative power, computational simplicity and tolerance against monotonic gray-scale changes. The original LBP operator forms labels for the image pixels by thresholding the 3×3 neighborhood with the center value and considering the result as a binary number. The histogram of these $2^8 = 256$ different labels can then be used as a image descriptor.

The original LBP operator was defined to only deal with the spatial information. Recently, it has been extended to a spatiotemporal representation for dynamic texture analysis (DT). This has yielded to so called Volume Local Binary Pattern operator (VLBP) [13]. The idea behind VLBP consists of looking at dynamic texture as a set of volumes in the (X,Y,T) space where X and Y denote the spatial coordinates and T denotes the frame index (time). The neighborhood of each pixel is thus defined in a three dimensional space. Then, similarly to basic LBP in spatial domain, volume textons can be defined and extracted into histograms. Therefore, VLBP combines motion and appearance into a dynamic texture description.

To make the VLBP computationally simple and easy to extend, the co-occurrences of the LBP on the three orthogonal planes (LBP-TOP) was also introduced [13]. LBP-TOP consists of the three orthogonal planes: XY, XT and YT, and concatenating local binary pattern co-occurrence statistics in these three directions. The circular neighborhoods are generalized to elliptical sampling to fit to the space-time statistics. The LBP codes are extracted from the XY, XT and YT planes, which are denoted as $XY - LBP$, $XT - LBP$ and $YT - LBP$, for all pixels, and statistics of the three different planes are obtained, and then concatenated into a single histogram. The procedure is shown in Fig. 1. In this representation, dynamic texture (DT) is encoded by the $XY - LBP$, $XT - LBP$ and $YT - LBP$.

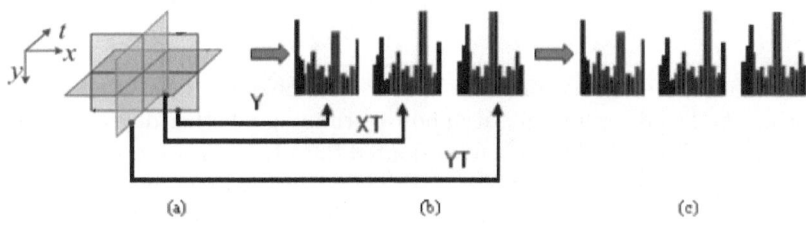

Fig. 1. (a) Three planes of dynamic texture; (b) LBP histogram from each plane; (c) Concatenated feature histogram [13]

Using equal radiuses for the time and spatial axes is not reasonable for dynamic textures [13] and therefore, in the XT and YT planes, different radii can be assigned to sample neighboring points in space and time. More generally, the radii in axes X, Y and T, and the number of neighboring points in the XY, XT and YT planes can also be different denoted by R_X, R_Y and R_T, P_{XY}, P_{XT} and P_{YT}. The corresponding feature is denoted as $LBP - TOP_{P_{XY},P_{XT},P_{YT},R_X,R_Y,R_T}$.

Let assume we are given an $X \times Y \times T$ dynamic texture ($x_c \in \{0, \cdots, X-1\}$, $y_c \in \{0, \cdots, Y-1\}$, $t_c \in \{0, \cdots, T-1\}$). A histogram of the DT can be defined as:

$$H_{i,j} = \sum_{x,y,t} I\{f_j(x,y,t) = i\}, \quad i = 0, \cdots, n_j - 1; j = 0, 1, 2 . \tag{1}$$

in which n_j is the number of different labels produced by the LBP operator in the jth plane ($j = 0 : XY$, $1 : XT$ and $2 : YT$) and $f_i(x,y,t)$ expresses the LBP code of central pixel (x,y,t) in the jth plane.

Similarly to the original LBP, the histograms must be normalized to get a coherent description for comparing the DTs:

$$N_{i,j} = \frac{H_{i,j}}{\sum_{k=0}^{n_j-1} H_{k,j}} . \tag{2}$$

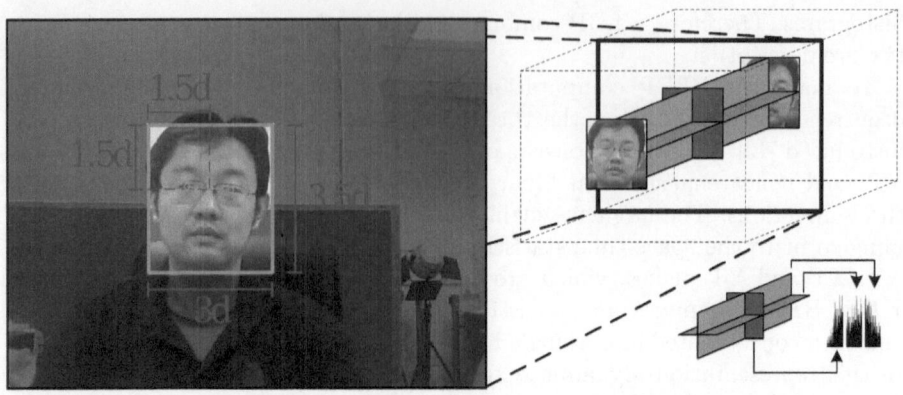

Fig. 2. Dynamic texture based face description

Due to its tolerance against monotonic gray-scale changes, LBP is adequate for measuring the facial texture quality and determining whether degradations due to spoofing medium are observed. We adopted the LBP based spoofing detection in spatiotemporal domain because LBP-TOP features have been successfully applied in describing dynamic events. Our key idea is use LBP-TOP features for detecting e.g. specific facial motion patterns or sudden characteristic reflections of planar spoofing media which might differentiate real faces from fake ones.

When deriving our proposed face liveness description, we aim to avoid scaling during geometric face normalization in order to keep all valuable information about the facial texture quality which is a crucial visual cue in spoofing detection. Simple head pose correction based on eye locations may also be too unstable between video frames, e.g. due to inaccurate eye detection, yielding performance degradation in dynamic texture analysis. To overcome these effects, LBP-TOP$_{8,8,8,1,2,2}$ operator is instead applied on each pixel and the dynamic LBP histogram for every frame is calculated over the volume bounded by the roughly normalized face location (see Fig. 2). Then, the histograms of 768 bins are accumulated over a period of two seconds (50 frames at 25fps) to form the final feature vector.

3   Experimental Analysis

To assess of the effectiveness of our proposed spatiotemporal face liveness description for spoofing detection, we performed a set of experiments on the CASIA Face Anti-Spoofing Database [7] and Print-Attack Database [12] from the Idiap Research Institute. We used Viola-Jones algorithm for face detection [16] while eye localization is performed using 2D Cascaded AdaBoost [17]. The relative eye positions from the first frame are used to refine the detected faces so that the face areas are roughly aligned in every frame. We also exploited the fact that the spoofing medium might be visible around the face, thus the height and width of the aligned face are set to 3.5d and 3.0d where d represents the distance between eyes (see Fig. 2). Once the face liveness description is derived, a homogeneous kernel map [18] is applied to obtain a five dimensional linear approximation of a χ^2 kernel. The approximated feature map is computed with VLFeat [19] and the final classification is performed using a linear SVM implementation of LIBLINEAR [20].

The SVM classifier is trained using a set of positive (genuine faces) and negative (fake faces) samples which are extracted from the provided training data. In order to get sufficient amount of data for building the model, the whole length of each training video is divided into several time windows with temporal overlap of one second over which the LBP-TOP features are computed. On the test sets, however, only the first two seconds from the beginning of each video sequence are used for determining whether a genuine face or a fake one is observed. The use of the whole video sequence may naturally lead to better detection results but at the cost of more computational time which could be an issue in real-life applications.

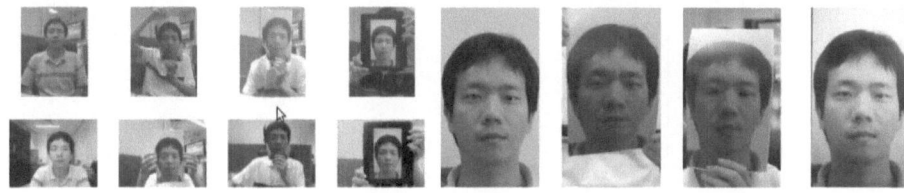

Fig. 3. Example images from the CASIA Face Anti-Spoofing Database [7]

3.1   Evaluation on the CASIA Face Anti-Spoofing Database

We first conducted extensive experiments on the CASIA Face Anti-Spoofing Database [7] and compared our results against those which are provided along with the database. The database includes significant improvements compared to previous databases, since it provides more variations in the collected data. The data set contains 50 real clients and the corresponding fake faces are captured with high quality from the original ones. The variety is achieved by introducing three imaging qualities (low, normal and high) and three fake face attacks which include warped photo, cut photo (eyeblink) and video attacks. Examples from the database can be seen in Fig. 3. Altogether the database consists of 600 video clips and the subjects are divided into subsets for training and testing (240 and 360, respectively). Results of a baseline system are also provided along the database for fair comparison. The baseline system considers the high frequency information in the facial region using multiple DoG features and SVM classifier and is inspired by the work of Tan *et al.* [6].

Since the main purpose of the database is to investigate the possible effects of different fake face types and imaging qualities, the test protocol consists of seven scenarios in which particular train and test samples are to be used. The quality test considers the three imaging qualities separately, low (1), normal (2) and high quality (3), and evaluates the overall spoofing detection perfomance under variety of attacks at the given imaging quality. Similarly, the fake face test assesses how robust the anti-spoofing measure is to specific fake face attacks, warped photo (4), cut photo (5) and video attacks (6), regardless of the imaging quality. In the overall test (7), all data is used to give a more general evaluation. The results of each scenario are reported as Detection-Error Trade-off (DET) curves and equal error rates (EER), which is the point where false acceptance rate (FAR) equals false rejection rate (FRR) on the DET curve.

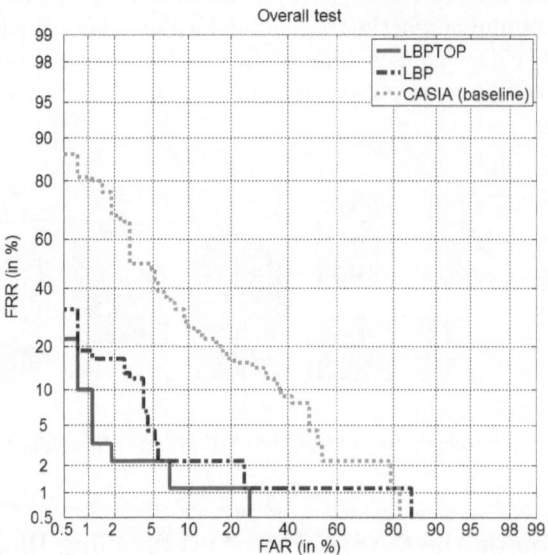

Fig. 4. Overall comparative results on the CASIA Face Anti-Spoofing Database

The results of the experiments are shown in Fig. 4 as DET curves for the overall test, i.e. including all scenarios. As it can be seen, the use of only facial appearance (LBP) leads to better results compared to the baseline method (CASIA baseline). Importantly, when the temporal planes XT and YT are also considered for spatio-temporal face description (LBP-TOP), a significant performance enhancement is obtained, thus confirming the benefits of encoding and exploiting not only the facial appearance but also the facial dynamics information.

More detailed results for each spoofing attack scenario are presented in Fig. 5 and in Table 1. The results indicate that the proposed LBP-TOP based face description yields best results in all configurations except at the highest imaging quality. The facial appearance description (LBP) works perfectly when the highest imaging quality is used because the skin texture of genuine faces looks

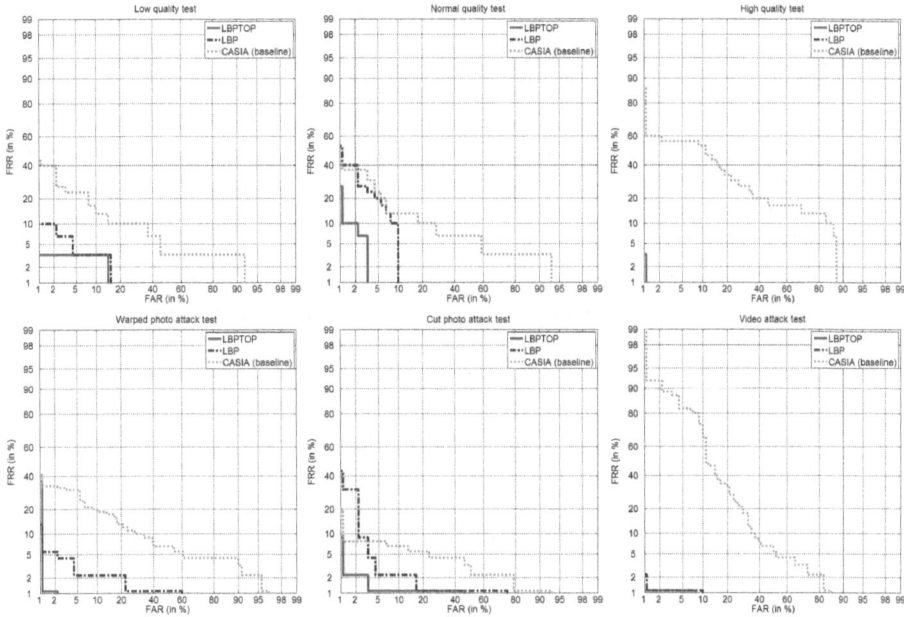

Fig. 5. DET curves under the different protocols of the CASIA Face Anti-Spoofing Database

Table 1. EER comparison between the baseline method, LBP and LBP-TOP on the CASIA Face Anti-Spoofing Database

Scenario	1	2	3	4	5	6	7
Baseline	0.13	0.13	0.26	0.16	0.06	0.24	0.17
LBP	0.04	0.10	0.00	0.04	0.04	0.01	0.04
LBPTOP	0.03	0.03	0.01	0.01	0.02	0.01	0.02

strikingly sharper compared to the fake ones. Thus, the measurement of facial texture quality seems to provide sufficient means to reveal whether degradation due to recapturing process is observed if the imaging quality is good enough to capture the fine details of a human face. However, the quality test shows that the use of facial dynamics enhances the spoofing detection results at lower imaging qualities without any significant performance drop when a high resolution camera is used for capturing the facial image. Furthermore, the fake face test indicates that adding temporal planes to the face description improves the robustness to different types of spoofing attacks, especially to warped and cut photo attacks, at various imaging qualities. The downsized resolution of the original high quality video spoofs (due to limited iPad screen resolution) [7] and the occasionally visible video screen frame around the fake faces also partially explain the less challenging nature of the video attack tests.

3.2   Evaluation on the Print-Attack Database

For extensive evaluation, we also conducted experiments on a second publicly available database namely Print-Attack Database [12] which was originally introduced within the IJCB 2011 Competition on Counter Measures to 2D Facial Spoofing Attacks [1]. The database consists of 200 real client accesses and 200 print-attack videos (50 clients) which were captured in controlled and uncontrolled lighting conditions using a webcam at 25fps with a resolution of 320×240 pixels. The print attacks were generated by taking high-resolution photographs of each client under the same conditions as in their authentication sessions and the captured images were printed in color on A4-sized paper. The spoofing attack attempts were performed with fixed or hand-held prints. Example images from the database are shown in Fig. 6. The database is divided into three sets, training, development and test data (see Table 2). Clients have been randomly divided for each subset so that the identities do not overlap between the subsets. The EER of development set is used for tuning the threshold which is applied for discriminating the test samples. For simplicity, we used the provided face locations for extracting the LBP-TOP representations.

Adverse   **Controlled**

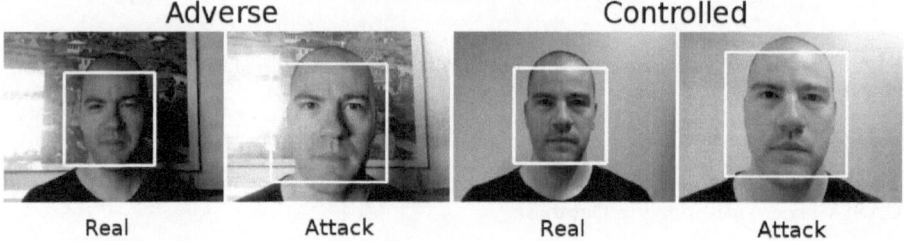

Real   Attack   Real   Attack

Fig. 6. Examples from Print-Attack Database [12] with the provided face locations. Note that the photo attacks suffer from apparent printing artefacts.

Table 2. The decomposition of the Print-Attack Database. The numbers indicate how many videos are included in each subset (the sums indicate the amount of hand-based and fixed-support attacks).

Type	Train	Devel.	Test	Total
Real	60	60	80	200
Attack	30+30	30+30	40+40	100+100
Total	120	120	160	400

Our dynamic texture based face description approach easily detected and characterized the printing artifacts and facial movements, e.g. eye blinking. Our approach yields perfect detection results (EER of 0%) on this database. Print attacks are perhaps less challenging to our method than the combination of different types of attacks as in the CASIA Face Anti-Spoofing Database. Table 3 shows a performance comparison between our proposed approach and the works of different research groups who participated in the IJCB 2011 Competition on Counter Measures to 2D Facial Spoofing Attacks [1]. It is worth mentioning that our proposed dynamic texture based face description performed very well even using only a single LBP-TOP feature vector which is easily extracted from the face area, whereas the other methods considered more complex analysis using multiple cues, e.g. fusion of separate motion and texture analysis, or relying on describing the strongly visible print defects which are quite obvious in the data set.

Table 3. Performance comparison between the proposed approach and the teams who participated in the IJCB 2011 Competition on Counter Measures to 2D Facial Spoofing Attacks [1]

Method	Development		Test		
	FAR	FRR	FAR	FRR	HTER
AMILAB [1]	0.00	0.00	0.00	1.25	0.63
CASIA [1]	1.67	1.67	0.00	0.00	0.00
IDIAP [1]	0.00	0.00	0.00	0.00	0.00
SIANI [1]	1.67	1.67	0.00	21.25	10.63
UNICAMP [1]	1.67	1.67	1.25	0.00	0.63
UOULU [1]	0.00	0.00	0.00	0.00	0.00
Proposed approach	0.00	0.00	0.00	0.00	0.00

4 Conclusion

Inspired by the recent progress in dynamic texture, we investigated the problem of face spoofing detection using spatiotemporal local binary patterns. To the best of our knowledge, this is the first work in the literature applying dynamic texture to the spoofing detection problem. The key idea of our proposed approach consists of analyzing the structure and the dynamics of the micro-textures in the facial regions using LBP-TOP features which provide an efficient and compact representation for face liveness description. Experiments on two publicly available databases showed excellent results under various fake face attacks, including video replay attacks, at different imaging qualities. The incorporation of facial dynamics significantly increased the robustness of the LBP based face description regardless of the imaging quality, especially under warped and cut photo attacks. Our obtained results can be used by the research community as a new reference on these spoofing databases for future research.

The excellent obtained results on these two publicly available databases suggest that more challenging databases (e.g. using 3D skin-like masks of very high quality and precision) should be designed, captured and made publically available for the research community in the near future. It would be then of great interest to evaluate our approach on such challenging data when available. We are currently incorporating the described anti-spoofing measure into our existing access control system for deployment in real-world applications. We plan to release the source code of our described anti-spoofing method for the research community after the publication of this work.

Acknowledgement. This work has been performed within the TABULA RASA project 7th Framework Research Programme of the European Union (EU), grant agreement number: 257289. The financial support of the Academy of Finland and Infotech Oulu Graduate School is also gratefully acknowledged.

References

1. Chakka, M.M., Anjos, A., Marcel, S., Tronci, R., Muntoni, D., Fadda, G., Pili, M., Sirena, N., Murgia, G., Ristori, M., Roli, F., Yan, J., Yi, D., Lei, Z., Zhang, Z., Li, S., Schwartz, W.R., Rocha, A., Pedrini, H., Lorenzo-Navarro, J., Castrillón-Santana, M., Määttä, J., Hadid, A., Pietikäinen, M.: Competition on counter measures to 2-d facial spoofing attacks. In: Proceedings of IAPR IEEE International Joint Conference on Biometrics (IJCB), Washington DC, USA (2011)
2. Pan, G., Wu, Z., Sun, L.: Liveness detection for face recognition. In: Delac, K., Grgic, M., Bartlett, M.S. (eds.) Recent Advances in Face Recognition, ch. 9. IN-TECH (2009)
3. Kollreider, K., Fronthaler, H., Bigun, J.: Non-intrusive liveness detection by face images. Image and Vision Computing 27, 233–244 (2009)
4. Bao, W., Li, H., Li, N., Jiang, W.: A liveness detection method for face recognition based on optical flow field. In: 2009 International Conference on Image Analysis and Signal Processing, pp. 233–236. IEEE (2009)
5. Li, J., Wang, Y., Tan, T., Jain, A.K.: Live face detection based on the analysis of fourier spectra. In: Biometric Technology for Human Identification, pp. 296–303 (2004)
6. Tan, X., Li, Y., Liu, J., Jiang, L.: Face Liveness Detection from a Single Image with Sparse Low Rank Bilinear Discriminative Model. In: Daniilidis, K., Maragos, P., Paragios, N. (eds.) ECCV 2010, Part VI. LNCS, vol. 6316, pp. 504–517. Springer, Heidelberg (2010)
7. Zhang, Z., Yan, J., Liu, S., Lei, Z., Yi, D., Li, S.Z.: A face antispoofing database with diverse attacks. In: Proceedings of 5th IAPR International Conference on Biometrics (ICB 2012), New Delhi, India (2012)
8. Bai, J., Ng, T.T., Gao, X., Shi, Y.Q.: Is physics-based liveness detection truly possible with a single image? In: IEEE International Symposium on Circuits and Systems (ISCAS), pp. 3425–3428 (2010)
9. Määttä, J., Hadid, A., Pietikäinen, M.: Face spoofing detection from single images using micro-texture analysis. In: Proceedings of IAPR IEEE International Joint Conference on Biometrics (IJCB), Washington DC, USA (2011)
10. Zhang, Z., Yi, D., Lei, Z., Li, S.Z.: Face liveness detection by learning multispectral reflectance distributions. In: International Conference on Face and Gesture, pp. 436–441 (2011)
11. Pietikäinen, M., Hadid, A., Zhao, G., Ahonen, T.: Computer Vision Using Local Binary Patterns. Springer (2011)

12. Anjos, A., Marcel, S.: Counter-measures to photo attacks in face recognition: a public database and a baseline. In: Proceedings of IAPR IEEE International Joint Conference on Biometrics (IJCB), Washington DC, USA (2011)
13. Zhao, G., Pietikäinen, M.: Dynamic texture recognition using local binary patterns with an application to facial expressions. IEEE Transactions on Pattern Analysis and Machine Intelligence 29, 915–928 (2007)
14. Ojala, T., Pietikäinen, M., Harwood, D.: A comparative study of texture measures with classification based on feature distributions. Pattern Recognition 29, 51–59 (1996)
15. Ojala, T., Pietikäinen, M., Mäenpää, T.: Multiresolution gray-scale and rotation invariant texture classification with local binary patterns. IEEE Trans. on PAMI 24 (2002)
16. Viola, P.A., Jones, M.J.: Rapid object detection using a boosted cascade of simple features. In: Proceedings of the IEEE Conference on Computer Vision and Pattern Recognition, pp. 511–518 (2001)
17. Niu, Z., Shan, S., Yan, S., Chen, X., Gao, W.: 2d cascaded adaboost for eye localization. In: Proc. of the 18th International Conference on Pattern Recognition (2006)
18. Vedaldi, A., Zisserman, A.: Efficient additive kernels via explicit feature maps. In: Proceedings of the IEEE Conference on Computer Vision and Pattern Recognition (2010)
19. Vedaldi, A., Fulkerson, B.: VLFeat: An open and portable library of computer vision algorithms (2008)
20. Fan, R.E., Chang, K.W., Hsieh, C.J., Wang, X.R., Lin, C.J.: LIBLINEAR: A library for large linear classification. Journal of Machine Learning Research 9, 1871–1874 (2008)

Class-Specified Segmentation with Multi-scale Superpixels

Han Liu[1], Yanyun Qu[1,*], Yang Wu[2], and Hanzi Wang[3]

[1] Computer Science Department, Xiamen University, China
[2] Academic Center for Computing and Media Studies, Kyoto University, Japan
[3] Center for Pattern Analysis and Machine Intelligence, Xiamen University, China

Abstract. This paper proposes a class-specified segmentation method, which can not only segment foreground objects from background at pixel level, but also parse images. Such class-specified segmentation is very helpful to many other computer vision tasks including computational photography. The novelty of our method is that we use multi-scale superpixels to effectively extract object-level regions instead of using only single scale superpixels. The contextual information across scales and the spatial coherency of neighboring superpixels in the same scale are represented and integrated via a Conditional Random Field model on multi-scale superpixels. Compared with the other methods that have ever used multi-scale superpixel extraction together with across-scale contextual information modeling, our method not only has fewer free parameters but also is simpler and effective. The superiority of our method, compared with related approaches, is demonstrated on the two widely used datasets of Graz02 and MSRC.

1 Introduction

This paper aims to segment an image into semantic objects. As a special case, we can extract foreground region from background region at pixel level. More generally, we can segment an image according to the class labels of its components as well; namely, label all objects in the image. Such a task is referred to as class-specified image segmentation in this paper.

Class-specified image segmentation is quite different from the unsupervised bottom-up image segmentation. A single region generated by the bottom-up image segmentation rarely represents a physical object, which is usually troublesome when used for higher level vision tasks. Moreover, bottom-up image segmentation is likely to be sensitive to the model parameters and the image data itself. Different choices of the parameters in a particular bottom-up image segmentation algorithm could generate segments with different quality on the same image. Therefore, the class-specified segmentation is proposed to overcome such problems. It segments an image according to the semantic information of the objects within it, which is expected to be consistent with humans perceptions.

---

* Corresponding author.

J.-I. Park and J. Kim (Eds.): ACCV 2012 Workshops, Part I, LNCS 7728, pp. 158–169, 2013.

The development of object localization has shed some light on class-specified segmentation. Dalal et al. [1] implemented a sliding window scheme combined with SVM classifiers to detect pedestrians. However, that method is time consuming. In order to solve this problem, Lampert et al. [2] proposed an efficient subwindow search method, which is based on the branch-and-bound scheme, to detect the generic object. Blaschko et al. [3] treated the problem of object localization as a regression problem, in which the objects location is an output of a learned objective function. However, the above-mentioned methods are all conditioned on the existence of an object template, which is hard to be made robust to the change of object appearance, such as rotation, illumination changes, occlusion, etc. Utilizing multiple templates might somewhat ease the problem, but for many objects in the unconstrained real images, such a strategy may lead to a significant increase of the total number of the required templates. Another flaw of those methods is that they only extract an object with a bounding box, thus being unable to provide accurate segmentation at pixel level.

Recent success in pixel-level categorization has shown a promise for object localization, in which one can label image pixels with the corresponding classes instead of roughly bounding an object with only a rectangle. Shotton et al. [4] constructed semantic texton forests (STF) to learn the local representation. They used a grid with small cells as the input to STF. However, their method is sensitive to the size of the cells and its accuracy decreases as it meets a higher speed demand. Fulkerson et al. [5] used superpixels instead of the regular patch grid for representation. They represented the local image information in an adaptive domain rather than in a fixed window and adopted Conditional Random Field (CRF) [6] to extract the object-level regions. Tighe et al. [7] proposed a similar method as [5]. The difference is that they used superpixel matching instead of classifying to compute the likelihood score for each class, while their commonness is to base themselves on the superpixels of a single scale. Therefore, both of their methods can only capture the context of neighboring superpixels, but not cover the across scale context of the informative superpixels of multiple levels in the scale space. Their performances are thus sensitive to the scale of superpixels and the range of superpixel neighbors, which results in a relative unstable object-level segmentation. Kohli et al. [8] proposed an image parsing method based on both pixels and unlabeled segments, encouraging pixels in the same segment to share the same label. Similar to [5][7], this method does not take into account the scale space context as well. The latest work that explored both the spatial coherency of neighboring superpixels in the same scale and the contextual information across scales was presented by Lubor et al. [9], in which a hierarchical CRF model was performed. However, this work has two shortcomings which limit its effectiveness and applicability. One is that its performance depends much on the goodness of the initial unsupervised segmentation, and the other is that it has many free parameters to be predefined, which is not a trivial task.

In this paper, we propose a new approach for class-specified segmentation which inherits the virtues of the existing methods while at the same time avoids

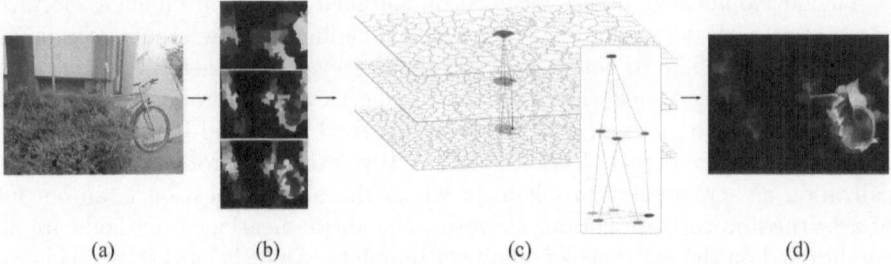

(a)   (b)   (c)   (d)

Fig. 1. The framework of the class-specified segmentation method. a) the original image; b) the classification of the segments at three scales where the red color means a high probability of the corresponding superpixels belonging to the bike and the blue color means a high probability of the corresponding superpixels belonging to the background; c) graph construction on the multi-scale superpixels; d) the obtained confidence map. (The figure is best viewed in color.)

their shortages. The proposed approach follows the idea of using CRF to integrate both the spatial coherency and across-scale consistency of multi-scale superpixels, but in a simpler and more effective way than the one presented in [9]. More precisely, instead of using appearance for representing the across-scale contextual information, we use the overlapping ratio which is proved to be more efficient and more effective. Our model has only one single free parameter: the number of scales, which is not sensitive to the input data, as to be witnessed in our experiments. All the other parameters of our model can be learned in the training stage. Besides its applicability, its superiority in terms of segmentation performance will be demonstrated in this paper, especially when it is compared with the most related method [5].

The rest of the paper is organized as follows. In section 2, we give the details about our method. In section 3, the experimental results are given to show the performance of our method. Conclusions are given in the section 4.

2   Class-Specified Segmentation

The framework of our method is shown in Figure 1. We firstly obtain the superpixels at multiple scales by changing the number of segments at each scale. Then an adaboost classifier for the foreground object is learned on the labeled training data. After that, the confidence values of superpixels are computed using the classifier. We employ the CRF model [6] to enforce the spatial consistency between the superpixels and their neighbors both in the same scale and in the consecutive levels in the scale space. Finally, we obtain the class-specified segmentation of an image, as shown in Figure 1(d).

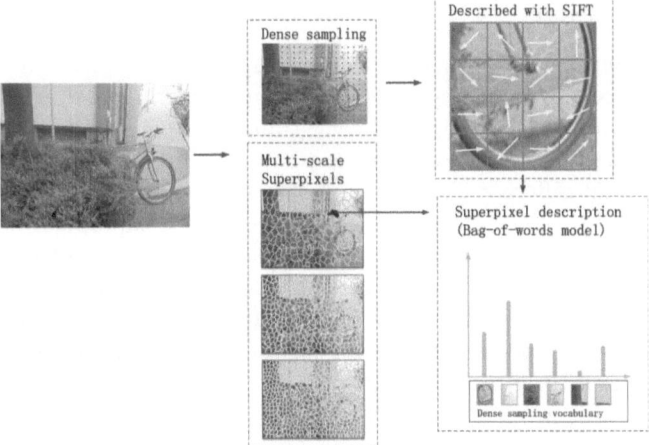

Fig. 2. The flowchart of superpixel's description with bag of words model. We get the SIFT descriptors for all the pixels. With the vocabulary of visual words, we describe each superpixel by the word frequency of the pixels within it.

2.1 Superpixel and Description

The motivation of using multi-scale superpixels is to capture the context of the superpixels of multiple levels in the scale space which may be critical for stable object-level segmentation. In our method, we firstly use SLIC superpixels [10] to oversegment an image with different numbers of segments and obtain the multi-scale superpixels. We have evaluated the following image segmentations in our framework: graph-based-segmentation [11], quickshift [12] and SLIC [10], and we found that the superpixels obtained by SLIC achieved the best performance with our model. It probably dues to the fact that the parameters in quickshift segmentation and graph-based-segmentation are relatively less sensitive to the change of color, which results in many repeated superpixels.

We employ the bag-of-words model (BOW) to describe these superpixels. Since sparse sampling may end up with a representation of superpixels which is not informative and stable enough, we use the dense description instead and describe each pixel by a SIFT descriptor [13] as shown in Figure 2, which is similar to [5,7]. These descriptors are then mapped to a vocabulary of visual words which are computed using vector quantization based on the K-means scheme. Before representing a superpixel, we dilate each superpixel region by four pixels in order to enforce the boundary information to the superpixel descriptor following [5,7]. To represent superpixels, we build a histogram of word frequency for each superpixel with the vocabulary. Moreover, we use the color cue as well. We compute the average color for each superpixel in the Lab color space for its high discriminative ability on colors. Finally we simply concatenate the histogram of visual words and the average color to form a high dimension feature vector for each superpixel.

2.2   Classification

In order to compute the confidence value of superpixels, we learn adaboost classifiers [14] based on the superpixels contained in the labeled object regions in
the training datasets. The label of each training superpixel is decided by the
labels of the pixels in the superpixel. If the pixels in the region of a superpixel
belong to several different classes, the label of the superpixel is determined by
the label shared by the largest number of pixels in the superpixel region. In the
case of binary segmentation, we learn a single binary adaboost classifier using
the labeled training data. For the case of the multi-class segmentation, we learn
the adaboost classifiers with multiple weak learners trained in a one-vs-rest way,
and the confidence of predicted label for each superpixel is decided by calculating
the votes from all these weak learners.

2.3   Graph Construction

Considering the spatial consistency between superpixels, we construct an three-
dimensional adjacency graph $G(S, E)$ to encode the spatial constraints, in which
S is the set of nodes, indicating all the superpixels from all scales, while E is the
set of edges connecting pairs of superpixels (s_i, s_j) being adjacent either spatially
in the same scale or across consecutive scales. As shown in Figure 1(c), we define
these two types of edges as horizontal edges and vertical edges. We connect the
pairwise superpixels in the same scale of an image with a horizontal edge if
they share a boundary, which represents the spatial context. And we connect
the pairwise superpixels in the multiple levels in the scale space with a vertical
edge if they share pixels, which stands for the scale context. Compared with
[5], we add the vertical edges which enable our method to capture the context
of multiple levels in the scale space and extract a stable object-level region. In
contrast, the performance of [5] is sensitive to the size of superpixels.

2.4   Inferring with CRF

We introduce CRF to carry out inference on the graph we built. Let $P(c|G, \omega, \nu)$
be the conditional probability of predicting label $\{c_1, \cdots, c_n\} \in C$ given the
adjacent graph $G(S, E)$ and the weights ω and ν:

$$-log(P(c|G, \omega, \nu)) = \sum_{s_i \in S} \psi(c_i|s_i) + \omega \sum_{(s_i, s_j) \in E_h} \phi(c_i, c_j|s_i, s_j) + \nu \sum_{(s_i, s_j) \in E_v} \varphi(c_i, c_j|s_i, s_j)$$

(1)

where E_h is the set of the horizontal edges, and E_v is the set of the vertical edges.
Moreover, ψ is the unary potential and ϕ is the horizontal pairwise potential,
while φ is the vertical pairwise potential. There are two weights ω and ν used
in our model: ω is the tradeoff parameter between the unary potentials and
the horizontal edge potentials, and ν is the tradeoff parameter between the
unary potentials and the vertical edge potentials. Since each graph may contain
more than one thousand nodes and thousands of edges, it could take several
days to train the parameters if we use gradient decent scheme. Alternatively, we

use an approximate scheme called stochastic gradient descent [15] to train the parameters ω and ν. For each iteration t, this scheme randomly selects a sample which contains about 5 to 20 batches of points, and computing its gradient by optimizing the maximum-likelihood estimation of C with $P(c|G, \omega, \nu)$. Then update the current parameters with the gradient by a small step. Repeat this process until it converges or iterates sufficient times. It is very fast and efficient.

We define the unary potential $\psi(c_i|s_i)$ by the confidence value obtained from Adaboost which is operated on the superpixels obtained in subsection 2.2. The horizontal pairwise edge potential ϕ is defined as:

$$\phi(c_i, c_j|s_i, s_j) = \frac{1}{1 + \|s_i - s_j\|} \cdot [c_i \neq c_j] \qquad (s_i, s_j) \in E_h \qquad (2)$$

and the vertical pairwise edge potential φ is defined as:

$$\varphi(c_i, c_j|s_i, s_j) = \frac{|s_i \cap s_j|}{|s_i \cup s_j|} \cdot [c_i \neq c_j] \qquad (s_i, s_j) \in E_v \qquad (3)$$

where $[c_i \neq c_j]$ is the zero-one indicator function. $\|s_i - s_j\|$ is the norm of the color distance between superpixels in Lab color space. The vertical pairwise edge potential is the ratio of the intersection area $|s_i \cap s_j|$ and the union area $|s_i \cup s_j|$ of the pairwise superpixels.

In our experiments we find that the vertical edges have contributed more than horizontal ones, because ν/ω is greater than 1 in most cases. It indicates that the context across scales is more important for object segmentation. As we mentioned before, the total number of the nodes in the graph is usually over a thousand, thus an exact inference is intractable. Therefore, we carry out approximate inference by employing the loopy belief propagation (LBP) [16], which is simple and efficient.

2.5 Across-Scale Label Confidence Integration

For each test image, we get the superpixels and the corresponding descriptions as mentioned in section 2.1. And then all the superpixels are tested through the Adaboost classifier obtained in section 2.2. After that, the CRF inference is carried out with the graph constructed in section 2.3, and the confidence value of each superpixel is obtained. Based on the CRF inference result, we can construct a pixel-wise confidence map for each category by averaging the class-specified confidence values from all the corresponding superpixels, that is, the confidence map is an image whose dimension is equal to the number of classes. Finally, a pixel is labeled according to its the maximum value of the labels, as shown in Figure 3(d) and 5(d).

3 Experimental Results

We evaluate our method on two publically available databases: Graz-02 and MSRC, and all of our results have been released on our website [1].

---

[1] https://sites.google.com/site/hanliupers/research/image-parsing

3.1   Graz-02

There are three categories in the Graz-02 dataset: car(300), bike(300) and person(300). It is a challenging dataset because the objects significantly vary with rotation, occlusion, scales, etc. We mainly compare our work with Fulkerson et al.'s work [5] because it is most related to our method, and we use the same training and testing data (i.e. the oddly indexed images are used as the training set, while the evenly indexed ones are used as the testing set). Some representative results are shown in Figure 3(d). As a result, our method has achieved better performance than Fulkerson et al.'s work (see Figure 3(c) and Table 1). In their work, different dilate sizes have been applied to superpixels, which results in different segmentation accuracies. On the contrast, in our work we dilate each superpixel with four pixels. In Table 1, we compare our segmentation accuracy with the best result of [5]. We present the results of our method with various numbers of scales (NS) to test the influence of NS on the performance. The results show that our approach work best when the number of scales is equal to 5. Compared with [5], our method achieves 11% higher accuracy on car, while the accuracy improvements on bike and person are 7% and 9%, respectively.

We have tried several different vocabulary sizes K=[100, 200, 400, 600, 800, 1000] in our experiment, and the results show that larger K tends to result in a better performance. However, when K gets greater than 200, it has little effect on the performance improvement. Therefore, we select $K = 600$ in our method.

Table 1. The comparison results in terms of the recall=precision points between [5] and the proposed method with different numbers of scales on the Graz-02 dataset

		Car	Bike	Person
The method in [5]		72.2%	72.2%	66.3%
The proposed method with different numbers of scales.	NS=3	79.5%	76.4%	72.2%
	NS=4	**83.5%**	77.1%	72.9%
	NS=5	81.4%	**79.4%**	**75.1%**
	NS=6	81.2%	78.2%	74.8%

3.2   MSRC

The MSRC dataset contains twenty-three categories. Similar to [4], we discard two categories: horse and mountain because they have too few samples. In this dataset, misclassifications usually happen between similar categories. For example, some parts of a cow can be misclassified as those of a sheep. Shotton [4] suggested to use an image-level prior (ILP) to solve this problem, considering that one may have some prior knowledge about what an image possibly contains before image parsing. To evaluate the ILP in our experiment, we simply describe each trained image with the bag of words model of Spatial Pyramid scheme [17], and learn a classifier from the training images. For each test image, we compute the prior probability(ILP) $P(c)$ on twenty-one categories with the

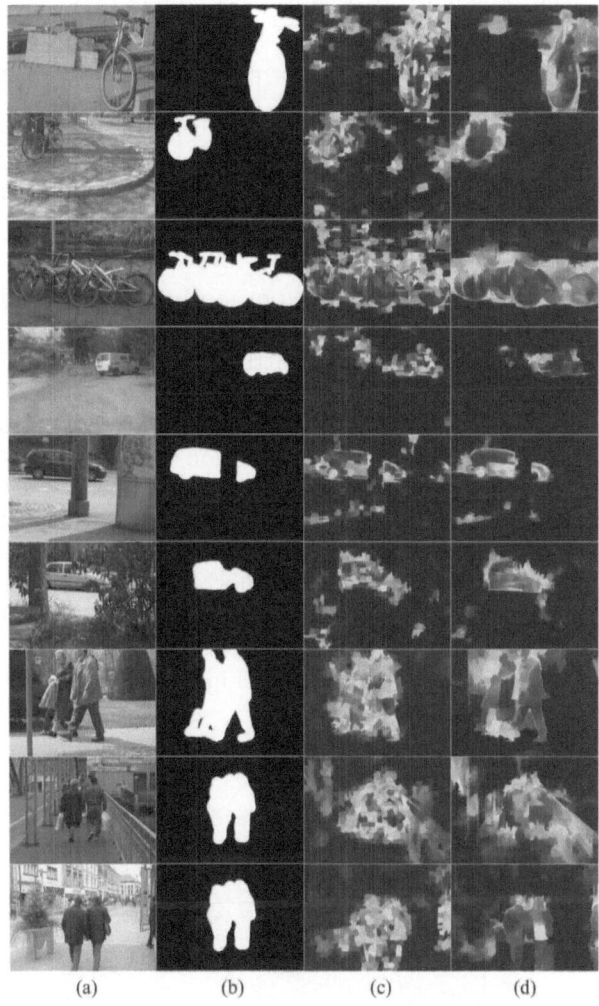

(a)   (b)   (c)   (d)

Fig. 3. Representative results on the Graz-02 dataset in which the red pixels represent the predicted foreground region; a) original images; b) ground-truth labels; c) the results of [5]; d) our results. (They are best viewed in color.)

trained classifier, and then multiply $P(c|G, \omega, \nu)$ by the posterior probability $P(c)$ as:

$$P'(c|G) = P(c|G, \omega, \nu) \cdot P(c)^{\alpha} \tag{4}$$

where α is used to soften the prior probability. $P'(c|G)$ is the final confidence value on each category. In section 2.4 we mentioned that the vertical edge is more important than the horizontal edge. To prove it, we show the value of

Table 2. The comparison of the tradeoff parameter ratios between vertical pairwise potential and horizontal pairwise potential for the 21 categories on the MSRC dataset

	building	grass	tree	cow	sheep	sky	airplane	water	face	car	bicycle	flower	sign	bird	book	chair	road	cat	dog	body	boat
ν/ω	2.0	3.4	2.2	0.3	4.8	1.4	1.3	4.3	2.1	1.2	1.3	3.4	2.1	5.7	1.3	3.3	1.8	6.7	1.5	3.0	1.8

the tradeoff ν/ω for each category trained by stochastic gradient descent [15] in Table 2. We have $\nu/\omega > 1$ for all the 21 categories except cow.

We present the pixel-level confusion matrix of our method on MSRC dataset in Figure 4 and show some representative segmentation results in Figure 5. Since Fulkerson et al. [5] did not experiment on the MSRC dataset, to compare with it, we test their method with our own implementation and show its results in Figure 5(c) and Table 3. Besides of that, we also compare our method with [4] and [18] in terms of segmentation accuracy. As shown in Table 3, our method achieves both the highest global accuracy (total proportion of correctly predicted pixels) of 75% and the highest averaged accuracy of 68%, and performs better than the other methods on 11 categories (more than half of 21). In terms of efficiency, our method takes an average of 8 seconds to process an image, while [18] required 3 minutes per image. Unlike [4] relies on learning from a large pool of features, our method only uses quite simple ones.

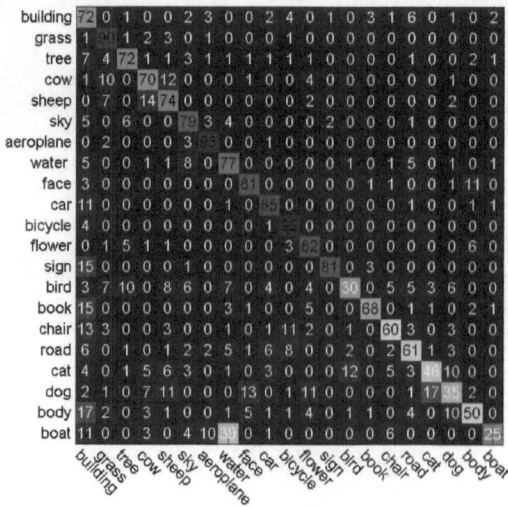

Fig. 4. The pixel-level confusion matrix of our method on the MSRC dataset

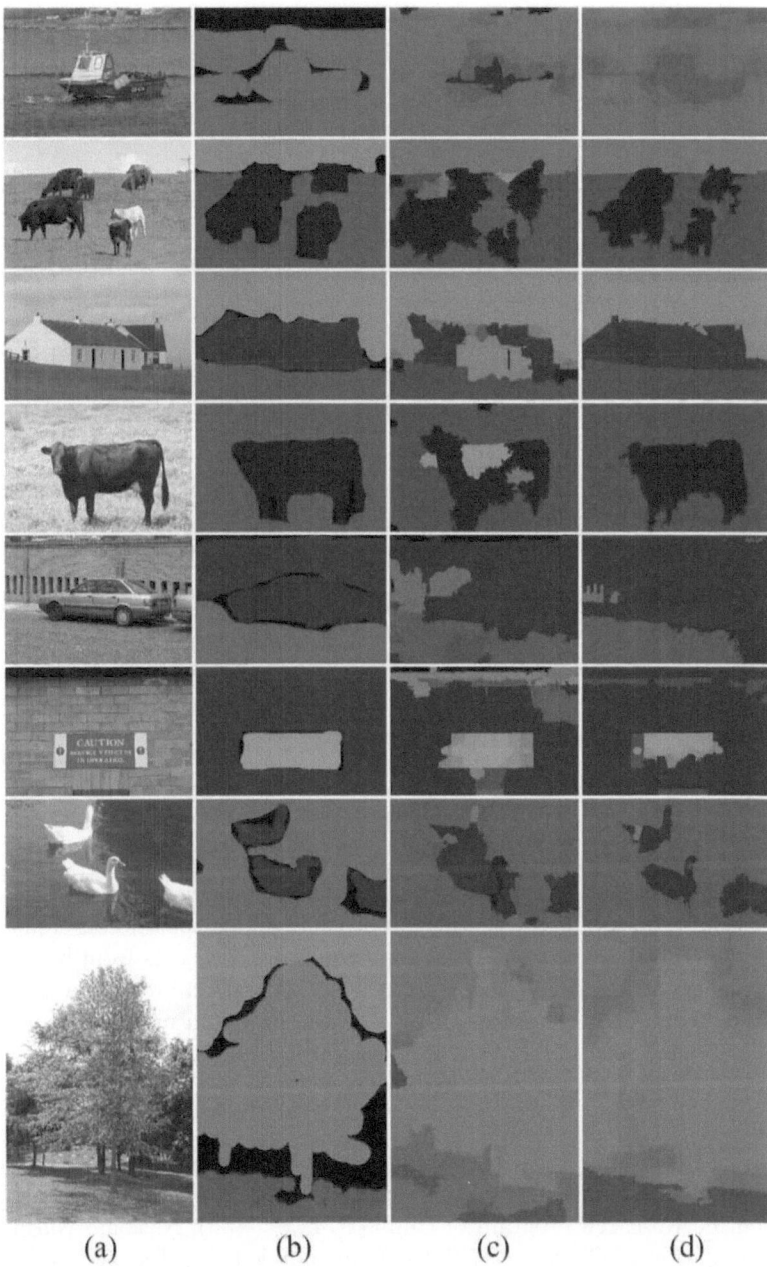

(a)   (b)   (c)   (d)

Fig. 5. The results of segmentation and classification for the MSRC dataset; a) original images; b) ground-truth labels; c) the results of [5]; d) our results, each map shows the most confident class at pixel level. (The figure is best viewed in color.)

Table 3. Segmentation accuracy (in percentage) for each class on the MSRC dataset

	Ours	[5]	[18]	[4]
building	**72**	54	62	49
grass	90	73	**98**	88
tree	72	66	**86**	79
cow	70	65	58	**97**
sheep	74	68	50	**97**
sky	79	**89**	83	78
airplane	**93**	90	60	82
water	**77**	65	53	54
face	81	75	74	**87**
car	**85**	76	63	74
bicycle	**95**	89	75	72
flower	**82**	69	63	74
sign	**81**	78	35	36
bird	**30**	24	19	24
book	68	50	92	**93**
chair	**60**	59	15	51
road	61	46	**86**	78
cat	46	53	54	**75**
dog	**35**	31	19	**35**
body	50	55	62	**66**
boat	**25**	23	7	18
Global	**75**	65	71	72
Average	**68**	62	58	67

4   Conclusions

We propose a class-specified segmentation method, which utilizes CRF to integrate the information of multi-scale superpixels under spatial constraints. The proposed method can be used to segment foreground objects from background, and it can also be used for image parsing. The experimental results on the widely used Graz02 and MSRC datasets show that the proposed method is superior to the related methods [4,5,18] and is more simpler and more effecient than the work [9].

Acknowledgement. This research work was supported by Fundamental Research Funds for the Central Universities (2010121067), National Defence Basic Scientific Research program of China(B1420****55), National Natural Science Foundation of China (61170179), Xiamen Science & Technology Planning Project Fund (3502Z20116005) of China, and "R&D Program for Implementation of Anti-Crime and Anti-Terrorism Technologies for a Safe and Secure Society", Special Coordination Fund for Promoting Science and Technology of the Ministry of Education, Culture, Sports, Science and Technology, the Japanese Government.

References

1. Dalal, N., Triggs, B.: Histograms of oriented gradients for human detection. In: Computer Vision and Pattern Recognition, vol. 1, pp. 886–893 (2005)
2. Lampert, C., Blaschko, M., Hofmann, T.: Beyond sliding windows: Object localization by efficient subwindow search. In: Computer Vision and Pattern Recognition, pp. 1–8 (2008)
3. Blaschko, M.B., Lampert, C.H.: Learning to Localize Objects with Structured Output Regression. In: Forsyth, D., Torr, P., Zisserman, A. (eds.) ECCV 2008, Part I. LNCS, vol. 5302, pp. 2–15. Springer, Heidelberg (2008)
4. Johnson, M., Shotton, J.: Semantic Texton Forests. In: Cipolla, R., Battiato, S., Farinella, G.M. (eds.) Computer Vision. SCI, pp. 173–203. Springer, Heidelberg (2010)
5. Fulkerson, B., Vedaldi, A., Soatto, S.: Class segmentation and object localization with superpixel neighborhoods. In: Computer Vision and Pattern Recognition, pp. 670–677 (2009)
6. Sutton, C., Mccallum, A.: An Introduction to Conditional Random Fields for Relational Learning. In: Getoor, L., Taskar, B. (eds.) Introduction to Statistical Relational Learning, MIT Press (2006)
7. Tighe, J., Lazebnik, S.: Superparsing: Scalable nonparametric image parsing with superpixels. International Journal of Computer Vision (2012)
8. Kohli, P., Ladický, L., Torr, P.H.: Robust higher order potentials for enforcing label consistency. Int. J. Comput. Vision 82, 302–324 (2009)
9. Ladicky, L., Russell, C., Kohli, P., Torr, P.H.S.: Associative hierarchical crfs for object class image segmentation. In: ICCV 2009, pp. 739–746 (2009)
10. Achanta, R., Shaji, A., Smith, K., Lucchi, A., Fua, P., Süsstrunk, S.: SLIC Superpixels Compared to State-of-the-art Superpixel Methods. Pattern Analysis and Machine Intelligence (2012)
11. Felzenszwalb, P.F., Huttenlocher, D.P.: Efficient Graph-Based Image Segmentation. International Journal of Computer Vision 59, 167–181 (2004)
12. Vedaldi, A., Soatto, S.: Quick Shift and Kernel Methods for Mode Seeking. In: Forsyth, D., Torr, P., Zisserman, A. (eds.) ECCV 2008, Part IV. LNCS, vol. 5305, pp. 705–718. Springer, Heidelberg (2008)
13. Lowe, D.G.: Distinctive Image Features from Scale-Invariant Keypoints. International Journal of Computer Vision 60, 91–110 (2004)
14. Friedman, J., Hastie, T., Tibshirani, R.: Additive Logistic Regression: a Statistical View of Boosting. The Annals of Statistics 38 (2000)
15. Vishwanathan, S.V.N., Schraudolph, N.N., Schmidt, M.W., Murphy, K.P.: Accelerated training of conditional random fields with stochastic gradient methods. In: International Conference on Machine learning, ICML 2006, pp. 969–976. ACM, New York (2006)
16. Murphy, K.P., Weiss, Y., Jordan, M.I.: Loopy belief propagation for approximate inference: an empirical study. In: Proceedings of the Fifteenth Conference on Uncertainty in Artificial Intelligence, UAI 1999, pp. 467–475. Morgan Kaufmann Publishers Inc., San Francisco (1999)
17. Lazebnik, S., Schmid, C., Ponce, J.: Beyond Bags of Features: Spatial Pyramid Matching for Recognizing Natural Scene Categories. In: Computer Vision and Pattern Recognition, vol. 2, pp. 2169–2178 (2006)
18. Shotton, J., Winn, J., Rother, C., Criminisi, A.: Textonboost for image understanding: Multi-class object recognition and segmentation by jointly modeling texture, layout, and context. International Journal of Computer Vision 81, 2–23 (2009)

A Flexible Auto White Balance
Based on Histogram Overlap

Tao Jiang, Duong Nguyen, and K.-D. Kuhnert

Institute of Real Time Learning Systems, University of Siegen, Siegen, Germany
{jiang,kuhnert}@fb12.uni-siegen.de

Abstract. Auto white-balance plays a very important role in computer vision, and also is a prerequisite of color processing algorithms. For keeping the color constancy in the real-time outdoor environment, a simple and flexible auto white balance algorithm based on the color histogram overlap of the image is presented in this paper. After looking at a numerous images under different illuminance, an essential characteristic of the white-balance, the color histogram coincidence, is generalized as the basic criterion. Furthermore the overlap area of the color histogram directly reflects this coincidence, namely, when the overlap area of the color histogram reaches the maximum, the respective gain coefficients of color channels can be derived to achieve the white-balance of the camera. Through the subjective and objective evaluations based on the processing of real world images, the proposed histogram overlap algorithm can not only flexibly implement the auto white-balance of the camera but also achieve the outstanding performance in the real-time outdoor condition.

1   Introduction

The formation of an image captured by the digital camera depends on the physical content of the scene, the sensitive property of the camera and the illumination of the environment. In the case of the different illumination from various light sources, the image will appear the different color against the same scene. For instance, a white object shotted still keeps its white color under the normal daylight, but it will present the reddish color when it's captured under the incandescent bulb, similarly the bluish color will appear under the fluorescent light source[13]. This color cast more easily occurs to the vision system of an autonomous vehicle running outdoor. Therefore, it is necessary to correct the white balance of cameras against the various illuminants with the specified temperature so that the autonomous vehicle can perceive the objects in the same scene as the non-deviation color objects and recognize them under the different illumination.

The auto white balance(AWB), a basic function on the digital camera, is to remove the color case as the camera imaging, keep the color constancy in the different illumination. In the past decades years, many researchers proposed a number of AWB algorithms in many manners. The Gray-World is an oldest and best-known assumption, it assumes that the average reflectance in a scene under

J.-I. Park and J. Kim (Eds.): ACCV 2012 Workshops, Part I, LNCS 7728, pp. 170–181, 2013.

a neutral light source is achromatic[3,12], that is, all the pixel values of R, G and B channels are nearly equivalent and the color in each sensor channel averages to gray over the entire image[1,2,14]. However, when there are some big areas with single color or some predominant color in a scene, such as face, grassland and sea, the Gray-World algorithm will cause obviously errors. The White-Patch algorithm also called the perfect reflectance, assuming the observed pixels with greatest intensity must correspond to a color-neutral surface patch in the scene, dedicates to look for the maximum values for the R, G and B channels in the image, and obtain the chromaticity of the light source, so as to compensate the color cast efficiently[12,10,6,5]. Although the White-Patch algorithm amended the defect of the Gray-World method, it will inevitably encounter the failure when there are greatly bright pixels in plenty in the image. Instead of only using statistics of pixel values for determining the AWB, more sophisticated and complex methods were developed, which exploited information and features captured in a learning phase. One of them was the gamut mapping by Forsyth[9], which assumed that one may observe only a limited number of colors for a given illuminant in real-world images[12]. The gamut mapping algorithms derive the results quit well[7,8], but the computationally expensive cost limits their application in real-time environment such as autonomous robot. Most of methods are more or less based on some assumptions, which makes them limited to apply in the common environment. For the sake of algorithms' robustness, some researchers combined different approaches together[15,4]. Obviously this combination made a large increase of the performance comparing to those state-of-the-art single algorithms.

Although many AWB also knows as color constancy algorithms achieved numerous successful applications in various fields, there is still no full fitted algorithm to be consider as universal. Especially in the context of the autonomous mobile vehicle, there is still a need to develop a real-time and robust AWB algorithm for it's vision system. In this paper, a simple and flexible AWB algorithm based on the coincidence of the color histogram is proposed for deriving the stable and efficient performance of the AWB at low computational cost. Furthermore the experiment results indicate that the proposed algorithm present a good performance of removing color cast and decreasing computational cost.

2   White Balance Approach

As the above reviewed, there are various kinds of methods to implement the AWB at present. In general, aiming at the single input image for determining the AWB, most of the AWB methods can obviously be divided into two procedures. Firstly, the color temperature of the input image should be determined according to the features and properties of the input image, i.e. deducing the color temperature of the unknown light source. After that, the color cast correction can be obtained from the mapping between the unknown light source and the canonical light source.

2.1   Color Temperature Determination

According to the theory of the digital image formation, the simple and practical assumption model is the Lambertian model that only considering the major and key factors: the color of input light, the surface reflectance and the camera sensitivity.

Based on the Lambertian assumption, a color image is represented as follows[12],

$$E_k(x, y, \lambda) = \int_\omega R(x, y, \lambda) L(\lambda) S_k(\lambda) d\lambda \tag{1}$$

where, the $R(x, y, \lambda)$ denotes the surface reflectance, the $L(\lambda)$ is the illumination property, and the $S_k(\lambda)$ is the sensor characteristic, all the three variables are as a function of the wavelength λ, over the visible spectrumω. Every channel of the sensor is represented by the subscript $k(k = R, G, B)$, and the $E_k(x, y, \lambda)$ is the image corresponding to the kth channel. Generally, under the assumption that the surface reflectance is a constant and the sensor's property is also known, the color appearance of the image will directly response to the change of illuminance in a scene. Thus the color temperature of the input image (or more accurately, the color temperature of unknown light) can be derived from the features of input image. For example, the Grey-World method fully utilizes the RGB mean, the statistic feature, through the whole or parts of the image.

2.2   Color Correction

Once deriving the color temperature of the image, the color correction can be implemented through transforming the input image shot under an known light source to the image without color cast. Considering the application in the vision of the autonomous robot, the linear diagonal transformation also called von kries model[12], can be used to simplify the complication of processing and reduce the computational cost, instead of the complicate others such as the linearized Bradford, CIECST02 or affine transformation[12]. Most of the algorithms mentioned previously adopted to this linear model exactly.

$$\begin{pmatrix} R_c \\ G_c \\ B_c \end{pmatrix} = \begin{pmatrix} k_R & 0 & 0 \\ 0 & k_G & 0 \\ 0 & 0 & k_B \end{pmatrix} \cdot \begin{pmatrix} R_u \\ G_u \\ B_u \end{pmatrix} \tag{2}$$

From the (2), the $R_c, G_c and B_c$ represent three color values from RGB channels respectively, which were taken under an unknown light source, correspondingly the R_u, G_u and B_u describe the color values under canonical light source. the linear and diagonal transform matrix maps the color that were taken under the unknown light source to their corresponding color under the reference light source.

2.3   Color Histogram and White Balance

Histograms exactly show us the 256 different brightness levels from black to white, here the level 0 represents the pure black and the level 255 denotes pure white, So this makes that the histogram can give an indication of how many dark, middle tone, and bright pixels there are, and whether any shadow or highlight detail is lost. Ideally luminosity histogram can be applied to monitor the camera's

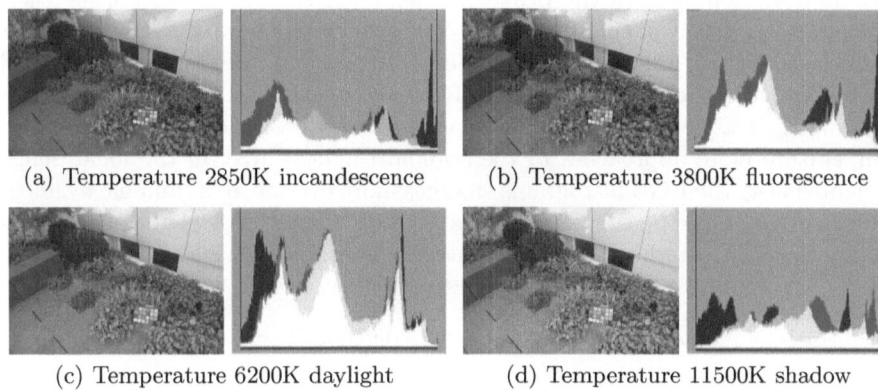

(a) Temperature 2850K incandescence   (b) Temperature 3800K fluorescence

(c) Temperature 6200K daylight   (d) Temperature 11500K shadow

Fig. 1. Images under different illuminance and their respective histograms

exposure and assess the changes in brightness or contrast. However, we're more interested in analyzing and modifying how tones the entire image distributes, obviously a color histogram of the image can really reflect those features instead of luminosity. The color histogram provides a compact summarization of the color distribution in an image. The following equation reflects the essence of the color histogram in depth,

$$h^c[n] = \frac{1}{XY} \sum_{x \in X} \sum_{y \in Y} \begin{cases} 1 & if\ I^c(x,y) = n \\ 0 & otherwise. \end{cases} \quad n \in [0, 255] \quad (3)$$

The h is the statistic amount of pixels on every grey level n , the superscript c denotes each channel in the RGB color space. After accumulating the frequency of every color value $I^c(x,y)$ throughout the entire image with the size of width X and height Y, the color histogram h attained can represent the color distribution of the image. Generally, the color histogram of an image has the relative invariance with the translation and rotation regarding the viewing axis. So it becomes an important basic method in the field of pattern recognition. Furthermore, the color histogram still presents high sensitive to the variation of illumination in the RGB color space. Considering the (1) and the (2) together, the color histogram also describes the changes of the color temperature, namely the illuminance in the shot scene. Based on the significant feature of the color histogram's high lightness sensitivity, determining the accurate change of the gain regarding the white balance becomes more simply and easier in an application to the real-time autonomous vehicles. How to derive the parameters of the white balance from the color histogram of an image? From the research of a large number of images shot under different color temperatures, we know that the coincide of the color histogram reflects the change of color temperature. As the Fig.1 shows this relationship, when there is obviously color cast in the 1(a) and 1(d), i.e. severely bluish and reddish, accordingly their respective color histograms show the big deviation at both the peak and valley position, in the meantime their figures also appear extreme dissimilar. Their color temperatures are rather far from the 6200K of the standard light source. In the 1(b), the color deviation still exists, but the respective histogram of R, G and B channels has comparable figure, and their extreme points are quite closed in position, additionally the color temperature more approaches to the standard light source. In the ideal image 1(c), obviously the RGB histograms emerges more overlap areas and have almost the same figure except the part of blue area on the left side that is caused by dominated color. As the Table 1 presented, when the deviation of color temper-

Table 1. Overlap areas under different illuminants

illuminant	Area(pixels)	Ratio(%)
Incandescence	123586	45.33
Fluorescence	178440	65.45
Daylight	188272	69.06
Shadow	131475	48.22

ature from normal daylight increases hardly, correspondingly the overlap area decreases obviously. In the case of daylight, the color constant reaches in the image, simultaneously the overlap area also presents greatest value comparing to others.

In the other hand, the assumptions of the Gray-World and the White-Patch similarly embody the principle of the coincide of the histogram. The Gray-World assume that the average intensities of the Red, Green and Blue channels should be equal when plenty of colors exist in an image. When the average intensities reach the equality, their color histogram coincide very well, i.e. the number of overlapping pixels reaches the most amount. The White-Patch assume that the greatest intensity must correspond to a color-neutral surface patch in the scene, this color-neutral also means the corresponding color histogram has the greatest coincidence and its intensities of pixels in the RGB channels are nearly equal.

In view of the above analysis, a novel white-balance method is proposed, which utilizes the coincidence of the color histogram to estimate the white-balance of images. That is, according to the RGB histograms' overlap, the related position and figure of color histogram represents the different action in different light source. Only determining the max measurement of overlap, and then the corresponding illumination is easy to obtain.

3 Proposal Algorithm

The purpose of white-balance algorithm is to convert the image shot under the unknown light source to the image shot under the canonical light source. From the previous analysis,we know that the histogram represents the color distribution, only if the histogram keep the max overlap between three color channels, the image must be considered as being shot under canonical light. In this case, the color of the image from the camera should be consistent with that from human eyes.

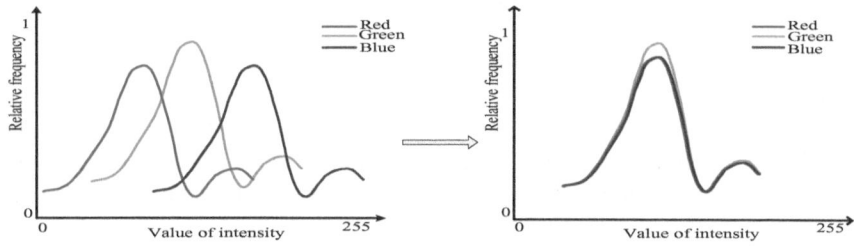

Fig. 2. White balance principle chart

As Fig.2 shown, the RGB channels' histograms appear some deviation each other in the left graph, that is, the overlap rate is quite low. The goal of obtaining the white-balance of the image through its histogram is to determine the related coefficients of the gains when making three channels' histograms greatest overlapping, like as the right graph. The Fig.3 is the implementing diagram of

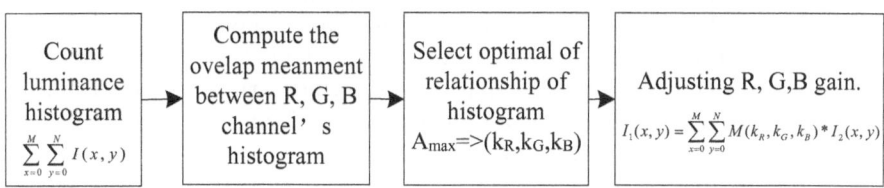

Fig. 3. The block diagram of implementing Auto white balance

the proposed method, it implements auto white-balance in four steps. Firstly, the color histogram,the statistic measurement that describes the distribution of the color of an image, is counted through the whole image or its down-sampling image. Then the proposed algorithm measures the common overlap of the color histogram between three color channels. Afterwards it compares the different overlap area regarding the gains of the different R, G and B channels, and determine the optimal gain of channels, which really is the expected gain coefficients of the image. In the end the color cast image can be convert to the ideal color image by color correction matrix, looking like that is shot under the standard light source.

3.1   Obtain the Overlap of the Color Channel Histogram

For deriving the overlap of the color histogram, the minimum value will be kept at every grey level and the overlap area can reach through the sum from level 0 to level 255. the (4) generalize the process simply.

$$H(h^R, h^G, h^B) = \frac{\sum_{j=1}^{n} min(h_j^R, h_j^G, h_j^B)}{\sum_{j=1}^{n} h_j^G} \tag{4}$$

where, the $h_j^R, h_j^G,$ and h_j^B are the value of the jth grey level in the R,G and B channel respectively.

The overlap area can be obtained between three channels from the (4). Here the H indicates the overlap area of the R, G and B channels' histograms, which derives from the accumulated minimum on all grey levels regarding three color channels.

3.2   Find Optimal Coincide

Since the overlap area of the color histogram reflects the variation of the color temperature of images, and the max overlap area is responsible for the standard color temperature, the corrected coefficients can be derived when the overlap area reaches the maximum.

$$arg \max_{k_R, k_G, k_B} H(h^R(k_R), h^G(k_G), h^B(k_B)) \tag{5}$$

As the (5) shown, if the overlap area H reaches the maximum, the gain coefficients k_R, k_G and k_B become the desired values. That can also be considered that the optimal status is obtained when the histograms of the R,G and B channels have closest features and smallest dissimilarity. Furthermore, the optimal parameters are applied directly to the linear and diagonal transform matrix to correct the gains of the R, G and B color channels.

4   Experiment and Evaluation

For verifying the performance of the proposed algorithm, the related experiments require the tested image with a scene illuminant known or standard marker as ground truth. The P.V. Gehler's database set [**??**] including 258 images with a wide variety of indoor and outdoor shots, which placed a MacBeth Color Checker in the scene as the ground truth of images, is applied to effectively evaluate the performance of algorithms. In the meantime, a number of real-world outdoor shots with diverse illumination and complicate scenes also are adopted to examine the proposed algorithm.

4.1   Experiment

In general, the outdoor images always include some noise that is caused by many factors such as the variation of illuminance. So the Gaussian filter is adopted to reduce the affection of the noise prior to other procedures. Subsequently the counting process of the color histogram starts in the image that has been eliminated the noise and still kept original color distribution. The metric of coincidence in the histograms is computing the overlap area, That is, by (4), the diverse overlap areas is easy to determine corresponding to the variable gain coefficients. Note that the overlap area of the color histogram shouldn't include the grey level 0 and 255 when summing the number of all pixels on the every grey level, because when an image shot in the condition of under-exposure or over-exposure, there are a lot of pixels accumulate on the grey level 0 or 255, which have lost abundant color information of the scene. Actually this situation will also arise similarly when searching the optimal gain coefficients in the image with normal exposure, if the varying ranges of gain parameters is too big, the extreme situation will be encountered easily. In this case, those pixels are useless to compute the overlap area. On the contrary, it not only severely affects the accuracy of the results but also causes the failure of the proposed method. The Fig.4 illustrates the exceptions. When the range of the gain parameters is

(a) Original image   (b) Exception with level 0   (c) Exception with level 255

Fig. 4. The case of exception with grey level 0 and 255

[0.2,1.2], as the Fig.4(b) shown, the image corrected looks dimmer than the original image. On the contrary, many too bright areas arise in Fig.4(c) within the range [0.6,5]. Therefore, the actual ranges of R and B channels are set in [0.4,2] and the results are quit satisfied. In practice, the variation of the Green channel should be kept in small range, such as [0.8,1.2]. Because if the Green channel change too more, which contribute more than the Red and Blue channels to the intensity, the intensity of the image will change hardly, this consequence will become unacceptable for the purpose of the AWB algorithm. Actually the algorithm holds the green channel unadjusted to sustain the intensity of the image with color cast, simultaneously the process also reduces the running time of the algorithm and complexity.

After deriving the max overlap area and its corresponding optimal parameters, the (2) can fast and linearly correct the color cast of images, namely implement color constancy. But there is still a notable matter that the value over 255 need be set to 255 during mapping the image shot under the unknown light source into the canonical light source using the diagonal matrix composed of the optimal parameters. Otherwise, the corrected image may arise some discontinuous pixels obviously.

(a) Original images

(b) Images corrected

Fig. 5. The results of the proposed algorithm

The results of the experiment using real-world images and database images were shown in the Fig.5. Obviously in the Fig.5(a), the three original images had the different color cast, while in the corresponding Fig.5(b), the color of images looked vivid and the color cast was eliminated after the processing of the proposed method even the image with a complicate scene. Hence the proposed algorithm implemented the AWB efficiently.

4.2   Subjective and Objective Evaluation

For comparing the proposed histogram overlap method with others, we use the Gray-World, White-Patch and Gamut mapping to correct the same original image with some color cast, and then listed the results in the Fig.6 as comparison. From their respective color histogram, all methods of the AWB had increased the coincidence of their color histograms contrasting with original image. However, the best coincidence of the color histogram derived from the proposed method. Meanwhile, the white car also reached the purest white and the blue parking sign, yellow barrier gate and lawn appeared more vivid. No doubt, this achieved greatest overlap area that leaded to best white-balance, and made the corrected

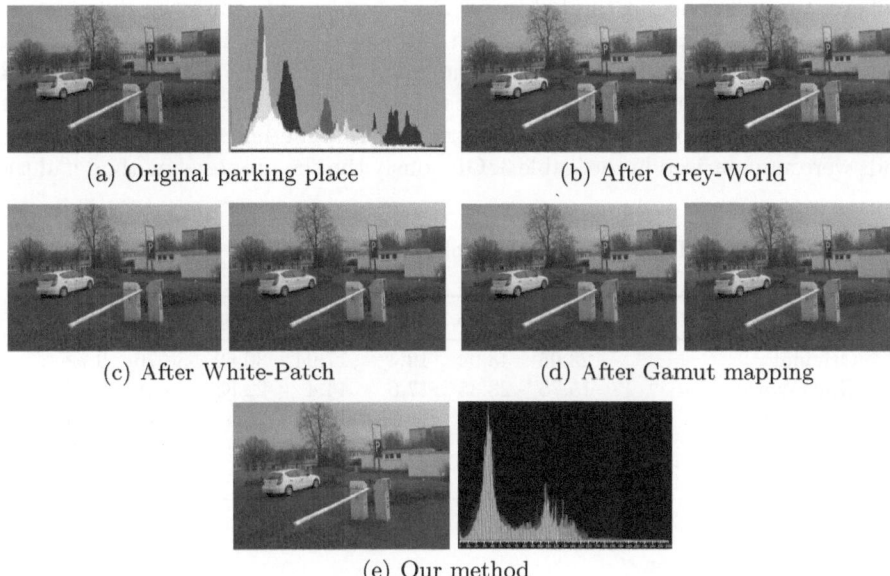

(a) Original parking place   (b) After Grey-World

(c) After White-Patch   (d) After Gamut mapping

(e) Our method

Fig. 6. Images after different AWB algorithms and their respective histograms

Table 2. Overlap areas comparing between algorithms

Algorithm	Area(pixels)	Ratio(%)
Original	119506	43.94
Grey-World	157594	57.94
White-Patch	121899	44.82
Gamut mapping	131412	48.31
Histogram overlap	189301	69.60

image look more consistent with real-world perceived by human eyes. Since the quantitative analysis is more objective to evaluate the performance of the proposed AWB algorithm , we summed the number of pixels in the overlap area of the color histogram after every AWB algorithm. From the Table 2, obviously the histogram overlap method obtained the greatest number of pixels in the overlap area, which also reached consistence with the above subjective evaluation. On the other hand, based on the color checker in the scene, the Euclidean distance (ΔE_{ab}^{*}) between the reference white and the white-balanced colors in the color space CIE L*a*b, provides the relative perceptual differences directly. The computer equation is just as the (6) shown.

$$\Delta E_{ab}^{*} = \sqrt{(\Delta L^*)^2 + (\Delta a^*)^2 + (\Delta b^*)^2} \tag{6}$$

where, the difference of the lightness: $\Delta L^* = L_2^* - L_1^*$, and the difference of the chromaticity :$\Delta a^* = a_2^* - a_1^*$ and $\Delta b^* = b_2^* - b_1^*$. In the equation, the L_1^*, a_1^* and b_1^* indicate the luminance and color values of the color checker , while the L_2^*, a_2^* and b_2^* indicate the luminance and color values of the captured image. The results of the presses by the histogram overlap method and the other methods were summarized in the Table 3. Obviously the the proposed method had the

Table 3. The average of ΔE_{ab} of images under different methods

Algorithm	image1	image2	image3	image4	image5	image6	image7
Original	23.95	43.94	19.4	17.94	30.52	38.76	9.72
Grey-World	15.75	23.35	17.6	14.4	22.83	23.58	9.02
White-Patch	12.18	11.82	18.95	12.51	27.5	22.14	8.86
Gamut mapping	13.14	9.81	16.96	8.23	21.3	19.91	7.98
Histogram overlap	11.23	8.36	16.92	7.36	17.33	18.84	7.76

lest Euclidean distance comparing to the Grey-World, White-Patch and Gamut mapping methods in the same scene under the same illuminance, namely this histogram overlap is a highly effective algorithm and its color is more consistent with the perception of human eyes.

Although the proposed method had outperformed others in numerous real scenes, it sometime lacked of strength for the cases of dominant color. In the case of the scenes with the extreme dominant color such as grassland and red wall, the overlap area of color histogram rapidly decreased and became more sensitive to the dominant channel rather than non-dominant channels, that's likely to result the local optimal values.

5   Conclusions

In this paper, a simple and practical AWB algorithm is proposed based on the coincidence of the color histogram of an input image, which dynamically measures the overlap area of the histograms in the R,G and B channels and derives the optimal parameters when the overlap area reaches or approaches the maximum. According to the determined adjusting coefficients, the camera can automatically adjust its gains of the R,G and B channels to eliminate the color case of the images and implement the white-balance. This proposed histogram overlap method utilizes the essential characteristic, the coincidence of each basic color hue, to achieve the color consistent, it not only reaches the same goal by different route instead of the classic assumptions like the gray world and white patch, but also overcomes their shortcoming and obtains the better results. Through the test of hundreds of images, the proposed algorithm can very well implement the auto white balance of the camera, furthermore, it simplifies and improves the performance of the AWB without training images beforehand.

References

1. Agarwal, V., Abidi, B., Koschan, A., Abidi, M.: An Overview of Color Constancy Algorithms. J. of Pattern Recognition Research 1, 42–54 (2006)
2. Barnard, K., Martin, L., Coath, A., Funt, B.: A comparison of computational color constancy algorithms, part ii: Experiments with image data. IEEE Transactions on Image Processing 11(9), 985–996 (2002)
3. Buchsbaum, G.: A Spatial Processor Model for Object Color Perception. J. of Franklin Institute 310, 1–26 (1980)
4. Cardei, V., Funt, B.: Committee-based color constancy. In: IS&T/SID's Color Imaging Conference, pp. 311–313 (1999)
5. Chikane, V., Fuh, C.: Automatic white balance for digital still camera. J. of Information Sciene and Engineering 22, 497–509 (2006)
6. Ebner, M.: Color constancy based on local space average color. J. of Machine Vision and Applications 20(5), 283–301 (2009)
7. Finlayson, G., Hordley, S.: Improving gamut mapping color constancy. IEEE Transactions on Image Processing 9, 1774–1783 (2000)
8. Finlayson, G.D., Hordley, S.D., Hubel, P.M.: Color by correlation: a simple, unifying framework for color constancy. IEEE Transactions on Pattern Analysis and Machine Intelligence 23(11), 1209–1221 (2001)
9. Forsyth, D.A.: A novel algorithm for color constancy. J. of Computer Vision 5(1), 5–36 (1990)
10. Gasparini, F., Schettini, R.: Color balancing of digital photos using simple image statistics. J. of Pattern Recognition 37, 1201–1217 (2004)
11. Gasparini, F., SchettiniColor, R.: Color balancing of digital photos using simple image statistics. J. of Pattern Recognition 37, 1201–1217 (2004)
12. Gijsenij, A., Gevers, T., Weijer, J.: Computational Color Constancy: Survey and Experiments. IEEE Trans. on Image Processing 20(9), 2475–2489 (2011)
13. Huo, J., Chang, Y., Wang, J., Wei, X.: Robust Automatic White Balance Algorithm using Gray Color Points in Images. IEEE Transactions on Consumer Electronics 52, 541–546 (2006)
14. Kim, Y., Lee, H.S., Morales, A.W.: A video camera system with enhanced zoom tracking and auto white balance. IEEE Transactons on Consumer Electron 48(3), 428–434 (2002)
15. Lam, E.: Combining Gray World and Retinex Theory for Automatic White Balance in Digital Photography. In: The Ninth International Symposium on Consumer Electronics, pp. 134–139 (2005)

Region Segmentation and Object Extraction Based on Virtual Edge and Global Features

Fumihiko Mori[1] and Terunori Mori[2]

[1] Tamagawa University, Tokyo
morif@lab.tamagawa.ac.jp
[2] Fuji Visual Science Laboratory, Kanagawa, Japan
spx82su9@lagoon.ocn.ne.jp

Abstract. We have developed a robust statistical edge detection method by combining the ideas of Kundus method, in which the region segmentation of local area is used, and Fukuis method, in which a statistic evaluation value separability is used for edge extraction and also have developed a region segmentation method based on the global features like the statistics of the region. A new region segmentation method is developed by combining these two methods, in which the edge extraction method is used instead of the first step of region segmentation method. We obtained the almost same results as the ones of previous region segmentation method. The proposed one has some advantages because we are able to introduce a new conspicuity degree including a clear contrast value with the adjacent regions, a envelopment degree based on clear edge and so on without much difficulty and it will contribute to develop a further unification algorithm and a new feature extraction method for scene recognition.

1 Introduction

Edge, where a sudden spatial change of a characteristic (such as brightness or color) occurs, is a fundamental feature for understanding images. The conventional methods of edge extraction can be classified into three broad categories: spatial differentiation-based methods [2, 6, 7, 11, 16–18], model-based methods [8], and statistics-based methods [4, 9, 20]. The spatial differentiation-based methods use a bottom-up approach, the model-based methods use a top-down approach, and the statistics-based methods use either a top-down or a bottom-up approach. In the statistics-based methods for edge extraction, a local area is divided into two sub-regions, R_1 and R_2, according to the top-down or the bottom-up criteria, and a statistical evaluation of edge existence is conducted.

Yakimovsky [20] used top-down approach for a local area segmentation and equation (1) for statistical evaluation.

$$\eta_0 = \frac{\sigma_0^{2(N_1+N_2)}}{\sigma_1^{2N_1}\sigma_2^{2N_2}},$$ (1)

where σ_0^2 is the variance of the feature in the local area, and σ_i^2 and N_i are respectively the variance in and the number of pixels belonging to R_i. The value

J.-I. Park and J. Kim (Eds.): ACCV 2012 Workshops, Part I, LNCS 7728, pp. 182–193, 2013.
© Springer-Verlag Berlin Heidelberg 2013

of η_0 is approximately 1 in a homogeneous feature area and it is infinite at a step-type edge.

Kundu [9] divided the local area according to the mean value μ_0 of the k-th maximum and k-th minimum brightness in the area (bottom-up approach). The evaluation was based on the distance between the centers of the two sub-regions R_1 and R_2, respectively (μ_{1x}, μ_{1y}) and (μ_{2x}, μ_{2y}), and on the size difference $|N_1 - N_2|$.

Fukui [4, 5, 19] proposed one rectangle separated into two equal-sized small rectangles and obtained four rectangles by rotating them in 45-degree steps. The evaluation was carried out using separability η_1 defined by equations (2)-(4) mentioned by Otsu [15].

$$\eta_1 = \frac{\sigma_b^2}{\sigma_T^2}, \tag{2}$$

$$\sigma_b^2 = N_1(\mu_1 - \mu_0)^2 + N_2(\mu_2 - \mu_0)^2, \tag{3}$$

$$\sigma_T^2 = \sum (I_j - \mu_0)^2, \tag{4}$$

where I_j is the brightness of the j-th pixel, and μ_0, μ_1, and μ_2 are the mean brightness in the original rectangle, one small rectangle, and the other small rectangle, respectively. The value η_1 is almost 1 in a step-type edge, approximately 0 in a homogeneous feature area, and approximately 0.75 (more precisely, $0.75/(1 - 1/n^2)$, where n is rectangle size) in a homogeneous gradient area (i.e., smooth gradation area). However, the value for a step-type corner edge is about 0.33, less than rather than greater than that for a homogenous gradient area.

Unfortunately, these conventional methods have the following problems:

(1) spatial differentiation-based methods often
require the use of smoothing filters for the suppression of noise and so produce thick edges,

(2) Kundus [9] method is ineffective under massive impulse-type noisy conditions [6], and

(3) Fukuis [4] method is ineffective for a corner or curved edge, as mentioned above.

We have ever proposed a method to solve the above mentioned problems (including the critical deficit of Canny operator) of the conventional method [14]. Herein, an application of the new method is presented.

Section 2 describes the fundamental concept of the statistical method [14], and Section 3 presents the simulation system combined the edge extraction method [14] and the region segmentation method [13] and the simulation results.

2   Statistical Method of Edge Extraction

A color image consisting of n × n-sized blocks is set as the first step to solve the problems mentioned in Section 1. At each block, the mean values

$(\mu_{0R}, \mu_{0G}, and \mu_{0B})$ for color components R, G, and B, and the variances $(\sigma_{0R}^2, \sigma_{0G}^2, and \sigma_{0B}^2)$ are calculated, and the color α with the highest variance is extracted. The block which has the standard deviation lower than a threshold λ_σ $(\sigma_0 = \sqrt{\sigma_{0R}^2 + \sigma_{0G}^2 + \sigma_{0B}^2} < \lambda_\sigma)$ is designated the homogeneous feature area, and it is not divided into two sub-regions. Other non-homogeneous block is divided into the two sub-regions R_1 and R_2 according to the mean value of the color α $(\mu_{0\alpha})$ like R_1 is composed of pixels whose color value α is greater than $\mu_{0\alpha}$. In the dividing process, the number N_i of pixels in region R_i, the mean values $\mu_{i\xi}$ of the colors and locations, and the variances $\sigma_{i\xi}^2$ are calculated, where $i \ni \{1, 2\}$, $\xi \ni \{x, y, R, G, B\}$. This process contributes to solve the problems of the conventional methods mentioned in Section 1. When $N_i < n$, the edge extraction process is stopped for the block, to ensure stability, because the edge might be near the block end and the stability might be lost. However, a boundary edge near the block end can be handled in images consisting of half-size $(n/2)$ shifted blocks. The extracted edges in the four shifted images are superimposed. In this way, the problem that a boundary edge that is near the block end might not be extracted is solved. Then, the proposed statistical evaluation value η, the edge location (e_x, e_y), the edge orientation (θ_x, θ_y), the mean color difference $Delta$, and the distance d between sub-regions are calculated. We call a set of these extracted features the virtual edge. The real edges are the boundary between sub-regions R_i. This concept is shown in Fig.1. The proposed statistical evaluation value η for the homogeneous block is defined as zero.

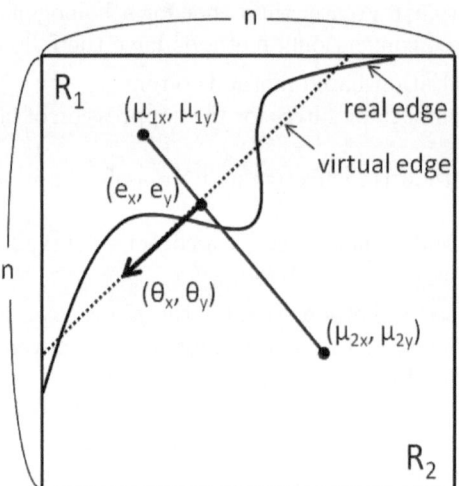

Fig. 1. Virtual edge location and orientation

$$\eta = 1 - \frac{N_1 \times \sigma_1 + N_2 \times \sigma_2}{n \times n \times \sigma_0}, \qquad (5)$$

$$(e_x, e_y) = \frac{N_1 \times (\mu_{2x}, \mu_{2y}) + N_2 \times (\mu_{1x}, \mu_{1y})}{n \times n}, \tag{6}$$

$$(\theta_x, \theta_y) = \frac{(-\mu_{2y} + \mu_{1y}, \mu_{2x} - \mu_{1x})}{|(-\mu_{2y} + \mu_{1y}, \mu_{2x} - \mu_{1x})|}, \tag{7}$$

$$\Delta = |(\mu_{2R} - \mu_{1R}, \mu_{2G} - \mu_{1G}, \mu_{2B} - \mu_{1B})|, \tag{8}$$

$$d = |(\mu_{2x} - \mu_{1x}, \mu_{2y} - \mu_{1y})|, \tag{9}$$

$$\eta(k) = 1 - \frac{N_1 \times \sigma_1^k + N_2 \times \sigma_2^k}{n \times n \times \sigma_0^k}, \tag{10}$$

One possible statistical evaluation measure of edge existence (separability) is $\eta(k)$, expressed by equation (10), where $\eta_1 = \eta(2)$ and $\eta = \eta(1)$. In equation (5), the separability for a step type, ramp type, linear gradation type and flat type is 1.0, 0.75, 0.5 $(= 1 - 0.5((n^2 - 4)/(n^2 - 1))^{1/2})$, and 0.0, respectively.

The distance d and the area of overlap of the sub-regions will be useful cues for discriminating a peak type, a valley type, and a convex/concave roof type from a step type and a ramp type. This process will be developed in future research.

3   Simulation System and Results Applied to Color Image [12]

A flow chart for extracting virtual edges is given as Fig.2.

To test the universality of the parameters, a fixed set of parameters ($n = 8, \lambda_\eta = 0.7, \lambda_\Delta = 30$) was applied to many color images. In Fig.3, the results for the proposed method are shown in the center column.

Experiments were executed to compare the separabilities η_0, η_1 , and η , where the block segmentation proposed herein was used to compare these under identical conditions. The results from using separability η_0, defined by Yakimovsky [20], are shown in the right-hand column of Fig.3. Noticeable are the lack of important edges and the addition of many noise edges. The results from using separability η_1, defined by Fukui [4], and the proposed separability η under their respective threshold values, $\lambda_{\eta 1}$ and λ_η (where $\lambda_{\eta 1} = 1.5\lambda_\eta$) , are shown in Fig.4. Separability η_1 had the drawback that it omitted more edges than the proposed separability, η, as shown in Fig.4. Also, more noise edges sometimes appeared for the threshold $\lambda_{\eta 1} = 0.66$ than for the threshold $\lambda_\eta = 0.5$.

The threshold of the parameter for determining a homogenously colored region was varied for both the proposed method (λ_Δ) and that of Kundu [9] (γ) and the performance compared. The results are shown in Fig.5. The number of detected edges dramatically changed in the case of Kundus method, but was

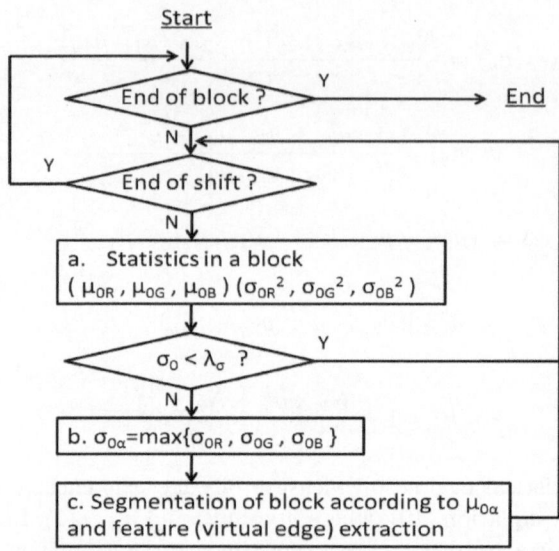

Fig. 2. Process of extracting a virtual edge [14]

almost constant for the proposed method. This property of the proposed method may be due to the introduction of separability.

Many applications of the proposed method [14] can be considered like a plausible true edge extraction, a base of disparity extraction, a base of motion extraction, a base of region segmentation and so on.

We present here some results that the virtual edge extraction method [14] was applied to color image segmentation and conspicuous objects extraction [13] as the first step of the processing. In the application of [14], small sub-regions obtained in the process of presented method are used as the start of segmentation process mentioned in [13] in which the region segmentation is done according to global statistics of location and feature similarity among small blocks. Although a method based on global features has been reported [3], the method has some following defects: (1)the number of regions to be extracted is fixed, (2)the large background may be split, and (3)the region boundaries sometimes do not follow object boundaries exactly, even when the object boundary is visually quite apparent(over-unification) [3]. We have proposed a completely different method using global features, in which the above mentioned defects are removed. [13] The proposed method is also completely different from the region segmentation method based on the closed line obtained by connecting edges [1, 10, 20], in which the closed line is not always obtained.

Some results of region segmentation and the most conspicuous object-regions extracted in the system are shown in Fig.6 and Fig.7 respectively.

The extracted conspicuous regions for 21 images exhibited high correspondence $(90 \pm 16\%)$ with the regions reported on the touch panel screen by the 60

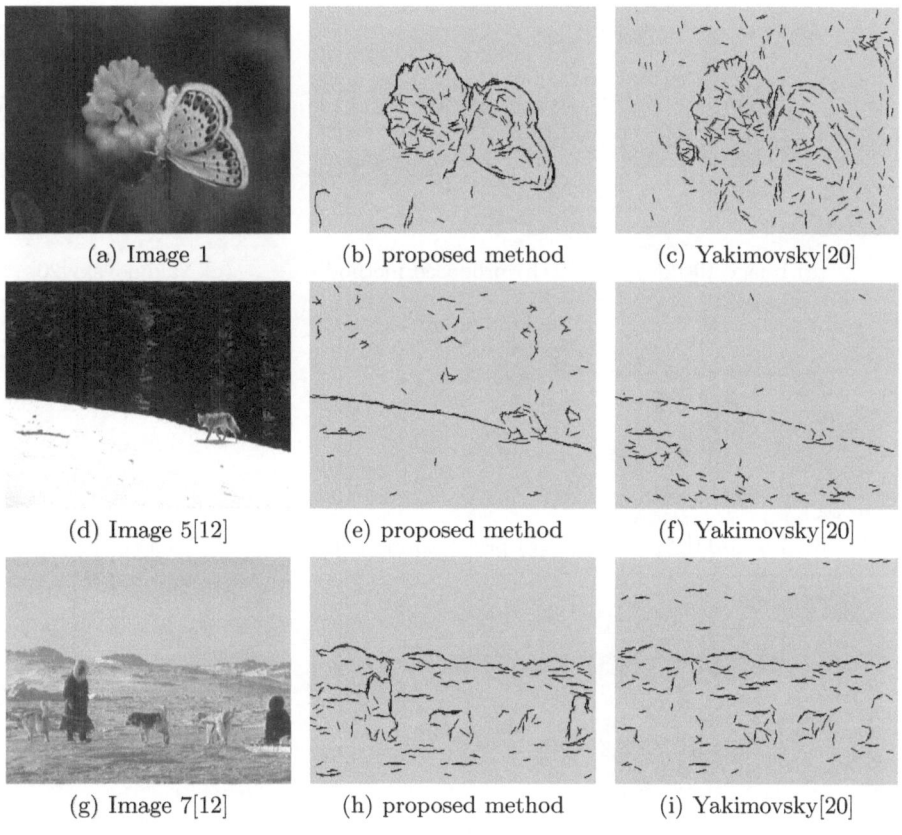

(a) Image 1   (b) proposed method   (c) Yakimovsky[20]

(d) Image 5[12]   (e) proposed method   (f) Yakimovsky[20]

(g) Image 7[12]   (h) proposed method   (i) Yakimovsky[20]

Fig. 3. Virtual edges using a fixed set of parameters and edges using the proposed method [14] ($\lambda_\eta = 0.46, 8 \times 8, \lambda_\Delta = 30$) and that of Yakimovsky[20] ($\lambda_{\eta 0} = 400, \lambda_\Delta = 3$)

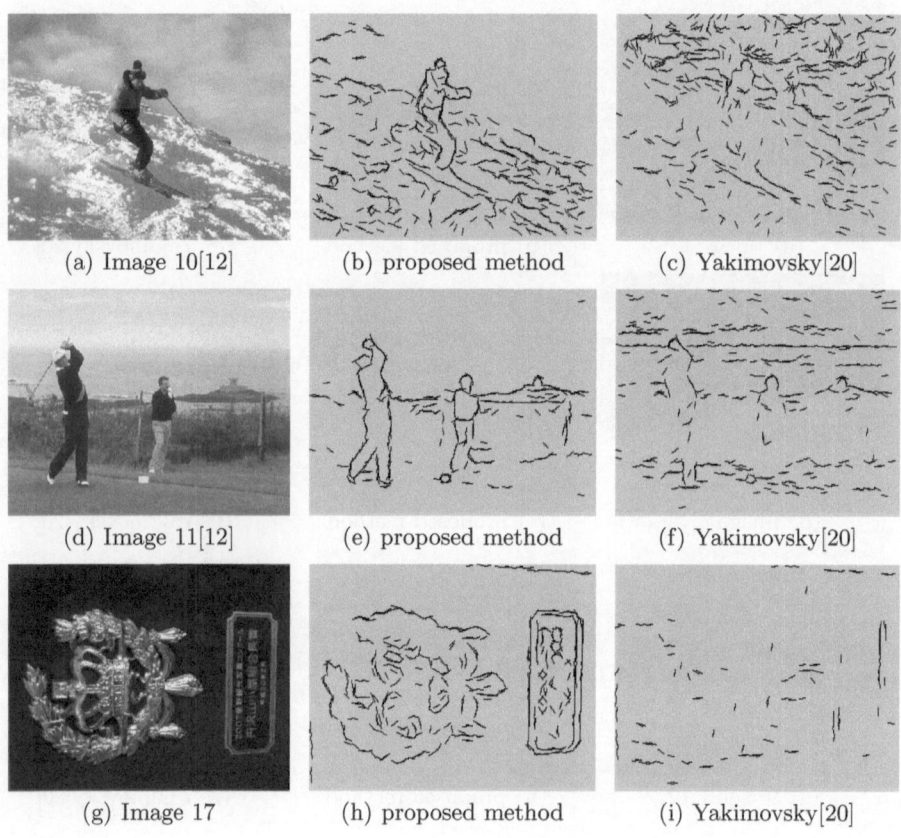

(a) Image 10[12]   (b) proposed method   (c) Yakimovsky[20]

(d) Image 11[12]   (e) proposed method   (f) Yakimovsky[20]

(g) Image 17   (h) proposed method   (i) Yakimovsky[20]

Fig. 3. It Continued

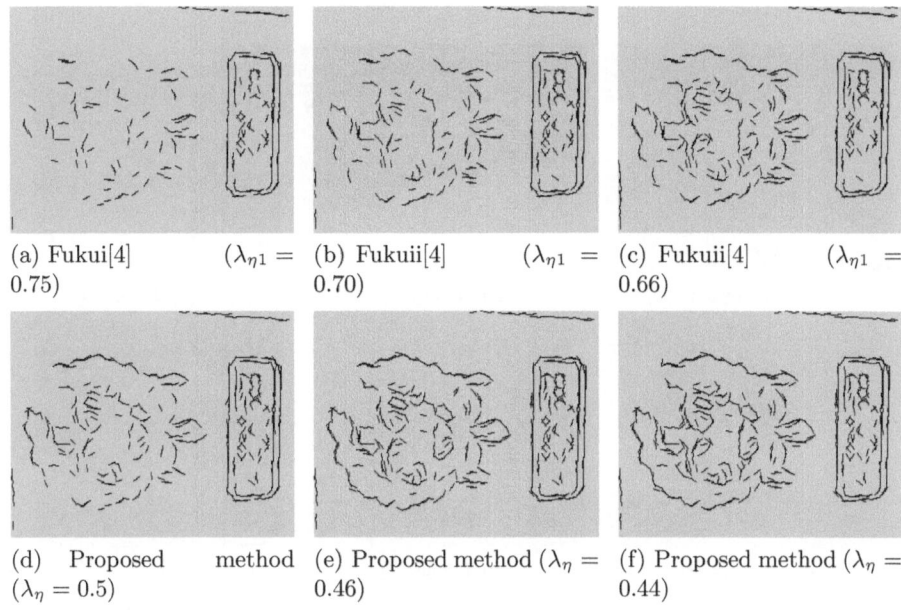

(a) Fukui[4] $(\lambda_{\eta1} = 0.75)$

(b) Fukuii[4] $(\lambda_{\eta1} = 0.70)$

(c) Fukuii[4] $(\lambda_{\eta1} = 0.66)$

(d) Proposed method $(\lambda_\eta = 0.5)$

(e) Proposed method $(\lambda_\eta = 0.46)$

(f) Proposed method $(\lambda_\eta = 0.44)$

Fig. 4. Results from using the method of Fukui [4] and the proposed method [14] $(\lambda_{\eta1} = 1.5\lambda_\eta)$

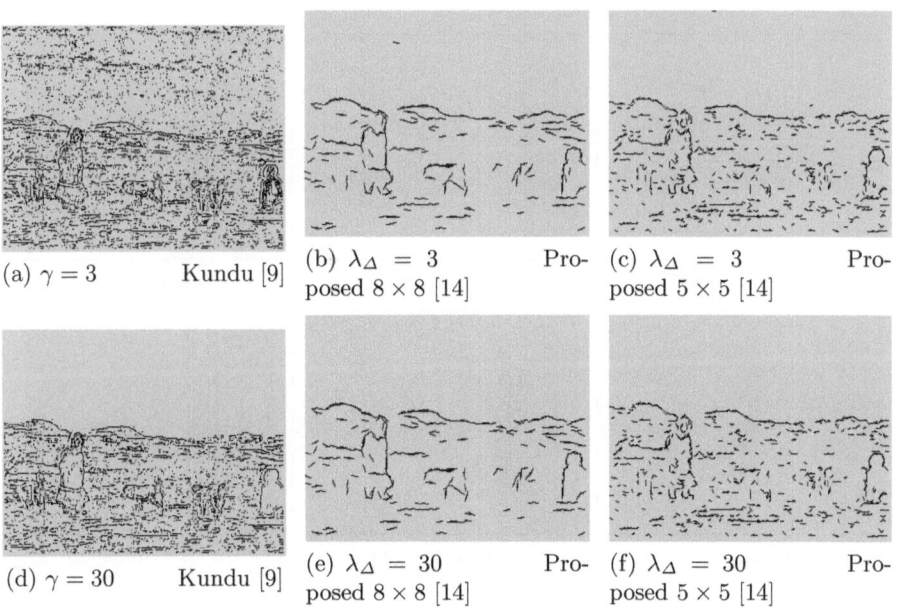

(a) $\gamma = 3$ Kundu [9]

(b) $\lambda_\Delta = 3$ Proposed 8×8 [14]

(c) $\lambda_\Delta = 3$ Proposed 5×5 [14]

(d) $\gamma = 30$ Kundu [9]

(e) $\lambda_\Delta = 30$ Proposed 8×8 [14]

(f) $\lambda_\Delta = 30$ Proposed 5×5 [14]

Fig. 5. Variation in edge detection for changes in threshold (γ and λ_Δ)

(a) Image 1   (b) Image 5   (c) Image 7

(d) Image 10   (e) Image 11   (f) Image 17

Fig. 6. Result of image segmentation

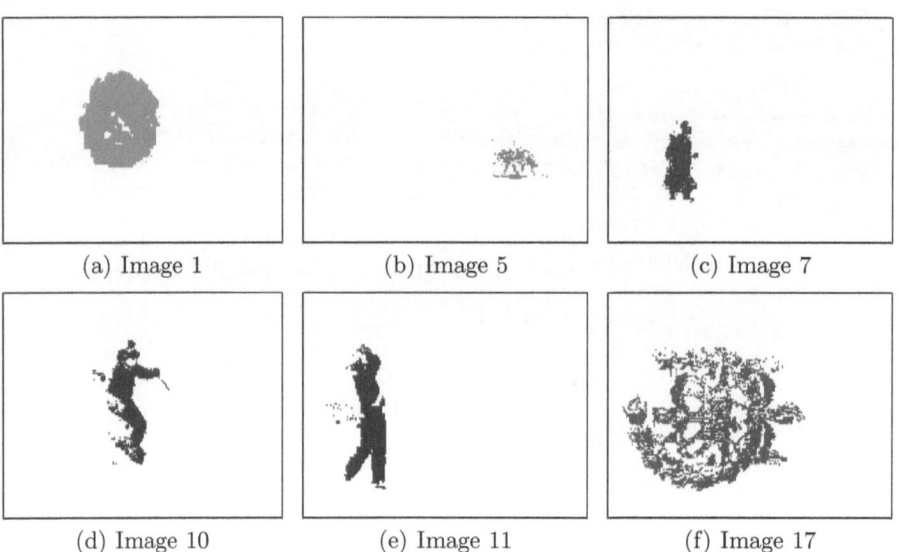

(a) Image 1   (b) Image 5   (c) Image 7

(d) Image 10   (e) Image 11   (f) Image 17

Fig. 7. Result of extracted most conspicuous objects

subjects (The reaction times of the touch were 1.9 ± 0.5 sec). These results are about equals to those of paper [14]. However, as the current system has the precise boundary information with the adjacent region, more adequate conspicuity value of the segmented region will be obtained. The boundary information may be useful for the integration of adjacent regions.

The process of image segmentation in the paper [14] is shown in Fig.8 for reference. The sub-regions R_1 and R_2 are used instead of the object block or pixels in the non-object blocks in paper [14](See processes c and d in Fig.8).

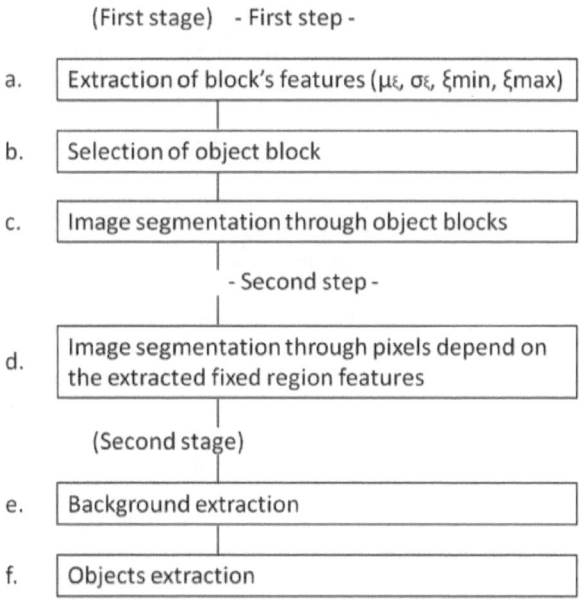

Fig. 8. Image segmentation process [13]

4   Conclusion

We have developed a robust statistical edge detection method [14] by combining the ideas of Kundus method, in which the region segmentation of local area is used, and Fukuis method, in which a statistical evaluation referred to as separability is used for edge extraction and also developed a region segmentation method [13] based on the global features like the statistics of the region. A new region segmentation method is developed by combining these two methods, in which the sub-regions in the edge extraction method [14] is used instead of object blocks or pixels in the first step of region segmentation method [13]. We obtained the almost same results as the ones of previous region segmentation method [13]. This means that the proposed method worked effectively. A new conspicuity degree including a clear contrast value with the adjacent regions, a envelopment degree by clear edge and so on will be included in the next version of the proposed method without much difficulty and it will contribute to develop a further unification algorithm and a new feature extraction method for scene recognition.

References

1. Arbetaez, P., Maire, M., Fowlkes, C., Malik, J.: Contour detection and image segmentation resources. IEEE Trans. PAMI 33(5), 898–916 (2011)
2. Canny, J.: A Computational approach to edge detection. IEEE Trans. PAMI PAMI-8(6), 679–698 (1986)
3. Carson, C., Belongie, S., Greenspan, H., Malik, J.: Blobworld:Image segmentation using expectation-maximization and its application to image querying. IEEE Trans. PAMI 24(8), 1026–1038 (2002)
4. Fukui, K.: Edge extraction method based on separability of image features. IEICE Trans. Inf. & Syst. E78-D(12), 1533–1538 (1995)
5. Fukui, K.: Contour extraction method based on separability of image features. Trans. IEICE J80-D-II(6), 1406–1414 (1997)
6. Hou, Z., Koh, T.S.: Robust edge detection. Pattern Recognition 36, 2083–2091 (2003)
7. Hubel, D.H., Wiesel, T.N.: Receptive fields, binocular interaction and fundamental architecture in the cats visual cortex. J. Physiology 160, 106–154 (1962)
8. Hueckel, M.: An operator which locates edges in digitized pictures. J. ACM 18(1), 113–125 (1971)
9. Kundu, A.: Robust edge detection. Pattern Recognition 23(5), 423–440 (1990)
10. Mairal, J., Leordeanu, M., Bach, F., Hebert, M., Ponce, J.: Discriminative Sparse Image Models for Class-Specific Edge Detection and Image Interpretation. In: Forsyth, D., Torr, P., Zisserman, A. (eds.) ECCV 2008, Part III. LNCS, vol. 5304, pp. 43–56. Springer, Heidelberg (2008)
11. Marr, D., Hildreth, E.: Theory of edge detection. Proc. Royal Soc. London B-207, 187–217 (1980)
12. Martin, D., Fowlkes, C., Tal, D., Malik, J.: A database of human segmented natural images and its application to evaluating segmentation algorithms and measuring ecological statistics. In: Proc. 8th ICCV, vol. 2, pp. 416–423 (2001)
13. Mori, F., Yamada, H., Mizuno, M., Sugano, N.: Color image segmentation based on statistics of location and feature similarity. IEEJ Trans. Electronics, Information and Systems 131(11), 2022–2029 (2012)
14. Mori, F., Yamada, H., Mizuno, M., Sugano, N.: Virtual edge extraction method based on new separability. IEICE Trans. on Information and System (Japanese Edition) J94-D(12), 2105–2113 (2012)
15. Otsu, N.: A threshold selection method from gray-level histogram. IEEE Trans. SMC SMC-9, 62 (1979)
16. Robinson, G.S.: Edge detection by compass gradient masks. Computer Graphics and Image Processing 6, 492–501 (1977)
17. Smith, S.M., Brady, J.M.: SUSAN-a new approach to low level image processing. Proc. IJCV 23(1), 45–78 (1997)

18. Sobel, I.: An isotropic 3x3 image gradient operater. In: Freeman, H. (ed.) Machine Vision for Three-Dimensional Scenes, pp. 376–379. Academic Press (1990)
19. Wakasugi, T., Nishiura, M., Yamaguchi, O., Fukui, K.: Lip contour extraction using separability of color distributions. Trans. IEICE J89-D(9), 2025–2032 (2006)
20. Yakimovsky, Y.: Boundary and object detection in real world images. J. ACM 23(4), 599–618 (1976)

Adaptive Sampling for Low Latency Vision Processing

David Gibson[1], Henk Muller[2], Neill Campbell[1], and David Bull[1]

[1] Computer Science, University of Bristol, UK
[2] XMOS Ltd, Bristol, UK

Abstract. In this paper we describe a close-to-sensor low latency visual processing system. We show that by adaptively sampling visual information, low level tracking can be achieved at high temporal frequencies with no increase in bandwidth and using very little memory. By having close-to-sensor processing, image regions can be captured and processed at millisecond sub-frame rates. If spatiotemporal regions have little useful information in them they can be discarded without further processing. Spatiotemporal regions that contain 'interesting' changes are further processed to determine what the interesting changes are. Close-to-sensor processing enables low latency programming of the image sensor such that interesting parts of a scene are sampled more often than less interesting parts. Using a small set of low level rules to define what is interesting, early visual processing proceeds autonomously. We demonstrate system performance with two applications. Firstly, to test the absolute performance of the system, we show low level visual tracking at millisecond rates and secondly a more general recursive Baysian tracker.

1 Introduction

There is increasing interest in low cost computer vision systems with a wide range of applications including gesture based user interfaces, surveillance, automotive systems and robotics. As the complexity of consumer, sensing and military systems increase the demands on energy resources becomes critical for high-level computing performance. Vision systems are proving to be extremely valuable across a range of applications and to be able to efficiently process visual information offers a huge advantage in the functionality of such systems.

Traditional computer vision systems typically consist of a camera continually capturing and transmitting images at a fixed frame rate and resolution with a host computer sequentially processing them to obtain a result such as the trajectory of a moving object. A major drawback of this pipeline is that large amounts of memory are required to store the image data before it is processed, especially as frame rate and image resolution increase. Additionally large amounts of the image data is transmitted to the host for processing regardless of the amount of information contained in this data. In the case of object tracking, computer vision algorithms work towards creating a concise description such as, 'a group of pixels at a certain location is moving in a particular way'. Often the object is

J.-I. Park and J. Kim (Eds.): ACCV 2012 Workshops, Part I, LNCS 7728, pp. 194–205, 2013.
© Springer-Verlag Berlin Heidelberg 2013

relatively small compared to the whole image and the background maybe static. In cases like this, the traditional computer vision processing pipeline could be considered as being highly inefficient as large amounts of image data are being captured, transmitted to the host, stored in memory and being processed on a per-pixel basis while most of the visual information comes from a small number of changing pixels. In such cases most of the image data is discarded as it contains no useful information.

In the case of a scene with an object moving across a static background most of the image data changes very little while some pixel areas might change rapidly or move a different speeds. The fixed temporal sampling rate of standard camera systems cannot take this into account and artifacts such as motion blur and temporal incoherence are introduced. These artifacts consequently confound down stream processing necessitating ever more complex computer vision algorithms to overcome these imaging effects.

A significant problem with digital video capture is that of temporal quantization. Given a finite set of resources, digital video capture proceeds by sequentially sampling frames at *fixed* temporal rates and spatial resolution within a range of luminance in a non-interactive passive manner. Biological systems proceed very differently; unable to sequentially process entire views, selective scene sampling is performed using a combination of eye movements. In Rucci et.al. [1,2,3,4,5] it is shown that a number of strategies exist for visual sampling in human vision depending on the task being carried out. A human eye is constantly moving in order to avoid fading, the loss of sight due to a lack of change on the retina. As well as head movement, eye movements include saccades, micro-saccades and drift. These movements enhance and stabilize the binocular view allowing the process of foveation to create a highly detailed perception capable of difficult tasks such as threading a needle.

Modern high speed cameras are capable of capturing images at thousands of frames a second and can have dedicated processing modules close to the sensor. In [6] the wing of a fly was tracked using regions of interest (ROI) at 6000Hz. The system used edge detection to analyse the shape and motion of the fly's wing via feedback from a tracker which predicted the next ROI. One problem with such systems is they are rigid in their FPGA based design, are task specific and highly implementation and environment dependent.

In order to increase the flexibility and efficiency of high-level downstream processing, image sensor design companies are developing devices that can compute interest points and local descriptors in silicon [7]. Other silicon devices include the artificial retina [8]. The artificial retina is being investigated in a number of contexts, one of which is 'event-based stereo vision in real-time with a 3D representation of moving objects' [9]. The low latency of this device is of particular interest. However, there is no illumination detail provided. This is overcome in a hybrid system that includes a traditional digital camera system to investigate selective attention or saliency for real-time active vision [10].

In a keynote speech Ed Dowski [11], lead for new technologies at OmniVision CDM Optics, Inc. suggested that:"An important class of future imaging systems, in our view, will be Application-Specific Imaging Systems (ASIS). These imaging systems will not be general purpose in that they are meant primarily for human viewing, but will be specialized systems that capture and communicate specific types of information depending on the particular application."

The central hypothesis of this paper is that low latency visual sampling can provide a framework for solving many challenging real world vision problems. The proposed system has characteristics similar to those of several compressive sensing methods [12]. Non-linear visual sampling in the spatial and temporal domains followed by image reconstruction of whole image sequences is a popular research avenue. However, current compressive sensing techniques generally involve highly specialised and expensive components with the results being reconstituted using time consuming and computationally demanding algorithms rendering them difficult to apply to practical real-world problems.

In this paper we are particularly interested in exploiting spatiotemporal redundancy **and** the low latency control offered by close-to-sensor processing through the use of non-linear visual sampling and **piecewise visual processing**. By exploiting spatiotemporal redundancy, high speed imaging can be accomplished without increasing bandwidth while reducing errors introduced by temporal quantization. Low latency enables software pipelines that can be switched so as to adapt to changes in visual input and be posed as a **functional visual sampling** problem. The proposed system and associated algorithms are strictly real-time in the sense that the capture and image processing relationship is directly linked and interdependent. An advantage of the proposed systems is that a broad range of traditionally hard or impossible vision based processing tasks can be addressed within a novel, cost and energy efficient framework. We propose a highly programmable visual sampling approach for application specific imaging pipelines (ASIP), that can provide output for machine vision systems as well as human observers.

1.1   Hardware System

Central to the system design is the XMOS microprocessor and the ability to reprogram image sensor parameters with very low latency. The processor allows a direct connection to an image sensor, has no operating system, does not use interrupts and supports concurrent threads within a parallel architecture. The XMOS[1] XCore is a multi-threaded processing component with instruction set support for communication, I/O and timing. Thread execution is deterministic and the time taken to execute a sequence of instructions can be accurately predicted. This makes it possible for software executing on an XCore to perform many functions normally performed by hardware, especially DSP and I/O.

To investigate the exploitation of spatiotemporal redundancy via piecewise visual processing a development board has been designed and built. Figure 1 shows the layout of the latest visual processing system. The design is such that pixel information is read in from an image sensor a one end of the pipeline and

---

[1] www.xmos.com

then processed into higher and higher representations as data passes through the system. Communications via ethernet and RAM is available at the far end of the pipeline. The pipeline can be configured in software and feedback to the image sensor control thread can be provided at any stage of processing. It should be noted that the original sampled pixel information need never be lost and in the simplest configuration the system behaves as a standard camera. With nine xcores in total 4500MIPS of processing across 36 concurrent threads is available for processing. XLinks provide fast inter core bi-directional communications and extensive GPIO is available.

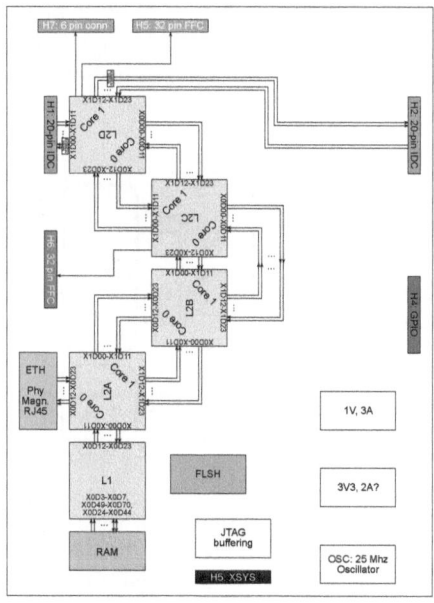

Fig. 1. Block diagram for the latest hardware design. It consists of four dual core and a single core processors, 128MB SDRAM, ethernet phy, approximately ten smaller chips (flash, buffers, oscillator, reset supervisors, etc), one FFC connector for the image sensor (H5 and H7), one FPC connector for the LCD (H6), male and female 16 pin IDC headers for XLinks (H1 and H2). XLinks enable multiple boards to be connected with each other and have a bandwidth of 320Mb/s. XLinks are also what form the backbone of the pipeline connecting the processor in series from sensor input at L2D to ethernet and RAM on L2A and L1 respectfully. Extensive GPIO is available on header H4.

Figure 2 shows an advanced system layout that could be implemented by the design in figure 1. For the work described in this paper less elaborate sub-system designs have been used. The minimal configuration consists of an image sensor connected to an XMOS processor and software running on a single xcore; one thread being used to read in pixels and control the sensor and a further

two threads to run a UDP ethernet transmitter. The architecture is highly programmable; if several xcores are linked a wide range of processing and feedback
designs can be implemented.

The underlying imaging used in this paper is, rather than capture an entire
image and then process the pixels at a standard frame rate, regions of interest
(ROI) are captured and processed at high frame rates. An advantage gained
using this approach is that an ROI can be processed in the time it takes to
expose the next ROI. The key advantage of this approach is that areas of an
image that have interesting changes occurring in then can be sampled more often
than in image areas where no changes are occurring. Throughout this paper the
pixel depth was set to 8 bits and ROI were set to 64 by 40 pixels, the sensor
resolution was set to 640 by 480 pixels and 2 by 2 pixel binning was used to
give an effective image resolution of 320 by 240 pixels. Each xcore has 64KB of
on chip memory, all of the follow experiments use only this memory and were
carried out using two quad-core XMOS processors.

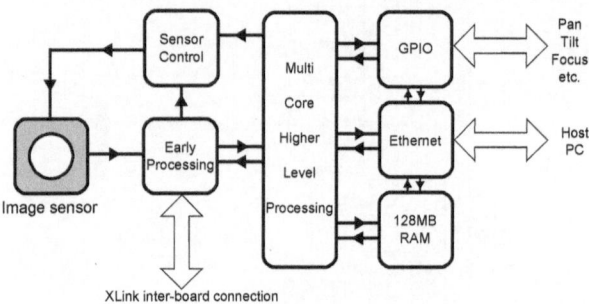

Fig. 2. A system diagram for a fully functioning processing pipeline. Early processing
and sensor control execute in separate dedicated threads close to the sensor. Higher
level processing and communications occurs in many threads over several cores and
processors.

In the following section two applications for low level visual tracking are
described. They are based on the system layout of figure 2. As such the two
trackers can be seen as running in parallel with each other. If there are no interesting changes with respect to the millisecond tracker pixel data is passed
on to the higher level Baysian tracker. The Baysian tracker then directs the re-
programming of the image sensor according to what it determines as the next
interesting ROI to sample. However, if the millisecond tracker does detect pixels
that are interesting to it, it now overrides higher level processing and proceeds
of its own accord. When the millisecond tracker no longer detects interesting
changes control of the sensor re-programming is handed back to the higher level
processing system. These two application are now described in more detail.

1.2   Experiments and Results

After initialization the system proceeds by repeating the following; program the ROI position on the sensor, exposing the ROI, read in ROI pixels from sensor, process ROI pixels. After a short period of time temporal differences can be computed, i.e. the difference between two spatially coherent but temporally offset ROI. This temporal differencing provides the basis for low level event detection and tracking. The sensor can be run in two modes; free running or triggered. The design in figure 1 allows for both. However, the hardware used in this paper could only function in free running mode. To understand the performance and limitation of the system the sensor was run at full speed, table 1 describes the low level timings and identifies when and how processor cycles are being used.

Table 1. Timing information in CPU cycles at 100MHz with frame and row delay at zero and sensor clock at 25MHz. One frame is dropped in the first three rows with overall frame time being $\sim$ 103000 cycles (frame grab plus frame delay, 964fps). The time between row read-ins is 816 cycles.

Frame Grab	Data Send	Frame Update	Frame Delay	FPS
42384	2042	46480	61296	964
42384 [a]	2042	32544	61296	964
42384 [b]	32	32538	61296	964
42384 [c]	32	4682	9455	1928
42384 [d]	32	34	9455	1928

[a] No display window update. OK if display columns remain constant.
[b] No data send.
[c] No window update, with display window update.
[d] No update at all.

In table 1 timing information in CPU cycles at 100MHz with frame and row delay at zero and sensor clock at 25MHz is given. The order of processing is as follows; the frame ready pin is pulled high (this is when timing starts), pixels are read in on a line by line basis, the pixel data is transmitted to the ethernet transmitter thread, the ROI position is updated via I2C and the next exposure begins. This cycle is repeated over the whole sensor surface, returning back to $(0,0)$ after each 320 by 240 composite image is read if the sensor windows are updated. From table 1 it is clear that transmitting the pixel data on to the next thread takes roughly two thousand clock cycles. The display window is the sensor width by ROI height region that the sensor exposes and the window update is the region of pixels that the sensor reads out. The time it takes to update the display window does not effect the FPS value, however updating the read out window involves programming over twice as many registers. Updating the read out window with or without updating the display window takes longer than the exposure time, consequently a whole frame is dropped before the frame ready pin is pulled high. With no window updates and no data send the absolute performance is demonstrated; it take 42384 cycles to read in the pixel data and 9455 cycles to expose the sensor giving an absolute frame rate of 1928FPS $(100000000/(42384+9455))$. If the sensor windowing updates take less than the exposure time then a frame is not dropped.

Millisecond Tracker. The above timings are with respect to a single thread running at 100MHz and with the delay between rows read-ins (816 cycles) there remain roughly 50000 cycles for processing the 64 by 40 ROI. To perform single ROI low level tracking a background model is built for each incoming ROI. This consists of two histograms, one for the maximum values of each row and one for the maximum of each column. If a single peak over a certain threshold exists a point of interested is considered as detected. If in the next ROI a similar point exists tracking begins and the x and y offsets between the two peaks are used to initialize a predictive tracker. The mean value of the previous and current prediction is used to estimate where the point in the next ROI will be. The position of the ROI on the sensor is updated and the process is iterated until the interest point is lost. Figure 3 show the laser point stimulus. The motion is so fast that at 30fps the light is smeared across the exposures. In figure 4 some example frames from the tracking result are shown, behind the text there is a bright point light. It should be noted that there is not a direct one to one match of fields of view or temporal synchronization between the images shown in Figures 3 and 4 as the different sensors are in different positions and are not fully aligned.

Fig. 3. The the laser point stimulus. The motion is so fast that at 30fps the light is smeared across the exposures. The point light is moved using a servo that has a maximum rotation speed of 0.16s for 60 degrees.

Fig. 4. Fast point tracker. Behind the text there is a point light. The top row of numbers shows the x and y positions within the ROI of the max pixel value. The second row shows the predeicted position of the point ligth according to the first order tracker. The third row shows the x value of the predictive tracker.

Baysian Tracker. If the above criteria for a point of interest is not met, ROI pixel data is transmitted on to a more complex tracking system. This is now described in detail. Given an image sensor surface, $\mathbb{S}$, with resolution $[X, Y]$ a spatiotemporal volume is described as $v(x, y, t)$ where $x \in [1, X]$ and $y \in [1, Y]$ are the row and column coordinates respectively and t is the temporal coordinate. A non-overlapping rectangular grid is defined as $g_o \in G \equiv \mathbb{S}$ and g_s is the set of all possible rectangles that belong to G. The image sensor is programmed by registers that can be set, on a sub-image by sub-image basis, enabling selective sampling of G with variable intervals of t. Initially, over all g_o we sample sub-images, $s(x_s, y_s)$, where x_s and y_s are the rows and columns of each sub-image, generally $x_s < (X/4)$ and $y_s < (Y/4)$. There are N_o non-overlapping sub-images in g_o that cover $\mathbb{S}$ and as each sub-image is captured some sparse feature vector representation, $\boldsymbol{f}$, of each sub-image is computed. After $t(N_o + 1)$ samples, sub-images s_1 and s_{N_o+1} are compared to determine if any changes have occurred. Any metric can be used to determine if and how the samples might have changed, the simplest is a difference, $d_{g_o(1)} = (\boldsymbol{f}_{N_o+1} - \boldsymbol{f}_1)$. When $t(2 \times N_o)$ samples have been captured a multi-modal distribution of differences across the extent of G is computed as:

$$p(d)_{t=0} = pdf(d) = \int_{g_o} \frac{1}{\sqrt{2\pi}} e^{-d^2/2} \tag{1}$$

Equation 1 initializes the system; if $p(d) = 0$, no changes in the pairs of sub-images have occurred otherwise $p(d)$ is proportional to the magnitude of change according to the feature description and metric used. $p(d)$ is updated with each new differential observation, d, in a manner similar to a large class of algorithms that include sequential Baysian filtering, Kalman filters, particle filters, HMMs, etc.

$$p(d)_t = \int_{g_o} \frac{1}{\sqrt{2\pi}} e^{-(p(d)_{t-1}+d_t)^2/2} \tag{2}$$

So far, x_s and y_s belong to g_o and δt is constant. The proposed algorithm now proceeds by re-sampling $p(d)$ such that a new set of sub-images, s_{g_p}, where $g_p \in g_s$, predict likely visual changes at some time in the future:

$$(g_p, t_{g_p}) \leftarrow p(d) \tag{3}$$

The algorithm proceeds to iterate over the Equation 2 and 3 effectively tracking visual changes that are 'interesting' according the feature set description and difference metric. The above description represents the simplest formulation of the proposed system, more complex formulations easily fit within the same framework. Equation 3 provides the basis for the hypothesis of this paper; δt_{g_p} and the number of sub-images, s_{g_p} are not fixed. There are several interesting consequences of this; firstly no whole images exist in the traditional sense, secondly there is no fixed frame rate, sub-images are captured at different spatial location and at different temporal rates depending on changes in the visual scene. An area of a scene where little or no change occurs gets sampled infrequently and the

δt between corresponding temporal sub-images will be relatively large. An area that changes a lot and rapidly will be sample frequently and δt will tend towards its minimum. In the current and proposed hardware design $\min(\delta t) \leq 0.5ms$. As fewer sub-images are sampled more frequently there is no significant difference in bandwidth compared to the bandwidth of standard frames rates and resolutions. It should be noted that the original sampled pixel information need never be lost and offers the potential for compressive sensing or other more standard techniques to be implemented further down the pipeline. Figure 5 shows the overall composition of the higher level tracking system. Figure 6 shows the individual components of the image processing pipeline and figure 7 shows selected frames of an object being tracked.

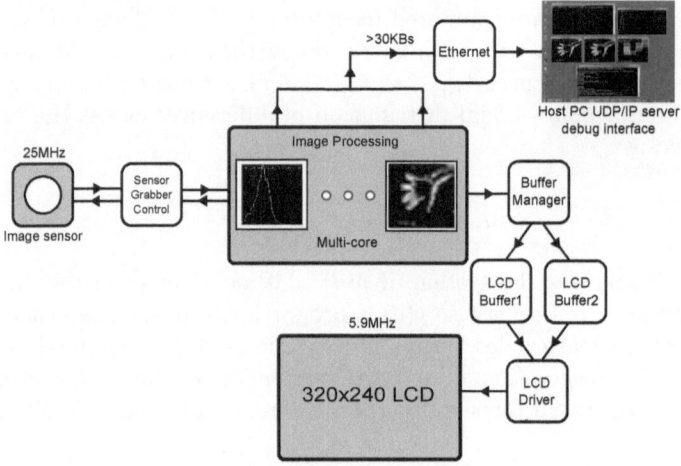

Fig. 5. The composition of the recursive Baysian tracker system. The full system is shown including a host PC that enables the internal states of the pipeline to be visualised.

1.3   Discussion

The work presented in this paper is in its infancy and the authors expect to be able to create much more advanced systems as dedicated hardware and higher quality image sensors become available. We will research and develop multiple pixel processing pipelines that implement low latency detection and tracking, autonomous stereo alignment and higher level vision processing. Multiple spatiotemporal resolutions will be used to direct focus of attention and stabilize vision algorithms. A major potential of the work is to investigate image sampling strategies given a particular stimulus and/or task. We will learn the underlying rules that enable the system to change its mode of operation. We will investigate autonomous pan, tilt and focus so as to provide continuously changing perspectives of any given visual stimulus. Sensor-processors will be able to change their

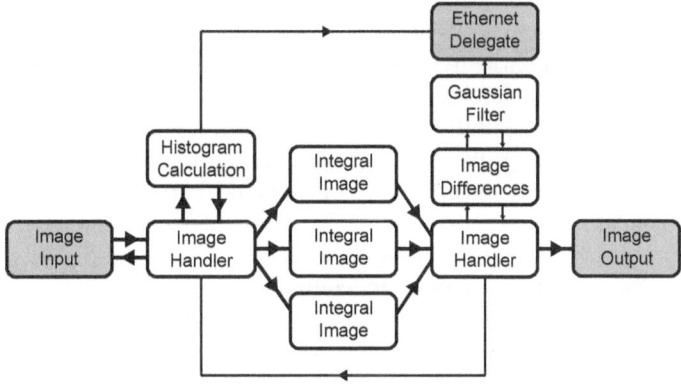

Fig. 6. The the individual components of the image processing pipeline. Each units represents a single concurrent thread running on a quadcore XMOS processor.

Fig. 7. Adaptive camera tracker, selected frames of an object being tracked. Blue rectangles are sensor surface regions that are momentarily being less frequently sampled.

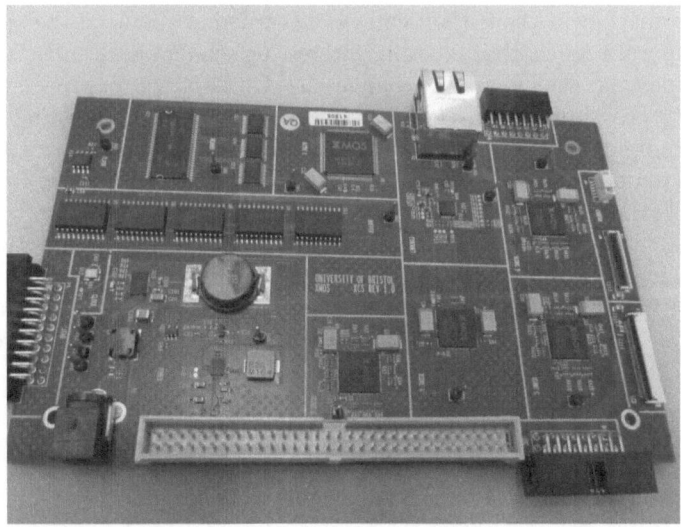

Fig. 8. The latest hardware design which is based on the block diagram layout described in figure 1

line of sight automatically based on low level rules for tasks such as; follow and focus on the largest moving object in a scene. To understand how a stereo pair of sensors might automatically configure themselves is particularly interesting, being able to move and focus independently allows a sensor pair to optically search over the depths of multiple objects within a scene.

Initial work on an adaptive processing pipeline for low level visual tracking has been presented, more advanced tasks could include; 'track a single object at a frame rate that minimizes motion blur', 'track the depth of the most interesting moving object in the scene', 'generate a super resolution snap shot of the most interesting object within a scene' or 'compute optical flow if the whole scene changes rapidly', etc. These tasks can be combined such that as a scene changes the most appropriate mode of pipeline operation is selected.

1.4   Conclusions

Traditional digital imaging is generally a passive process whereby images of fixed size and frame rates are captured regardless of what is occurring in the scene. Understanding motion cues is often made easier by increasing frame rate. However, this greatly increases the amount of data that needs to be transmitted and processed. Additionally, reprogramming a cameras image sensor often takes a number of frames leading to a latency between what an artificial system has processed and what the next image content might be. We have shown that by adaptively sampling visual information with respect to what is occurring in a scene, the performance of low level vision systems can be improved without increasing bandwidth. By having close-to-sensor processing, image regions can be captured and processed very rapidly. If spatiotemporal regions have little useful information in them they can be discarded without further processing. Spatiotemporal regions that contain 'interesting' changes are further processed to determine what the interesting changes are. Close-to-sensor processing enables low latency programming of the image sensor such that interesting parts of a scene are sampled more often than less interesting parts. Using a small set of low level rules to define what is interesting, early processing proceeds autonomously with very low latency.

The presented hardware design offers a cost effect high frame rate computational camera. As image processing is carried out in a piecewise manner a traditional frame store is not required. This in turn reduces the complexity of the system. The deterministic and parallel nature of the XMOS architecture allows for efficient and flexible visual processing pipeline designs.

1.5   Future Work

It should be noted that the default behaviour of the proposed system is that of a standard camera and original pixel information need never be lost. The proposed system is a computation camera capable a wide range of functionality. Multiple systems can be connected within the plug and play design to create multi-sensor systems. A four part system could consist of a monochrome stereo pair, a colour sensor and an IR sensor all with fast interconnections and 18000MIPS of parallel computing resources. With the existing design this would cost less than 1000USD, be the size of a small laptop and interface via a standard ethernet connection. In future work we plan to build a much smaller and more powerful design. One motivation for this is to be able to make the system more widely accessible to the vision and robotics community. Figure 8 shows the latest hardware design and roughly mirrors the layout described in figure 1.

References

1. Rucci, M., Iovin, R., Poletti, M., Santini, F.: Miniature eye movements enhance fine spatial detail. Nature (2007)
2. Santini, F., Nambisan, R., Rucci, M.: Active 3d vision through gaze relocation in a humanoid robot. International Journal of Humanoid Robotics (2009)
3. Ko, H., Poletti, M., Rucci, M.: Microsaccades precisely relocate gaze in a high visual acuity task. Nature Neuroscience (2010)
4. Poletti, M., Listorti, C., Rucci, M.: Stability of the visual world during eye drift. Journal of Neuroscience (2010)
5. Poletti, M., Rucci, M.: Eye movements under various conditions of image fading. Journal of Vision (2010)
6. Graetzel, C.F., Fry, S.N., Nelson, B.J.: A 6000 hz computer vision system for real-time wing beat analysis of drosophila. In: International Conference on Biomedical Robotics and Biomechatronics, vol. 1 (2006)
7. Kirsch, G.: Interest point and local descriptor generation in silicon. In: International Solid-State Circuits Conference (2012)
8. Lichtsteiner, P., Posch, C., Delbruck, T.: An 128x128 120db 15us-latency temporal contrast vision sensor. IEEE Journal Solid State Circuits (2007)
9. Shraml, S., Belbachir, A.N.: A spatio-temporal clustering method using real-time motion analysis on event-based 3d vision. In: International Conference on Computer Vision and Pattern Recognition (2010)
10. Sonnleithner, D., Indiveri, G.: A neuromorphic saliency-map based active vision system. In: International Conference on Information Sciences and Systems (2011)
11. Dowski, E.: A new paradigm for future application-specific imaging systems. In: International Conference on Computational Photography (2011)
12. Reddy, D., Veeraraghavan, A., Chellapa, R.: P2c2: Programmable pixel compressive camera for high speed imaging. In: International Conference on Computer Vision and Pattern Recognition (2011)

Colorimetric Correction
for Stereoscopic Camera Arrays

Clyde Mouffranc and Vincent Nozick

Gaspard Monge Institute, UMR 8049
Paris-Est Marne-la-Vallee University, France

Abstract. Colorimetric correction is a necessary task to generate comfortable stereoscopic images. This correction is usually performed with a 3D lookup table that can correct images in real-time and can deal with the non-independence of the colour channels. In this paper, we present a method to compute such 3D lookup table with a non-linear process that minimizes the colorimetric properties of the images. This lookup table is represented by a polynomial basis to reduce the number of required parameters. We also describe some optimizations to speedup the processing time.

1 Introduction

In recent years, stereoscopic technologies have been subject to an impressive growth and became incontrovertible in the movie maker industry. More recently, this technology has advanced from stereoscopic to autostereoscopic displays, involving more than two views and hence more than two cameras. The use of these multiview devices emphasizes technical issues in term of video stream synchronization, camera calibration, geometrical issues or colorimetric correction. This paper deals with this last problem, i.e. how to represent each object of the scene with a coherent colour in every view. Indeed, colorimetric inconsistencies in stereoscopic images may cause some perception troubles, as described in [1] as well as issues for multi-view video coding [2] or video-based rendering [3]. The goal of this paper is to get a uniform colour response among the camera and not to perform a colorimetric calibration [4] to get an absolute colour accuracy.

Most of the colorimetric inconsistencies mainly come from the camera sensors. Even if the cameras are the same model and come from the same factory, the sensor response is often quite different. Thus, selecting the same settings for each camera, i.e. gain, brightness or shutter speed may not solve the problem. Moreover, the camera response for two identical cameras may differ according to their respective position on the scene, where the illumination is not perceived the same or where the camera temperature is different. These colorimetric issues are clearly apparent on low cost cameras but are also visible with professional grade devices.

The usual requirement for a colorimetric correction technique is to be fast and accurate. Moreover, the process should be efficient on high resolution images

J.-I. Park and J. Kim (Eds.): ACCV 2012 Workshops, Part I, LNCS 7728, pp. 206–217, 2013.

for the movie industry as well as for daily applications such as teleconference running on consumer grade hardware. Most of the professionals claim that using separated 1D-LUT for each channel are not recommended since it is known that the colour channels are not independent. The standard approach for colorimetric corrections is to use 3D LookUp Tables (LUT) implemented on the GPU. Indeed, the GPU implementation is straightforward and the performances are excellent. Finally, for images with 256 colour levels per channel, a full size 3D-LUT would have a 256^3 voxel resolution. In practice, the 3D-LUT may have a much lower resolutions since the missing data are linearly interpolated by the GPU. As an example, a 32^3 3D-LUT is preformed enough and requires few memory (e.g. less memory than a RGB 200×200 image). The goal of this paper is to propose a 3D-LUT computation to perform a colorimetric correction between multiple cameras of a camera array.

2   Related Work

The problem of transferring the colorimetric properties of a source image to a target image, namely *colour transfer*, has been the starting point of numerous methods dealing more specifically with multiple view colorimetric correction. A survey of the related works for these two approaches is presented in the following parts.

2.1   Colour Transfer between Two Images

Reinhard et al. [5] present a method that matches the colour mean and variance of the target image to the source image. This operation is performed on the $l\alpha\beta$-colour space where the colour channels are not correlated. However this method is limited to linear transforms. Papadakis et al. [6] describe a variational formulation of the problem using cumulated histograms under colour conservation constraints, but provides 1-D transformations that is not suitable for our purpose.

Morovic and Sun [7] present a method to match the 3D colour histogram of the two images. Neumann and Neumann [8] have te same approach but also apply a smoothing and a contrast constraint to limit unexpected high gradients artifacts. Finally, Pitié et al. [9] matches the probability density function between the two images using a N-dimensional transfer function. These methodes are specially designed to perform colour transfer from images with very different colorimetric properties.

Finally, Abadpour and Kasaei [10] use a principal component analysis (PCA) to generate a new colour space where the channels are decorrelated. In [11], they use the PCA to compute a colour space from some specific image regions selected manually. PCA-based approaches will perform well on a static images, but can fail in video sequences where the variation of the colours may not match the initial colour space.

2.2   Colour Correction and Camera Array

Camera arrays dedicated to stereoscopic rendering are subject to several geometrical constraints (i.e. the cameras should be correctly aligned [12]) and hence the acquired images represent approximatively the same scene, with similar colorimetric properties.

Yamamoto et al. [13] extract SIFT correspondences [14] in order to handle scene occlusions and perform a multiview colorimetric correction. Yamamoto and Oi [15] use the same approach using an energy-minimization function on the 2D correspondences. Tehrani et al. [16] propose an iterative method based on an energy minimization of a nonlinearly weighted Gaussian-based kernel density function applied on SIFT corresponding feature points. The main drawback of these two methods is the fact that the colorimetric correction is performed on each RGB channels independently.

Shao et al. [17] distinguish the foreground and the background parts from a precomputed disparity map. They perform a PCA-based colorimetric correction only on the forground parts that are more likely to appear on each view. Shao et al. [18] also requires a precomputed disparity map to perform the correction using a linear operator on the YUV colour space.

Finally, Shao et al. [19] present a content adaptive method that performs a PCA on the data in order to select the relevant colours of the scene and generate a 3×3 correction matrix. This method does not require any disparity map but is limited to linear correction.

2.3   Outline of Our Method

As specified above, 1-dimension LUTs applied independently on each RGB channel are known for their limited colorimetric correction accuracy, whereas 3D-LUT based methods are much more accurate, still fast and easy to use. We propose a method to generate such 3D-LUT by a non-linear process that minimizes the colorimetric properties differences between each image. A 3D-LUT with full resolution would imply $3 \times 256^3 \simeq 5.10^7$ variables involving extremely long computation times. Even a 3D-LUT with a standard resolution of 32^3 would result in $3 \times 32^3 \simeq 10^4$ variables that still can not be computed in a reasonable time delay.

In this paper, we introduce a substitution of the 3D-LUT by an orthogonal basis functions that can represent the initial 3D-LUT with very few variables. We present a non-linear minimization process that finds optimal values for these variables such the recovered 3D-LUT generates corrected images with similar colorimetric properties. In regard to the related works, our method does not require any precomputed disparity map, can handle non-linear corrections, does not consider each channel independently, generates a set of 3D-LUT and is fully compatible with SIFT or other point correspondences approaches.

This paper is organized as follows: In section 3, we introduce the Chebyshev polynomial basis. Section 5 describes the non-linear minimization process used for the colorimetric correction. Section 6 presents some optimizations to speedup the process and section 7 shows some results.

3   3D-LUT and Basis Functions

3.1   Basis Functions for 3D-LUT

The purpose of the basis function is to decrease the number of variables representing the 3D-LUT. The basis function should be orthogonal to ensure the unicity of the LUT representation. We selected Chebyshev polynomial basis for several reasons. Indeed, the first order polynomials have soft variations, hence higher order polynomials can be ignored without a significant loose on the 3D-LUT description. Moreover, polynomial basis functions can represent the identity function used for the initialization. Some other well known basis such as discrete cosine transform can not unless they use all the functions of the basis. Finally, each Chebyshev polynomials are alternatively odd and even such the first polynomials have a specific signification in term of colour processing, as presented in Table 1.

Table 1. Polynomial basis for 3D-LUT: a signification for the first degrees

Degree	Effect
0	colour offset
1	identy function
2	brightness/gain
3	contrast

3.2   Chebyshev Polynomial Basis

The Chebyshev polynomials are a sequence of orthogonal functions defined for $x \in [-1, 1]$ as:

$$T_n(x) = \frac{(x - \sqrt{x^2 - 1})^n + (x + \sqrt{x^2 - 1})^n}{2}$$

They can also be expressed recursively with:

$$\begin{cases} T_0(x) = 1 \\ T_1(x) = x \\ T_{n+1}(x) = 2xT_n(x) - T_{n-1}(x) \end{cases}$$

Figure 1 depicts the first Chebyshev polynomials.

$T_0(x) = 1$
$T_1(x) = x$
$T_2(x) = 2x^2 - 1$
$T_3(x) = 4x^3 - 3x$
$T_4(x) = 8x^4 - 8x^2 + 1$
$T_5(x) = 16x^5 - 20x^3 + 5x$
$T_6(x) = 32x^6 - 48x^4 + 18x^2 - 1$
$T_7(x) = 64x^7 - 112x^5 + 56x^3 - 7x$
$T_8(x) = 128x^8 - 256x^6 + 160x^4 - 32x^2 + 1$
$T_9(x) = 256x^9 - 576x^7 + 432x^5 - 120x^3 + 9x$

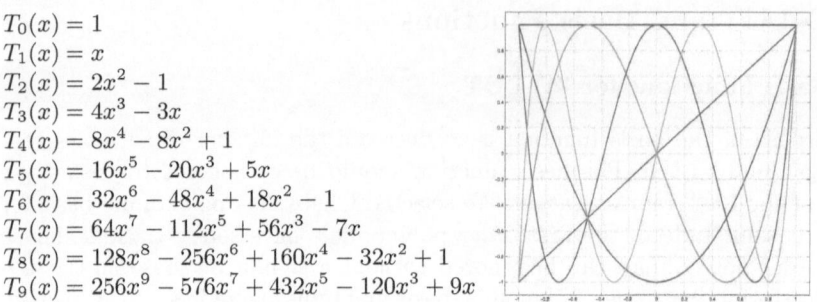

Fig. 1. Left: the first Chebyshev polynomials. Right: their graphical representation

3.3   3D-LUT Representation

Without loss of generality, we consider in the rest of this paper that the colour levels range from 0 to 1. We define a 3D-LUT f that transforms three input colours r, g and b into three output colours $(r', g', b')^\top = f(r, g, b)$ with the following formula:

$$\begin{pmatrix} r' \\ g' \\ b' \end{pmatrix} = \begin{pmatrix} \sum_{i=0}^{n} \sum_{j=0}^{n} \sum_{k=0}^{n} \alpha_{i,j,k}^{r} T_i(r).T_j(g).T_k(b) \\ \sum_{i=0}^{n} \sum_{j=0}^{n} \sum_{k=0}^{n} \alpha_{i,j,k}^{g} T_i(r).T_j(g).T_k(b) \\ \sum_{i=0}^{n} \sum_{j=0}^{n} \sum_{k=0}^{n} \alpha_{i,j,k}^{b} T_i(r).T_j(g).T_k(b) \end{pmatrix} \tag{1}$$

where n is the higher polynomial degree used and $\alpha_{i,j,k}^{c}$ is the coefficient associated to the polynomial $T_i(x) \times T_j(y) \times T_k(z)$, for the output colour channel c. An illustration of a 2-dimension Chebyshev basis functions is depicted in Figure 2.

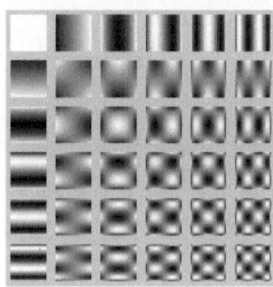

Fig. 2. A representation of the 2-dimensional Chebyshev basis with the 6 first levels, where each block $T_{ij}(x, y) = T_i(x) \times T_j(y)$. A coefficient $\alpha_{i,j}^{c}$ is associated to each block (in 3-dimension in our case) to represent a signal $g(x, y) = \sum_{i,j} \alpha_{i,j} T_{ij}(x, y)$.

The look-up table representation is defined by $3 \times n^3$ coefficients associated to the first n Chebyshev polynomials for the three channels r, g and b. A standard

3-dimension lookup table with n colour levels (i.e. 256 or 32) would involve $3 \times n^3$ variables. This representation also requires $3 \times n^3$ parameters, but the number n can be drastically diminished (e.g. $n = 7$). Moreover, the main response of the 3D-LUT is concentrated on the channel considered. Indeed, in (r, g, b), r is more significant than g and b to compute r' (respectively for g and b). Thus, it is possible to associate a higher accuracy for the channel considered rather than for the two other channels.

The conversion from the polynomial representation to the standard form is computed by applying equation (1) to all the lookup table elements.

4   Initialization

The 3D-LUT default initialization is the identity function, i.e. $(r, g, b)^\top = f(r, g, b)$. This configuration is obtained by using only the $L_1(x) = x$ polynomial for the channel related to the colour being processed. In term of coefficients, identity corresponds to:

$$\begin{cases} \alpha^r_{1,0,0} = \alpha^g_{0,1,0} = \alpha^b_{0,0,1} = 1 \\ \alpha^c_{i,j,k} = 0 \quad \text{otherwize} \end{cases} \tag{2}$$

It is also possible to convert an existing standard 3D-LUT to our model. Indeed, setting the initial function f with a good estimation of the expected lookup table will decrease the number of iterations required to reach convergence. Given a 1-dimensional lookup table $g(x)$ such that:

$$g(x) = \sum_{k=0}^{\infty} c_k L_k(x)$$

the coefficients c_k can be found by ([20], p.67):

$$c_k = \frac{4}{\pi} \int_{-1}^{1} \frac{g(x).T_k(x)}{\sqrt{1 - x^2}} dx$$

except for c_0 that should be divided by 2.
The discreet form with n discretization steps is:

$$c_k = \frac{4}{\pi n} \sum_{i=1}^{n-1} \frac{g(2\frac{i}{n} - 1).T_k(2\frac{i}{n} - 1)}{\sqrt{1 - (2\frac{i}{n} - 1)^2}}$$

Actually, a much faster estimation of c_k is given by:

$$c_k \simeq \frac{2}{n+1} \sum_{i=0}^{n} g\left(\cos \frac{\pi i}{n}\right).\cos \frac{\pi k i}{n} \tag{3}$$

still with c_0 divided by 2.

To perform this stage with a 3-dimensional lookup table, the previous method should be repeated on the three dimensions.

5   Non-Linear Process

5.1   Image Descriptors

Let $\{I_{i=1...k}\}$ be a set of k images and the vector $\mathbf{x} = D(I)$ a representation of the colorimetric properties of an image I. The vector $\mathbf{x}_i$ is a concatenation of measures on the image I_i. The purpose of the minimization process is to find the 3D-LUT such the transformed images provide similar vectors $\mathbf{x}$. As a minimal setup, we propose the following measures:

- **image average color**: returns a value (in $[0,1]$) for each r, g and b channel.
- **image saturation**: returns a single value (in $[0,1]$) corresponding to the average saturation per pixel. A pixel saturation is computed as the variance of the r, g and b channels.
- **image contrast**: returns a single value (in $[0,1]$) corresponding to the variance of the image 3D histogram.

In this paper, we mainly focus on these three measures, however any other measures satisfying distances properties can be added in the vector $\mathbf{x}$.

5.2   Minimization

The non-linear process consists in finding the parameters $\alpha_{i,j,k}^{c}$ representing a 3D-LUT that transforms the input images $I_{i=1...k}$ such the $\mathbf{x}_{i=1...k}$ become similar. Thus, this process is equivalent to minimize the cost function $M(\{\mathbf{x}_{i=1...k}\})$:

$$M(\{\mathbf{x}_{i=1...k}\}) = \|\sigma(\mathbf{x}_i)_{i=1...k}\| \tag{4}$$

Where $\sigma(\mathbf{x}_i)$ denotes the variance of the vectors $\mathbf{x}_i$. This approach makes the corrected images to have their descriptors converging to an average value $\hat{\mathbf{x}}$. Another possibility is to select a reference image I_r whose descriptor $\mathbf{x}_r$ will be considered as a target for the other images during the minimization process. The function $M_r(\{I_{i=1...k}\})$ becomes:

$$M_r(\{\mathbf{x}_{i=1...k}\}) = \sum_{\substack{i=1 \\ i \neq r}}^{k} \left(\mathbf{x}_i - \mathbf{x}_r \right)^2 \tag{5}$$

5.3   Point Correspondences

The minimization process can be performed on the whole images but can also be restricted on a set of selected areas. In that situation, point correspondences can be found using usual techniques such SIFT [14] or SURF [21]. Applying the minimization on a restricted set of areas on the images presents some advantages about robustness. Indeed, if a colour appears only on an image but not on the others, this colour will not be selected and hence will not contribute to the colorimetric correction. However, the risk of this method is to limit the diversity of colours encountered in the areas and hence to decrease the accuracy of the colorimetric correction.

6   Optimizations

The minimization process is sometimes long to compute, hence we propose in the following sections some optimizations to speedup the process to reach convergences. None of these methods affect the accuracy of the final results.

6.1   Initialization

As described in section 4, the initial parameters $\alpha_{i,j,k}^c$ are setup with equation (2) such the resulting 3D-LUT represents the identity function. An alternative is to start the iterative process from a solution that is fast to compute and not too far from the expected solution. In practice, we first compute an histogram equalization for each r, g and b channels, resulting in three 1-dimensional LUT. The corresponding coefficients $\alpha_{i,j,k}^c$ are extracted from these LUTs using equation (3).

6.2   Histogram Domain

Most of the descriptors presented in section 5.1 require the computation of a 3D histogram and the remaining descriptors can be computed from these histograms. Hence, the successive 3D-LUT computed during the non-linear process are directly applied on the histograms rather than to apply them on the images and then extract the histograms. Moreover, the 3D histogram data is stored on a 1D array with size equal to the number of the different colours appearing in the image. Thus, in the worth case (all pixels have different colours), this array has the same size as the image. Since a LUT is a surjective function, the size of the array will never increase during the iterative process. Finally, avoiding to apply the lookup tables to the images makes the computation time independent from the images'resolution and hence makes possible to work on high resolution images.

6.3   Pyramidal 3D Histograms

Finally, we use a pyramidal method on the iterative process. During the first iterations, the 3D histograms are quantized to decrease their size of 80%, involving a fast but inaccurate convergence. The quantization effect is progressively decreased during the iterations such that the last iterations become slower but use the image data with full details. The effect of this pyramidal method is first to speed up the computation during the first iterations and second to speed up total convergence.

7   Tests and Results

We implemented our minimization method in C++, with Levenberg-Marquardt minimization algorithm as described in [22] (p. 600). We tested our method on a set of images with different colorimetric properties and geometrically rectified

Table 2. Computation time with and without optimizations presented in section 6. The image resolution for the RGB and non-optimized version are reduced to 400×266 to get acceptable computation time, whereas the optimized version run on 2300×1533 resolution images. Moreover, using only 3×7^3 parameters for the RGB-LUT leads to unsatisfactory results.

number of variables	standard RGB 3D-LUT	without optimization	pyramidal histogram	pyramidal histogram + initialization
3×3^3	18 min	13 min	5 min	5 min
3×5^3	230 min	124 min	58 min	54 min
3×7^3	960 min	416 min	331 min	177 min

with [12]. Figure 4 depicts a result using the Chebyshev basis function with the degree 7, with a reference image as in equation (5). The resulting 3D-LUT for some images is shown in Figure 5 and clearly underline the non-independence between each channels.

The computational time is still long, Table 2 presents the computational time of our method, with and without the optimizations presented in section 6. As a comparison, our optimized method with 7 polynomials takes less than 3 hours to compute high resolution images when the direct RGB-3D-LUT computation

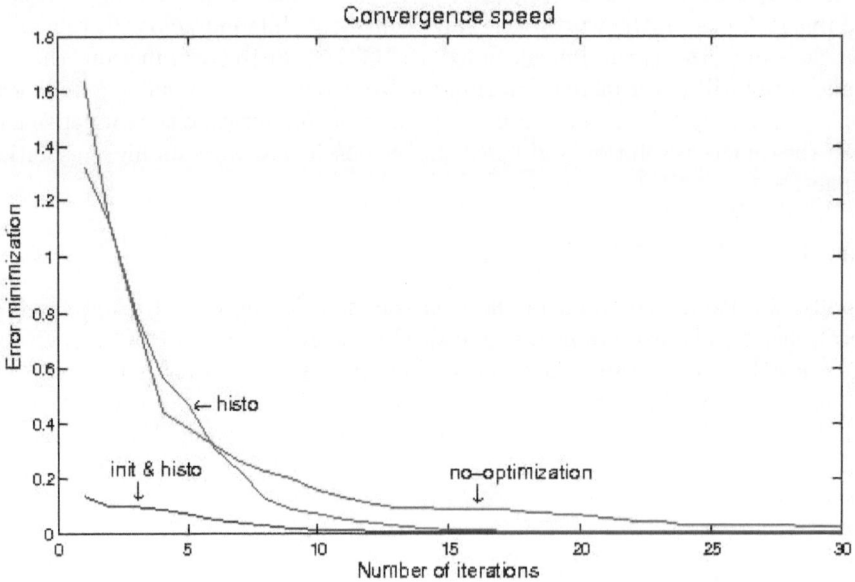

Fig. 3. Convergence speed with 3×7^3 polynomials for the methods using just the Chebyshev polynomial basis, the method with the 3D-histogram optimization and the method that also includes a 1D-LUT initialization

Fig. 4. Left: four input images with different colorimetric properties. Right: the corrected images with 7 polynomials using the top left image as a reference, with equation (5).

with 3×7^3 elements and low resolution images takes more than 16 hours, for very low quality results.

Figure 3 shows the minimization convergence speed comparison between our methods with or without the 3D histogram optimization and the initialization, for 3×7^3 polynomials. The use of the initialization drastically decreases the number of iterations required to reach convergence. The 3D-histogram optimization does not decrease the number of iterations, but reduces the computation time of an iteration. During the tests, we tried our method using L*a*b* colour space instead of rgb but we didn't noticed any changes in the results neither on the computation time.

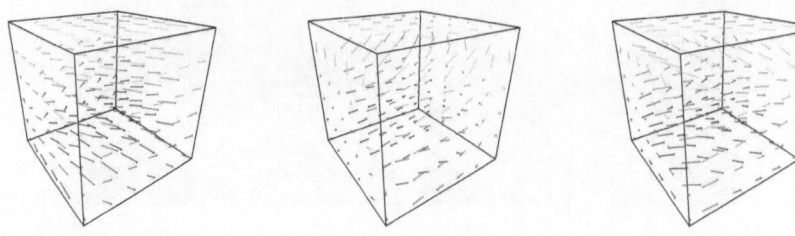

Fig. 5. 3D-LUT corresponding to the three corrected images of Figure 4

8   Conclusion

This paper presents a method to perform a colorimetric correction for a set of images captured with different cameras for stereoscopic purposes. The method produces a 3D lookup table that can be used in real-time on the GPU. This lookup table is represented by a basis function to reduce the number of required parameters. These parameters are computed by a non-linear method to minimize the difference of the colorimetric properties between the considered images.

In order to speedup the minimization process, we consider a compact form of the 3D histogram of the images rather than the images by them-self. This technique makes the process much faster and independent of the images resolution. The minimization process can start from the identity 3D lookup table or from any lookup table. Our tests show that using a fast 1D lookup table as an approximation of the results makes a very suitable starting point for our minimization process and produces a very fast convergence.

References

1. Lambooij, M.T.M., IJsselsteijn, W.A., Heynderickx, I.: Visual discomfort in stereoscopic displays: a review. Stereoscopic Displays and Virtual Reality Systems XIV 6490 (2007)
2. Smolic, A., Müller, K., Stefanoski, N., Ostermann, J., Gotchev, A., Akar, G.B., Triantafyllidis, G.A., Koz, A.: Coding algorithms for 3dtv - a survey. IEEE Trans. Circuits Syst. Video Techn. 17, 1606–1621 (2007)
3. Nozick, V., Saito, H.: Online Multiple View Computation for Autostereoscopic Display. In: Mery, D., Rueda, L. (eds.) PSIVT 2007. LNCS, vol. 4872, pp. 399–412. Springer, Heidelberg (2007)
4. Joshi, N., Wilburn, B., Vaish, V., Levoy, M., Horowitz, M.: Automatic color calibration for large camera arrays. Technical report, UCSD CSE Tech Report CS2005-0821 (2005)
5. Reinhard, E., Ashikhmin, M., Gooch, B., Shirley, P.: Color transfer between images. IEEE Comput. Graph. Appl. 21, 34–41 (2001)
6. Papadakis, N., Edoardo, P., Caselles, V.: A variational model for histogram transfer of color images. IEEE Transactions on Image Processing 20, 1682–1695 (2011)
7. Morovic, J., Sun, P.: Accurate 3d image color histogram transformation. Pattern Recognition Letters 24, 1725–1735 (2003)

8. Neumann, L., Neumann, A.: Color style transfer techniques using hue, lightness and saturation histogram matching. In: Proceedings of Computational Aesthetics in Graphics, Visualization and Imaging, pp. 111–122 (2005)
9. Pitié, F., Kokaram, A.C., Dahyot, R.: Automated colour grading using colour distribution transfer. Computer Vision and Image Understanding 107, 123–137 (2007)
10. Abadpour, A., Kasaei, S.: An efficient pca-based color transfer method. J. Visual Communication and Image Representation 18, 15–34 (2007)
11. Adabpour, A., Kasaei, S.: A fast and efficient fuzzy color transfer method. In: Proceedings of the IEEE Symposium on Signal Processing and Information Technology, pp. 491–494 (2004)
12. Nozick, V.: Multiple view image rectification. In: Proc. of IEEE-ISAS 2011, International Symposium on Access Spaces, Yokohama, Japan, pp. 277–282 (2011)
13. Yamamoto, K., Yendo, T., Fujii, T., Tanimoto, M., Suter, D.: Colour correction for multiple-camera system by using correspondences. The Journal of the Institute of Image Information and Television Engineers 61, 213–222 (2007)
14. Lowe, D.G.: Distinctive image features from scale-invariant keypoints. Int. J. Comput. Vision 60, 91–110 (2004)
15. Yamamoto, K., Oi, R.: Color correction for multi-view video using energy minimization of view networks. International Journal of Automation and Computing 5, 234–245 (2008)
16. Tehrani, M.P., Ishikawa, A., Sakazawa, S., Koike, A.: Iterative colour correction of multicamera systems using corresponding feature points. Journal of Visual Communication and Image Representation 21, 377–391 (2010)
17. Shao, F., Peng, Z., Yang, Y.: Color correction for multi-view video based on background segmentation and dominant color extraction. WSEAS Transactions on Computers 7, 1838–1847 (2008)
18. Shao, F., Jiang, G.Y., Yu, M., Ho, H.S.: Fast color correction for multi-view video by modeling spatio-temporal variation. Journal of Visual Communication and Image Representation 21, 392–403 (2010)
19. Shao, F., Jiang, G., Yu, M., Chen, K.: A content-adaptive multi-view video color correction algorithm. In: IEEE International Conference on Acoustics, Speech, and Signal Processing, vol. 1, pp. 969–972 (2007)
20. Gil, A., Segura, J., Temme, N.M.: Numerical Methods for Special Functions, 1st edn. Society for Industrial and Applied Mathematics, Philadelphia (2007)
21. Evans, C.: Notes on the opensurf library. Technical Report CSTR-09-001, University of Bristol (2009)
22. Hartley, R.I., Zisserman, A.: Multiple View Geometry in Computer Vision. 2nd edn. Cambridge University Press (2004) ISBN: 0521540518

Camera Calibration Using Vertical Lines

Jing Kong, Xianghua Ying*, Songtao Pu, Yongbo Hou,
Sheng Guan, Ganwen Wang, and Hongbin Zha

Key Laboratory of Machine Perception (Ministry of Eduction)
School of EECS, Peking University
Beijing 100871, China

Abstract. In this paper we present an easy method for multiple camera calibration with common field of view only from vertical lines. The locations of the vertical lines are known in advance. Compared to other calibration objects, the vertical lines have some good properties, since they can be easily built and can be visible by cameras in any direction simultaneously. Given 5 fixed vertical lines, an image containing them taken by a camera may provide 2 constraints in the intrinsic parameters of the camera, and extrinsic parameters can then be recovered. The calibration procedure consists of three main steps: Firstly, the image is rectified by a homography, which makes the projections of vertical lines parallel to u-axis in the rectified image. Secondly, for any vertical scan line in the rectified image, if we consider the scan line is taken by a virtual 1D camera, then we can calibrate the 1D camera. Finally, the intrinsic parameters of the original camera can be determined from the intrinsic parameters of the virtual 1D camera. By evaluating on both simulated and real data we demonstrate that our method is efficient and robust.

1 Introduction

Camera calibration is a necessary process of recovering 3D information from 2D images. Generally there are two classes of methods to calibrate cameras: the first class uses calibration objects, while the other one doesn't need to use any calibration objects. Since cameras are becoming cheaper and more precise, researchers have paid much attention to multi-camera system applications [1],[2],[3],[4],[5]. It's important to calibrate multiple cameras in a single coordinate frame. But traditional calibration methods may not meet this requirement. Some solutions using one-dimensional objects were proposed, but they all need to move or rotate the objects [6],[7]. The self-calibration method also had to capture a sequence of images [13].

Nowadays cameras are of good precision so that we can simplify the camera models. For the pinhole camera model, we can set the camera skew to zero, the pixel aspect ratio to one and also the principal point to the center of the image. So we usually care much more about the focal length than other intrinsic

---

* Corresponding Author; present address: No.2 Science Building, Department of EECS, Peking University, Beijing, China.

J.-I. Park and J. Kim (Eds.): ACCV 2012 Workshops, Part I, LNCS 7728, pp. 218–229, 2013.

parameters. In this paper we present a convenient method for camera calibration with a set of 5 fixed vertical lines, as shown in Fig.1, which are visible by cameras in any direction simultaneously. Lines can be easily set exactly vertical because of the gravity (such as plumb lines). From one image of vertical lines we can obtain two constraints on the intrinsic parameters and full information of the extrinsic parameters can then be determined. Our proposed calibration procedure consists of three steps. First rectify the observed lines with a homography matrix to make them parallel to the u-axis of image. After rectification, the problem turns into a virtual pure pan camera calibration, which can be solved by one-dimensional camera calibration. From the two intrinsic parameters of the virtual pure pan camera and the homography matrix, at last we obtain two constraints on the original camera's intrinsic parameters. By adding a fixed visible point on one of the vertical lines, we also can obtain the extrinsic parameters.

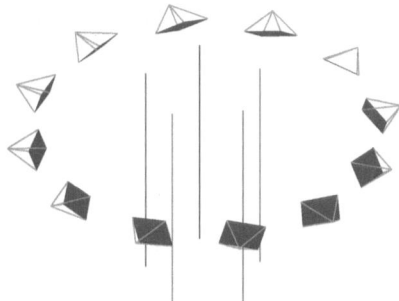

Fig. 1. When captured the five located vertical lines, we can get two constraints on the intrinsic parameters and then recover the extrinsic parameters

2 Related Work

There are a variety of calibration methods with different kinds of objects. A planar pattern viewed from at least three different orientations is used in [8]. Some other methods use spheres[9], [10], circles [11]. Most of these methods use the objections that are not visible by cameras in any direction simultaneously.

In some multi-camera systems [12], they use a moving plate for calibration. Svoboda et al. [13] calibrated their system of at least three cameras by obtaining a set of virtual 3D points made by waving a bright spot throughout the working volume. Baker and Aloimonos [14] proposed a calibration method for a multi-camera network which requires a planar pattern with a precise grid. Lee et al. [15] established a common coordinate frame for a sparse set of cameras so that all cameras observe a common dominant plane. They tracked objects moving in this plane and from their trajectories they estimated the external parameters of the cameras in a single coordinate system. Sinha et al. [16] calibrated multi-camera system using epipolar geometry constraints.

There are two studies that are most related to this paper. Wong et al. [17] obtain an approximate circular motion of the camera by using a string, a peg and a tripod. The string in the image approximately coincide with the vertical direction. They rectified the image by a homography induced by the string in the image, in order to allow the camera motion to be estimated using the circular motion algorithm. Ying et al. [18] propose a closed form solution for computing the camera pose from a set of three or more known parallel lines. But the solution is under the assumption that focal length is known, and the pose is determined up to a translation along the line direction.

As shown in several recent works [19],[20], knowledge about the vertical direction were used for reducing the minimum number of points for instantiating a hypothesis about the relative camera pose down to three or even only two in the perspective pose computation case. Most of these methods assume the intrinsic parameters are given in advance.

In [21], straight lines in the scene were used to provide constraints on the distortion parameters, and Kang [22] proposed radial distortion snakes to solve radial distortion problem. These methods can be used in conjunction with our method.

3   Notations

3.1   Two-Dimensional Pinhole Camera Model

Let $\mathbf{M} = (x, y, z, 1)^T$ be a world point and $m = (u, v, 1)^T$ be its image point, both in the homogeneous coordinates. Under pinhole camera model, they satisfy

$$\mu\mathbf{m} = \mathbf{PM} \tag{1}$$

where $\mathbf{P}$ is a 3×4 homogeneous projection matrix describing the perspective projection process. μ is an unknown scale factor. The projection matrix can be decomposed as

$$\mathbf{P} = \mathbf{K}\left[\mathbf{R}\,|\,\mathbf{t}\right], \tag{2}$$

where

$$\mathbf{K} = \begin{bmatrix} f_x & s & u_0 \\ 0 & f_y & v_0 \\ 0 & 0 & 1 \end{bmatrix}. \tag{3}$$

Here the upper triangular matrix $\mathbf{K}$ is the intrinsic parameter matrix, and $(\mathbf{R}, \mathbf{t})$ denotes a rigid transformation (i.e., $\mathbf{R}$ is a 3×3 rotation matrix and $\mathbf{t}$ is a 3×1 translation vector) which indicates the orientation and position of the camera with respect to the world coordinate system.

3.2   One-Dimensional Pinhole Camera Model

A one-dimensional pinhole camera project points from 2D to 1D [23]. So let $\bar{\mathbf{M}} = (y, z, 1)^T$ be a point of 2D and $\bar{\mathbf{m}} = (v, 1)^T$ be its image points of 1D. Similarly they satisfy

$$\mu\bar{\mathbf{m}} = \bar{P}\bar{\mathbf{M}}, \tag{4}$$

where $\bar{\mathbf{P}}$ is a 2×3 homogeneous projection matrix, and μ is an unknown scale factor. The projection matrix also can be decomposed as

$$\bar{\mathbf{P}} = \bar{\mathbf{K}} \left[\bar{\mathbf{R}} \,|\, \bar{\mathbf{t}} \right], \tag{5}$$

where

$$\bar{\mathbf{K}} = \begin{bmatrix} \bar{f} & \bar{v} \\ 0 & 1 \end{bmatrix}. \tag{6}$$

Here the upper triangular matrix $\bar{\mathbf{K}}$ is the intrinsic parameter matrix, and $\bar{\mathbf{R}}$ is a 2×2 rotation matrix and $\bar{\mathbf{t}}$ is a 2×1 translation vector.

4   Pure Pan Camera Calibration

For a pure pan camera, without loss of generality, we assume that its principal line is perpendicular to the vertical lines, while the u-axis of the image is parallel to them. The five vertical lines are represented as:

$$\mathbf{L}_i = \mathbf{M}_i + \lambda \mathbf{e}_x, \text{ for } \lambda \in R, i = 1, 2, ..., 5. \tag{7}$$

Here $\mathbf{M_i} = (0, y_i, z_i, 1)^T$ denotes a point of 2D on line $\mathbf{L}_i$, and $\mathbf{e}_x = (1, 0, 0, 0)^T$ is the direction of $\mathbf{L}_i$. We have

$$\mu l_i = \mathbf{P}\mathbf{L}_i, \tag{8}$$

where $\mathbf{P} = \mathbf{K} \left[\mathbf{R} \,|\, \mathbf{t} \right] = \mathbf{K} \left[\mathbf{R}_x \,|\, \mathbf{t} \right]$. $\mathbf{R}_x$ indicates the rotation matrix around x-axis.

Considering the v-axis of image plane as an image of a one-dimensional camera, we can rewrite equation (8) in the following form:

$$\mu \bar{\mathbf{m}}_i = \bar{\mathbf{P}}\bar{\mathbf{M}}_i, \tag{9}$$

in which we use $\bar{\mathbf{m}}_i = (v_i, 1)^T$ and $\bar{\mathbf{M}}_i = (y_i, z_i, 1)^T$ instead of $\mathbf{L}_i$ and l_i to express the one-dimensional projection process. The projection matrix $\mathbf{P}$ between $\mathbf{L}_i$ and l_i satisfies

$$\mathbf{P} = \begin{bmatrix} f_x & s & u_0 \\ 0 & f_y & v_0 \\ 0 & 0 & 1 \end{bmatrix} \begin{bmatrix} 1 & 0 & 0 & t_1 \\ 0 & \cos\varphi & \sin\varphi & t_2 \\ 0 & -\sin\varphi & \cos\varphi & t_3 \end{bmatrix}, \tag{10}$$

where φ indicates the rotation angel of the camera around x-axis.

The orientation and position of the one-dimensional camera with respect to the world coordinate system are the same with the two-dimensional camera. So we have

$$\bar{\mathbf{P}} = \begin{bmatrix} \bar{f} & \bar{v} \\ 0 & 1 \end{bmatrix} \begin{bmatrix} \cos\varphi & \sin\varphi & t_2 \\ -\sin\varphi & \cos\varphi & t_3 \end{bmatrix}. \tag{11}$$

Thus

$$\mu \begin{bmatrix} \lambda_{mi} \\ v_i \\ 1 \end{bmatrix} = \begin{bmatrix} f_x & s & u_0 \\ 0 & f_y & v_0 \\ 0 & 0 & 1 \end{bmatrix} \begin{bmatrix} 1 & 0 & 0 & t_1 \\ 0 & \cos\varphi & \sin\varphi & t_2 \\ 0 & -\sin\varphi & \cos\varphi & t_3 \end{bmatrix} \begin{bmatrix} \lambda_{Mi} \\ y_i \\ z_i \\ 1 \end{bmatrix}, \tag{12}$$

and

$$\mu \begin{bmatrix} v_i \\ 1 \end{bmatrix} = \begin{bmatrix} \bar{f} & \bar{v} \\ 0 & 1 \end{bmatrix} \begin{bmatrix} \cos\varphi & \sin\varphi & t_2 \\ -\sin\varphi & \cos\varphi & t_3 \end{bmatrix} \begin{bmatrix} y_i \\ z_i \\ 1 \end{bmatrix}. \tag{13}$$

From equation (12) and equation (13), we find constraints between $\mathbf{K}$ and $\bar{\mathbf{K}}$ that

$$\mathbf{K} = \begin{bmatrix} f_x & s & u_0 \\ 0 & \bar{f} & \bar{v} \\ 0 & 0 & 1 \end{bmatrix}. \tag{14}$$

That is to say $f_y = \bar{f}$ and $v_0 = \bar{v}$.

When given the location of vertical lines (i.e. $\bar{\mathbf{m}}_i$), we can solve $\bar{\mathbf{K}}$ through DLT solution. Then we can obtain two of the five parameters of $\mathbf{K}$.

As shown in equation (13), one vertical line provides one linear constraints on the parameters of the one-dimensional camera. So we have to capture at least 5 vertical lines to calibrate this pure pan camera.

5   Camera Calibration

In this section, we describe in detail how to solve the camera calibration problem from one or more observations of 5 vertical lines. Fig.2 shows the rectification procedure of the calibration.

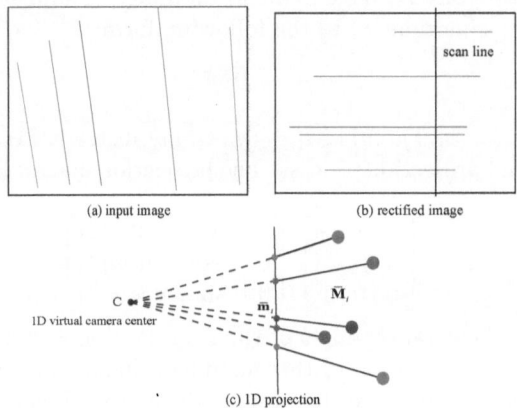

(a) input image   (b) rectified image

(c) 1D projection

Fig. 2. The procedure of calibration: (a) Capture an image of the five vertical lines. (b) After the rectification, image lines are parallel to the u-axis. (c) Calibrate the virtual 1D camera. Then from the parameters of 1D camera and the homography, we can calibrate the original camera.

Let $\mathbf{L}_i$ be the vertical lines. $\mathbf{l}_i$ are projected image lines, intersecting at the vanishing point, $\mathbf{m}_v = (u_v, v_v, 1)^T$. We use a projective transformation to make

l_i parallel,

$$\mathbf{m'}_v = \mathbf{H}\mathbf{m}_v = \begin{bmatrix} h_{11} & h_{12} & h_{13} \\ h_{21} & h_{22} & h_{23} \\ h_{31} & h_{32} & 1 \end{bmatrix} \begin{bmatrix} u_v \\ v_v \\ 1 \end{bmatrix}, \tag{15}$$

where $\mathbf{m'}_v$ is the transformed point on the line at infinity. And we have

$$\mathbf{m'}_v = (u'_v, v'_v, 0). \tag{16}$$

In order to simplify the solution, we set

$$\mathbf{H} = \begin{bmatrix} 1 & 0 & 0 \\ 0 & 1 & 0 \\ h_{31} & h_{32} & 1 \end{bmatrix}, \tag{17}$$

and $\mathbf{H}$ satisfies

$$h_{31}u_v + h_{32}v_v + 1 = 0. \tag{18}$$

After projective transformation, the parallel lines are l'_i. Rotate l'_i with a rotation matrix $\mathbf{R}_\theta$,

$$\mathbf{R}_\theta = \begin{bmatrix} \cos\theta & \sin\theta & 0 \\ -\sin\theta & \cos\theta & 0 \\ 0 & 0 & 1 \end{bmatrix}, \tag{19}$$

where θ indicates the angel between l'_i and u-axis of image plane. Then we obtain a set of image lines parallel with u-axis. We can assume that they are projected by a virtual pure pan camera.

Let $\mathbf{M} = (x, y, z, 1)^T$ be a world point. After projection and the 2D planar transformation, its image point is $\mathbf{m''} = (u, v, 1)^T$. The process can be presented as follows,

$$\mu\mathbf{m''} = \mathbf{R}_\theta\mathbf{H}\mathbf{K}[\mathbf{R}\,|\mathbf{t}\,]\mathbf{M}. \tag{20}$$

Let

$$\mathbf{A} = \mathbf{R}_\theta\mathbf{H}\mathbf{K}. \tag{21}$$

From equations (3), (17) and (19), we have

$$\mathbf{A} = \begin{bmatrix} f_x\cos\theta & s\cos\theta + f_y\sin\theta & u_0\cos\theta + v_0\sin\theta \\ -f_x\sin\theta & -s\sin\theta + f_y\cos\theta & -u_0\sin\theta + v_0\cos\theta \\ h_{31}f_x & h_{31}s + h_{32}f_y & h_{31}u_0 + h_{32}v_0 + 1 \end{bmatrix}. \tag{22}$$

After Schmidt decomposition, $\mathbf{A}$ becomes

$$\mathbf{A} = \mathbf{K'}\mathbf{R'}, \tag{23}$$

where $\mathbf{K'}$ is an upper triangular matrix and $\mathbf{R'}$ is an orthogonal matrix. $\mathbf{K'}$ is the intrinsic matrix of the virtual pure pan camera,

$$\mathbf{K'} = \begin{bmatrix} f'_x & s' & u'_0 \\ 0 & f'_y & v'_0 \\ 0 & 0 & 1 \end{bmatrix}. \tag{24}$$

We have had a solution under this condition in the previous section, f'_y and therefore we can compute the value of and . From (22) and (23), we have

$$v'_0 = (-f_x^2 h_{31} \sin\theta - s^2 h_{31} \sin\theta - f_y s h_{32} \sin\theta + f_y s h_{32} \cos\theta + f_y^2 h_{32} \cos\theta$$
$$- u_0^2 h_{31} \sin\theta - u_0 v_0 h_{32} \sin\theta - u_0 \sin\theta + u_0 v_0 h_{31} \cos\theta + v_0^2 h_{32} \cos\theta$$
$$+ v_0 \cos\theta)/(f_x^2 h_{31}^2 + h_{31}^2 s^2 + 2 f_y s h_{31} h_{32} + f_y^2 h_{32}^2 + u_0^2 h_{31}^2$$
$$+ 2 u_0 v_0 h_{31} h_{32} + 2 u_0 h_{31} + v_0^2 h_{32}^2 + 2 v_0 h_{32} + 1),$$

$$(25)$$

while the expression of f'_y is too complicated to be shown here.

At this point we have got 2 constraints of $\mathbf{K}$ from one image, and more images will produce more constraints.

In common we can set the camera skew to zero, the pixel aspect ratio to one and also the principal point to the center of the image. Then the intrinsic matrix $\mathbf{K}$ only has one degree of freedom,

$$\mathbf{K} = \begin{bmatrix} f & 0 & u_0 \\ 0 & f & v_0 \\ 0 & 0 & 1 \end{bmatrix}. \tag{26}$$

At this time, the expression of v'_0 becomes a rational fraction of 2 degrees and the expression of f'_y becomes a rational fraction of 6 degrees, so we have two equations for solving f. Both of the two equations have uniquely positive real solution.

From equation (20), (21) and (23), we have

$$\mu \mathbf{m}''_A = \mathbf{K}'[\mathbf{R}'\mathbf{R} \,|\, \mathbf{R}'\mathbf{t}\,]\mathbf{M}_A, \tag{27}$$

which presents the projection process of the pure pan camera. So it satisfies

$$\mathbf{R}'\mathbf{R} = \begin{bmatrix} 1 & \mathbf{0}^T \\ \mathbf{0} & \bar{\mathbf{R}} \end{bmatrix}, \text{ and } \mathbf{R}'\mathbf{t} = \begin{bmatrix} t_1 \\ \bar{\mathbf{t}} \end{bmatrix}. \tag{28}$$

where $\mathbf{0} = [0\ 0]^T$.

Given a control point $\mathbf{M}_f$ on $\mathbf{L}_i$, which can be detected in the image, we can compute the last unknown parameter t_i.

Without loss of generality, let $\mathbf{M}_f = (0, y_i, z_i, 1)^T$, and $\mathbf{m}_f = (u_f, v_f, 1)^T$ is its image point. Then we have

$$\mu \mathbf{m}_f = \mathbf{K}\,[\mathbf{R}\,|\,\mathbf{t}\,]\,\mathbf{M}_f. \tag{29}$$

From (28) and (29), we can solve t_i.

We need at least 5 vertical lines and a visible control point to calibrate the 8 parameters of the camera (2 intrinsic parameters and 6 extrinsic parameters), and additional vertical lines will not provide more constraints. We know that the vanishing point provides 2 constraints, each correspondence between the vertical lines and their image lines provides one constraint, the additional control point provides one constraint. If we have a sixth vertical line, from its location we can compute its image line in the rectified image (the virtual pure pan camera has already been calibrated using the other 5 vertical lines). So we can say that 5 or more vertical lines and a control point provides exact 8 constraints on the camera parameters.

6   Experiments

We have several tests on synthetic datasets (with various levels of noise and different solutions). We present an approximate solution for locating the vertical lines, and compared our calibrated focal length with that of Zhang's [8] method. We also test our method for 3D reconstruction on real datasets. In synthetic tests the virtual viewport size for the camera was [1000,1000] units, leading to focal length coordinates of comparable magnitude, while setting the principal point to the center of the image.

In real data experiments, we use a set of 5 colored electroluminescent wires as calibration object, which were straightened by heavy objects hanging below. So the gravity makes the lines exactly vertical. The whole cost of the electrolumines-cent wires is about \$ 30. Our calibration object is cheap and easy to build.

6.1   Synthetic Tests

For this series of tests, we calibrated synthetically generated ground-truth cameras. The camera's orientation and position were selected randomly but looking on ground-truth vertical lines. We also assume that the camera skew is zero, the pixel aspect ratio is one and also the principal point is the center of the image.

Properties of Different Solutions. Refer to (17) and (18), the homography matrix has two degrees of freedom but only one constraint. Fig.3 shows that $h31/h32$ might influence the calibration results. Let $\alpha = \arctan(h_{31}/h_{32})$,

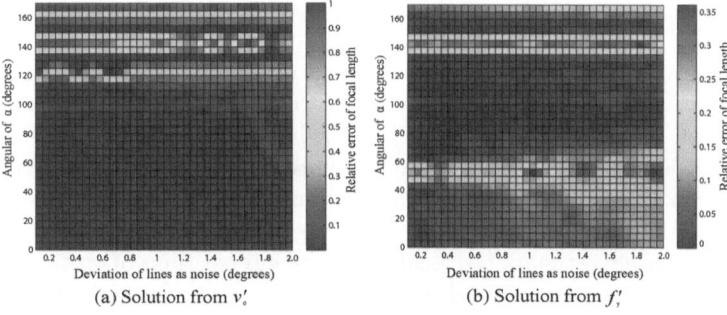

(a) Solution from v'_o (b) Solution from f'_y

Fig. 3. Relative error of focal length with different deviation of lines and α using two constrains: (a) Solve f from the expression of v'_0; (b) Solve f from the expression of f'_y

which varied from 0° to 180°. The noise was added randomly to the angular of the lines observed, with maximum from 0.1° to 2°. We took 200 estimates for each angular of α and each level of noise.

Evaluate the Extrinsic Parameters. In the real applications, the influence of our method generated by noise is mainly on the deviation of lines detected from the image. Therefore, we have tested the error of rotation and translation under different levels of lines' deviation. Hough Transform algorithm [24] and Least Square Method are widely used for line detection, and the accuracy can reach less than 0.1°. The line angular deviation varied from 0.1° to 1°, and for each level of line deviation we made 500 estimations.

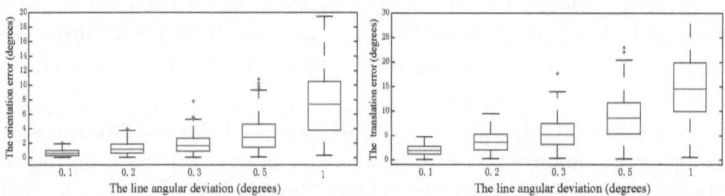

Fig. 4. The rotation error and the translation error caused by the deviation of line angular

Fig.4 shows the results of the estimations, which are represented by the MAT-LAB function boxplot. The blue box in Fig.4 represents values 25% to 75% quantile, and the red horizontal line in the middle of the box shows the median. The red crosses show data beyond 1.5 times the interquartile range. We can see that when the line angular deviation increased to 0.3°, the orientation error and translation error is still lower than 5°, but when the angular deviation reached 0.5° and even 1°, the error increased rapidly. We also noticed that the error of rotation is apparently lower than the translation error.

6.2   Real Data

We present an approximate solution for locating the vertical lines. As shown in Fig.5, the intersections of vertical lines and their shadows in the image are the image points of the intersections of vertical lines and ground. We assume that the ground is horizontal. If we place a square on the ground near the vertical lines, we can rectify the image through homography so that the square's image is perpendicular to the optical axis and in the middle of the viewport. Then we can approximately estimate the location of the intersections of vertical lines and ground (i.e. the location of vertical lines).

Table 1. shows the results of our method (using the estimated locations of vertical lines) and that of Zhang's method [8].

In order to illustrate the performance of our calibration results, we reconstruct a 3D model of a little boy, using shape-from-silhouette method. The experimental

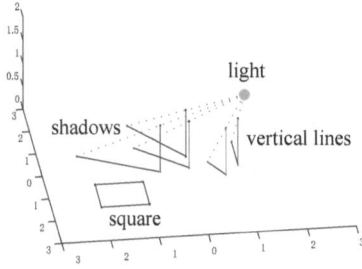

Fig. 5. Method for determining the coordinates of intersections between vertical lines and the ground plane (the horizontal plane)

Table 1. The focal length computed by our method and zhang's method [8]

Camera number	1	2	3	4	5	6
Our method	793.8629	816.6807	825.7224	859.7057	844.2413	841.1764
Zhang's method [8]	812.2360	815.8392	791.7388	840.8044	805.9483	821.3549

Camera number	7	8	9	10	11	12
Our method	834.0384	848.5016	839.4029	805.2660	790.1430	797.5329
Zhang's method [8]	813.7866	807.1752	796.1261	787.3946	778.2069	810.4973

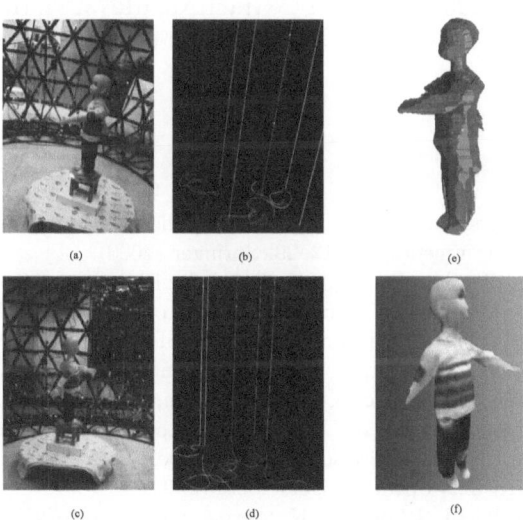

Fig. 6. (a) (c) are part of the 12 source images of the kid model, (b) (d) are corresponding images of 5 colored electroluminescent wires (the control point is on the red line) while the cameras remain fixed as (a) and (c). The projections of vertical lines can be easily extracted from such images. (e) is the reconstruction result, and (f) is the textured 3D model.

sequence consists of 12 images. The images was captured using 12 low resolution cameras (656×493), which are calibrated using one image of 5 vertical lines by our proposed method. Fig.6 shows the source images and the reconstruction result.

7   Conclusions

In this paper, we proposed a convenient method for multiple camera calibration only from vertical lines. Compared to other calibration objects, the vertical lines are easily built and are visible in any direction simultaneously. The most related study is proposed in [17], they use a string on the ground to generate an approximate circular motion, while we use plumb lines which are from a set of colored electroluminescent wires straightened by heavy objects hanging below. So the gravity makes the lines exactly vertical. It seems that we may provide better initial values for camera parameters. Given 5 fixed vertical lines and a fixed control point on one of the lines, one observation of them will provide 2 constraints for the intrinsic parameters and all the extrinsic parameters. We also present a convenient method to determine the locations of vertical lines. Both simulated and real data experiments were presented to evaluate the method we proposed, and the results demonstrate that our method is efficient and robust.

Acknowledgements. This work was supported in part by the NKBPRC 973 Grant No. 2011CB302202, the NNSFC Grant No. 61273283, the NNSFC Grant No. 61075034, the NNSFC Grant No. 91120004, and the NHTRDP 863 Grant No. 2009AA01Z329.

References

1. Brumitt, B., Meyers, B., Krumm, J., Kern, A., Shafer, S.: Easyliving: Technologies for intelligent environments, pp. 12–29. Springer (2000)
2. Bobick, A.F., Intille, S.S., Davis, J.W., Baird, F., Pinhanez, C.S., Campbell, L.W., Ivanov, Y.A., Schütte, A., Wilson, A.: The kidsroom: Perceptually-based interactive and immersive story environment. In: PRESENCE, pp. 367–391 (1999)
3. Khan, S., Javed, O., Rasheed, Z., Shah, M.: Human tracking in multiple cameras. In: International Conference on Computer Vision, pp. 331–336 (2001)
4. Svoboda, T., Hug, H., Van Gool, L.: ViRoom - Low Cost Synchronized Multicamera System and Its Self-calibration. In: Van Gool, L. (ed.) DAGM 2002. LNCS, vol. 2449, pp. 515–522. Springer, Heidelberg (2002)
5. Trivedi, M.M., Mikic, I., Bhonsle, S.K.: Active camera networks and semantic event databases for intelligent environments (2000)
6. Zhang, Z.: Camera Calibration with One-Dimensional Objects. In: Heyden, A., Sparr, G., Nielsen, M., Johansen, P. (eds.) ECCV 2002, Part IV. LNCS, vol. 2353, pp. 161–174. Springer, Heidelberg (2002)
7. Wu, F.C., Hu, Z.Y., Zhu, H.J.: Camera calibration with moving one-dimensional objects. Pattern Recogn. 38, 755–765 (2005)

8. Zhang, Z.: A flexible new technique for camera calibration. IEEE Trans. Pattern Anal. Mach. Intell. 22, 1330–1334 (2000)
9. Agrawal, M., Davis, L.S.: Complete camera calibration using spheres: A dual-space approach (2003)
10. Zhang, H., Zhang, G.: yee Kenneth Wong, K.: Camera calibration with spheres: Linear approaches (2005)
11. Jiang, G., Tsui, H.-T., Quan, L., Zisserman, A.: Single Axis Geometry by Fitting Conics. In: Heyden, A., Sparr, G., Nielsen, M., Johansen, P. (eds.) ECCV 2002, Part I. LNCS, vol. 2350, pp. 537–550. Springer, Heidelberg (2002)
12. Cheung, G.K.M., Baker, S., Simon, C., Kanade, T.: Shape-from-silhouette of articulated objects and its use for human body kinematics estimation and motion capture (2003)
13. Svoboda, T., Martinec, D., Pajdla, T.: A convenient multicamera self-calibration for virtual environments. Presence: Teleoper. Virtual Environ. 14, 407–422 (2005)
14. Baker, P.T., Aloimonos, Y.: Calibration of a multicamera network. In: Computer Vision and Pattern Recognition Workshop, vol. 7 (2003)
15. Lee, L., Romano, R., Stein, G.: Monitoring activities from multiple video streams: Establishing a common coordinate frame. IEEE Trans. Pattern Anal. Mach. Intell. 22, 758–767 (2000)
16. Sinha, S.N., Pollefeys, M., Mcmillan, L.: Camera network calibration from dynamic silhouettes. In: CVPR, pp. 195–202 (2004)
17. Wong, K.K.Y., Cipollat, R.: Reconstruction of sculpture from its profiles with unknown camera positions. IEEE Transactions on Image Processing 13, 381–389 (2004)
18. Ying, X., Zha, H.: Camera pose determination from a single view of parallel lines. In: ICIP, vol. (3), pp. 1056–1059 (2005)
19. Kukelova, Z., Bujnak, M., Pajdla, T.: Closed-Form Solutions to Minimal Absolute Pose Problems with Known Vertical Direction. In: Kimmel, R., Klette, R., Sugimoto, A. (eds.) ACCV 2010, Part II. LNCS, vol. 6493, pp. 216–229. Springer, Heidelberg (2011)
20. Fraundorfer, F., Tanskanen, P., Pollefeys, M.: A Minimal Case Solution to the Calibrated Relative Pose Problem for the Case of Two Known Orientation Angles. In: Daniilidis, K., Maragos, P., Paragios, N. (eds.) ECCV 2010, Part IV. LNCS, vol. 6314, pp. 269–282. Springer, Heidelberg (2010)
21. Kang, S.B.: Radial distortion snakes. IEICE Transactions on Information and Systems (2000)
22. Swaminathan, R., Nayar, S.: Non-metric calibration of wide-angle lenses and polycameras. In: IEEE Computer Society Conference on Computer Vision and Pattern Recognition, vol. 2, pp. 413–419 (1999)
23. Quan, L., Kanade, T.: Affine structure from line correspondences with uncalibrated affine cameras. IEEE Trans. Pattern Analysis and Machine Intelligence 19, 834–845 (1997)
24. Xu, L., Oja, E., Kultanen, P.: A new curve detection method: randomized hough transform (rht). Pattern Recogn. Lett. 11, 331–338 (1990)

Vehicle Localization Using Omnidirectional Camera with GPS Supporting in Wide Urban Area

My-Ha Le, Van-Dung Hoang, Andrey Vavilin, and Kang-Hyun Jo

Graduated School of Electrical Engineering, University of Ulsan, Ulsan, Korea
{lemyha,dungvanhoang,andy}@islab.ulsan.ac.kr, acejo@ulsan.ac.kr

Abstract. This paper proposes a method for long-range vehicle localization using fusion of omnidirectional camera and Global Positioning System (GPS) in wide urban environments. The main contributions are twofold: first, the positions estimated by visual sensor overcome the motion blur effects. The motion constrains of successive frames are obtained accurately under various scene structures and conditions. Second, the cumulative errors of visual odometry system are solved completely based on the fusion of local (visual odometry) and global positioning system. The visual odometry can yield the correct local position at short distance of movements but it will accumulate errors overtime, on the contrary, the GPS can yields the correct global positions but the local positions may be drifted. Moreover, the signals received from satellites are affected by multi-path and forward diffraction then the position errors increase when vehicles move in dense building regions or jump/miss in tunnels. To utilize the advantages of two sensors, the position information should be evaluated before fusion by Extended Kalman Filter (EKF) framework. This multiple sensor system can also compensate each other in the case of losing one of two. The simulation results demonstrate the accuracy of vehicle positions in long-range movements.

Keywords: Omni-directional camera, chamfer matching, visual odometry, GPS, cumulative error, EKF.

1 Introduction

Autonomous vehicle/robot navigation is an important research in various applications of localization, path planning and mapping. Although, the progress has been made in this field during the last few years, still there are no methods satisfying the requirement of high accurate as well as robustness in long distance of movements or different kind of conditions or environments.

In recent years, many methods have been developed for vehicle navigation, which can roughly be divided into several categories. Namely: methods using only electro-magnetic devices (e.g., Global Positioning System (GPS), Inertial Measurement Unit (IMU), wheel odometer, laser sensor), methods based on vision system only (e.g., monocular camera, stereo camera, catadioptric camera) and methods combined electro-magnetic devices and vision systems. In the first

J.-I. Park and J. Kim (Eds.): ACCV 2012 Workshops, Part I, LNCS 7728, pp. 230–241, 2013.
© Springer-Verlag Berlin Heidelberg 2013

group, the on-board GPS devices receive signals from satellites and then plot the absolute positions of vehicle on map. The accuracy of this method is often low. The signals from satellites are often drifted if it is compared with ground truth. Moreover, in the dense building region of urban scene these signals may be lost or jump in certain period. Therefore, vehicle may lose the direction information. The improvement of this method is supplementing other sensors. The system becomes multi-sensor system. With additional wheel odometer, for example, can improve the translation but it may work inaccurately if wheels slip or move on rough roads. Other device often considers is IMU. This can be used for acceleration and orientation measurement. However, the costs for this kind of devices are very expensive. Laser sensor is also a good choice in these cases, however, the signals may weak with objects appeared in far distance. Overall, the GPS signal is accurate in global shape of ego-motion but low accuracy in local position. This important characteristic will be utilized in this paper.

In the second group, vision systems are mounted on the vehicles and collect images/videos from scenes for process afterwards. Usually, the rotation and translation constraints of successive frames are first analyzed. Some robust features are considered as landmark, such as points, lines, planes. After finding out the correspondences of these features, the ego-motion will be achieved. Here, the scale ambiguity is worthy of remark. With the calibrated stereo systems (stereo perspective camera, stereo catadioptric camera), the scale of scene model and real scene is clearly known. However, with the monocular camera systems, this initial scale should be guessed or using additional device for calculation this scale (e.g., laser sensor). Overall, this kind of system can yield the accurate results in short distance of movements or in environment without GPS signals. The big problems in this kind of methods are accumulated errors. When the vehicles move in the large-scale scene, this cumulative error will be accumulated. Therefore the final trajectory will be diverged if compare with ground truth. However, one advantage of this approach is the local error in short distance of movements is small. This characteristic will be utilized in this paper.

In the third group, some combination of electro-magnetic devices and vision-based methods are proposed. These kinds of methods utilize the advantages of every sensor. The global signal from GPS can be used as the main orientation in each movement step of visual sensor. Moreover, the movement of visual systems can be performed easily in the rich texture of outdoor environment with many natural and human-made landmarks. This visual system may use the global direction to correct the main rotation direction. Besides, these sensors can compensate each other's in the cases of losing one of two signals. For example, when vehicles move in tunnels or very dense building city, the GPS signal may be lost. Then the visual system can work independently. In addition, the visual system can correct the GPS signals that are not often exact with standard GPS receiver devices. In this paper, the main idea also relies on that judgment.

This paper is organized into six sections. The next section will summarize the related works from former authors. Section 3 describes the local positioning system: using visual sensor for rotation estimation. Section 4 presents the precise evaluation and how combines the advantages of each positioning system. The experiments are shown in section 5. Finally, paper is finished with conclusions and point out future works discussed in section 6.

2   Related Works and Proposed Method

The combination of visual system and electro-magnetic devices are considered as a solution for accumulated error problem in recent years. The related works can be separate in three categories: the first group using only vision systems for localization, the second group use GPS combined magnetic devices, the third group combines vision and GPS devices. In the first group, the early research on vision based odometry using single conventional camera start by Nister et al. in [1] and Royer et al. [2]. Some authors propose using binocular camera as in [3], [4]. Because of limitation angle of view, some authors propose methods using omni-directional camera. Some typical Omni-vision based odometry can be listed as [5-8]. The basic principle of these approaches is using feature correspondence and epipolar geometry. The difficult point is: it is not easy to find the matching of features accurately. Most of them suffer from repeated textures, motion blur, illumination changes, high distortions due to the mirror, etc. Therefore, the results often incur the large error. The final trajectory may correct if vehicle/robot move in short distance. However, when they move in large-scale scene or work long time in outdoor environment, the trajectory will be diverge compared with ground truth. This is also the challenge in the increment methods of visual odometry or visual SLAM. In addition, the scale of trajectories is ambiguous if using only monocular vision system. In the second group, the multi-magnetic sensors are integrated in the system [9], [10]. Usually, the GPS is used for global position and IMU/wheel odometers are used as the local position estimation. These kinds of methods often yield correct result on the large-scale scene but the final trajectories are often drift. This is also the challenge of localization or navigation without vision systems. In recent years, the combination of two kinds of methods above are considers as the solution to overcome the disadvantages. Some authors, typically as in [11-13], propose method using stereo system and GPS. The results are significantly improved. However, these kinds of methods also rely on the feature correspondence and outlier removal in each process, so it is really time consuming.

According to the analysis above, the method based on Omni-directional camera and standard GPS is proposed in this paper. The rich information of environment is obtained by 3600 field of view camera. When the vehicle/robot rotates a large angle, the landmarks still tracked. With the orientation are guided by GPS, the final trajectories are both keep the local accuracy of Omni-vision and the global trajectory shape of standard GPS. The overview of this proposed method is shown in figure 1.

The detail of this method is described as follow: in the first frame, the direction of ego-motion is manually selected. Then the constraint of the second frame and the previous frame is computed. Here, the robust method using chamfer matching is applied instead of conventional point correspondence which is heavily

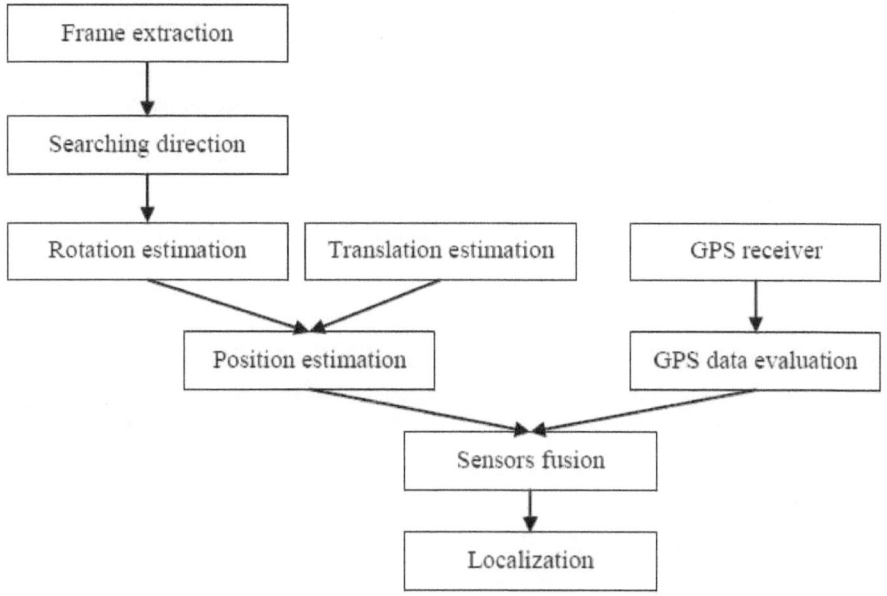

Fig. 1. General proposed scheme

suffered from motion blur affect. Besides, the translations of vehicle are absolute values, which are received from wheel odometry. Because of the ambiguous of monocular vision, this sensor is added to measure the true translation distances. In the next stage, the standard GPS receiver provides the global/absolute position of vehicle, which is extracted as the orientation guidance. However, the position from GPS may not correct in urban scene then the evaluation processes need to be performed. After obtaining the high accurate GPS position, the visual odometry and GPS position will be combined by using EKF. Especially, two these systems compensate together in the case of losing one of two signals, e.g., there are no GPS signals when vehicle move in tunnel or the direction estimation of vision system will be wrong if the tracked region is the very fast moving objects.

3   Omnidirectional Camera Based Localization

Main purpose of this step is finding the trajectory using only vision system. To do that, successive frame constrains must be computed. There are two main information must be achieved: translation and rotation. This is also the canonical topic in multiple view geometry last few decades. Some former authors propose using feature based method for calculating essential matrix whereas others proposing using appearance based methods. The absolute translation distance is typically calculated using magnetic devices, for instance, wheel encoder [14]. The rotation angles are computed exactly by using omnidirectional camera, which can cover 360-degree field of view.

3.1   Rotation Estimation

For the rotation estimation, the new approach is proposed. Note that the frames which are extracted from video often suffer from motion blur effects. Therefore, the feature point based correspondence methods yield low accurate matching position. This problem will lead to the wrong movement direction. Here, the edge based correspondence method is robust with this effect. The chamfer matching [15] is considered in this research but the application is used for omnidirectional images.

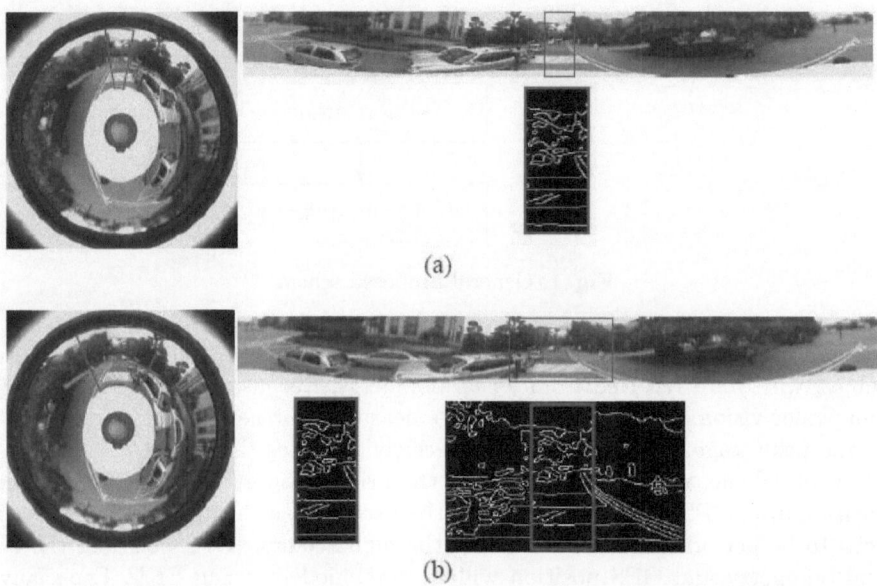

(a)

(b)

Fig. 2. Edge matching. (a) the region of forward direction. (b) the matching region in the successive frame.

The regions extracted in the forward direction of the first frame and second frame are considered as a template and region of interest (ROI). Illustration is shown in figure 2. Firstly, the edge of both template and ROI are extracted by applying edge detection algorithm. Here, Canny edge operator is applied. Secondly, the distance transformation of region of interest (DTROI) is computed. Initially, the pixels on the edge are assigned by zero and the remains are assigned by infinity. The Euclidean distance was used to compute the distance transformation.

$$
\begin{aligned}
D_{i,j}^k = \min(&D_{i-1,j}^{k-1}+1, D_{i-1,j-1}^{k-1}+\sqrt{2}, D_{i,j-1}^{k-1}+1, D_{i+1,j-1}^{k-1}+\sqrt{2}, D_{i+1,j}^{k-1}+1, \\
&D_{i+1,j+1}^{k-1}+\sqrt{2}, D_{i,j+1}^{k-1}+1, D_{i-1,j+1}^{k-1}+\sqrt{2}, D_{i,j}^{k-1})
\end{aligned}
\tag{1}
$$

where $D_{i,j}^k$ is the value of the pixel in the position (i, j) at iteration k. This process iterates until the value of each pixel is not change. The template is represented by binary image with the edge pixels are assigned one and zero for otherwise. Thirdly, the similarity measurement between template and sub-region of interest is computed. Then compute the sum of pixel value that the edge coincides on the DTROI. Slide template on DTROI to find the best matching which minimum the sum above. The perfect matching will be achieved when this sum reach to zero, i.e., the template superimpose on the sub-region of ROI. When the matching is found, the direction of the next frame also obtained as shown in figure 3.

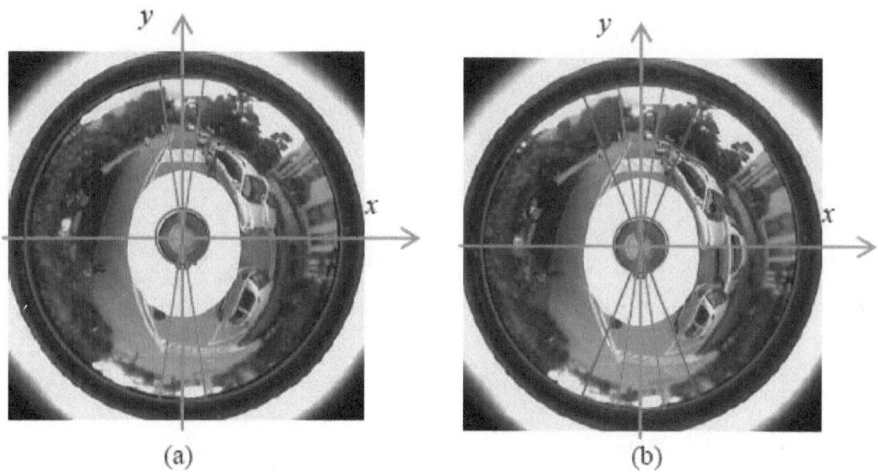

(a)   (b)

Fig. 3. Rotation direction (a) the direction of previous frame. (b) the computed direction in the successive frame.

After achieving the rotations and translations, the motion model can be constructed according to [16] without consider slip. If the position estimated from visual odometry is P(t) = (x, y, θ) and the error covariance matrix is $\Sigma_p(t)$, the kinematic equations which are formulated into EKF frame work are as follows:

$$P(t+\tau) = P(t) + \tau \begin{pmatrix} V(t)\cos(\theta(t)) \\ V(t)\cos(\theta(t)) \\ \omega(t) \end{pmatrix} \tag{2}$$

$$\Sigma_p(t+\tau) = J(t)\Sigma_p(t)J(t)^T + K(t)\Sigma_V K(t)^T + \Sigma_N \tag{3}$$

$$\Sigma_p(t) = \begin{pmatrix} \sigma_{xo}(t)^2 & \sigma_{xyo}(t) & \sigma_{x\theta o}(t) \\ \sigma_{xyo}(t) & \sigma_{yo}(t)^2 & \sigma_{y\theta o}(t) \\ \sigma_{x\theta o}(t) & \sigma_{y\theta o}(t) & \sigma_{\theta o}(t)^2 \end{pmatrix} \tag{4}$$

where τ is the sampling interval, $V(t)$, $\theta(t)$ and $\omega(t)$ represent vehicle velocity, orienta-tion and angular velocity respectively, $J(t)$ is the Jacobian of $P(t)$ with respect to x, y and θ, $K(t)$ is the Jacobian of $P(t)$ with respect to V and ω, Σ_V is the measurement error of odometry, Σ_N is the truncation error, Σ_V and Σ_N are determined accord-ing to the result of some autonomous motions because these parameters are different in each system.

4   Multiple Sensor Fusion

4.1   GPS Evaluation

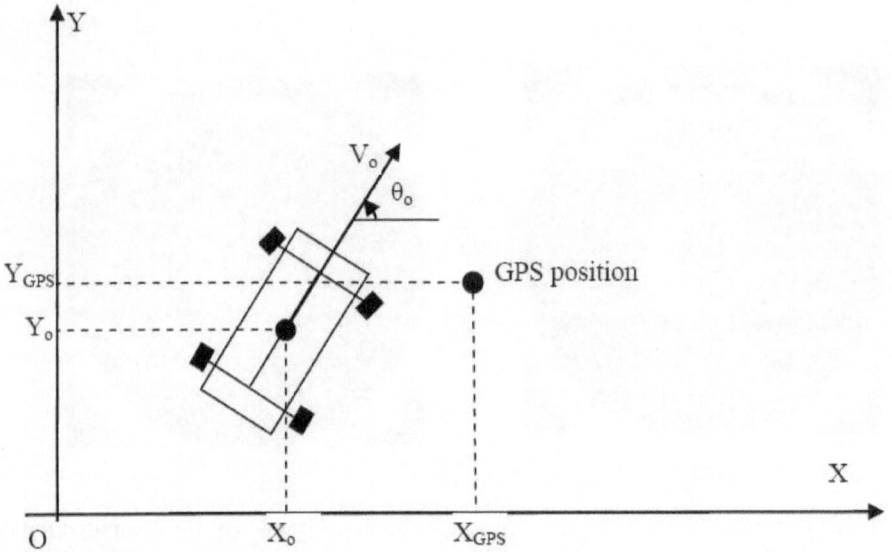

Fig. 4. Vehicle position estimated by visual odometer and GPS data

In most of long-range positioning system, the GPS is considered as the main orientation divide. However, to get the high accuracy position, it needs to pay the very expensive cost. With the standard GPS receiver, the GPS signal is uncertain. Therefore, the evaluation of the raw data should be performed (figure 4). Especially, when the vehicles move in the dense building regions or tunnels in urban scene the signal may be multi-path reflected, forward diffracted and even missed. The local signal may be wrong. The GPS position with small error will be extracted and use to correct the visual odometry. Here, authors assume that the trajectory of visual odometry is correct in short distance. Then the GPS with large error compared with visual odometry will be discarded [16]. The observation equation of GPS position is

$$P_{gps}(t) = \begin{pmatrix} x_{gps}(t) \\ y_{gps}(t) \\ \theta_{gps}(t) \end{pmatrix} \tag{5}$$

while the GPS observation $(x_{gps}(t), y_{gps}(t), \theta_{gps}(t))$ is provided by GPS position measurement. As mention above, the GPS measurement may be affected from a lot of noise, so the GPS position error can be described as follows:

$$W_{gps}^{-1} = \begin{pmatrix} \sigma_{xgps}^{2}{}^{-1} & 0 & 0 \\ 0 & \sigma_{ygps}^{2}{}^{-1} & 0 \\ 0 & 0 & \sigma_{\theta gps}^{2}{}^{-1} \end{pmatrix} \tag{6}$$

For compare the position distance of GPS and visual odometry, the Mahalanobis distance is analyzed. The position distance is computed as follows:

$$d_{xy}(t) = \sqrt{\frac{(X_{gps}(t) - X_o(t))^T (\Sigma_o(t) + \Sigma_{gps}(t))^{-1}(X_{gps}(t) - X_o(t))}{2}} \tag{7}$$

where

$$X_o(t) = (x_o(t), y_o(t)) \tag{8}$$

$$X_{gps}(t) = (x_{gps}(t), y_{gps}(t)) \tag{9}$$

$$\Sigma_o(t) = \begin{pmatrix} \sigma_{xo}(t)^2 & \sigma_{xyo}(t) \\ \sigma_{xyo}(t) & \sigma_{yo}(t)^2 \end{pmatrix} \tag{10}$$

$$\Sigma_{gps}(t) = \begin{pmatrix} \sigma_{xgps} & 0 \\ 0 & \sigma_{ygps} \end{pmatrix} \tag{11}$$

Similarly, the direction distance is computed as follows:

$$d_\theta(t) = \sqrt{\frac{(\theta_{gps}(t) - \theta_o(t))^T (\sigma_{\theta o}^2(t) + \sigma_{\theta gps}^2)^{-1}(\theta_{gps}(t) - \theta_o(t))}{2}} \tag{12}$$

Here, $X_o(t)$, $\Sigma_o(t)$, $\theta_o(t)$, $\sigma_{\theta o}(t)$ are the subset of P(t) and $\Sigma_p(t)$ which are obtained from the odometry. σ_{gps} and $\sigma_{\theta gps}$ are constant values which the authors determined from the observation of GPS measurement error. This distance implies the accuracy of GPS position and one threshold will be selected empirically. If the distance is smaller than the threshold, GPS should be used for visual trajectory correction or discarded on the contrary case.

4.2   Vision and GPS Sensor Fusion

There have several methods for multi sensors fusion. Here, the EKF is applied for this purpose. After evaluating the GPS position, the visual odometry trajectory will be corrected by using equation of maximum likelihood estimation in the EKF framework if it is small (less than threshold)

$$\hat{P}_f(t) = P(t) + \Sigma_f(t)W_{gps}^{-1}(P_{gps}(t) - P(t)) \tag{13}$$

$$\Sigma_f(t) = (\Sigma_p(t)^{-1} + W_{gps}^{-1})^{-1} \tag{14}$$

where $\hat{P}_f(t)$ and $\Sigma_f(t)$ represent corrected vehicle position and its error covariance matrix after fusion. For correcting the vehicle position which was estimated by odometry, each $P(t)$ and $\Sigma_p(t)$ is updated with $\hat{P}_f(t)$ and $\Sigma_f(t)$.

5   Experiments

In this section, the simulations are presented to evaluate effectiveness of proposed method. The experiments are performed in long-range movements covered a large region in urban environment. The videos and GPS positions are acquired by hyperbolic omnidirectional camera and GPS receiver mounted on a vehicle (figure 5).

In this experiment, the distance of movement is around 1.6km and 5716 frames

Fig. 5. Vehicle equipped with multiple sensors

are processed. The Omni-images are unwrapped into panoramic images and the direction of the car is defined on the first frame. To find the direction for the next frame, one region of size [50x240] on the forward direction of the first image is slide and match to the second frame. Here, to reduce the time consuming of appearance based matching method, the chamfer matching are performed. The edges of template region and panoramic image of the second frame are obtained by using Canny edge detection. The direction of car on the second frame is calculated based on the matching position the second frame. To reduce the direction error which occurs suddenly, one threshold of direction angle is defined. With the remark that even the road corner is 90 degree or smaller, the angle direction change regularly in the arc shape because of the mechanical structure and kinematic model of vehicle. With this assumption, some large error in direction cause by the near objects appeared in the images are discarded. In addition, the wheel encoder provides the translation distance in the absolute values. However, the result of only vision system and odometry often suffer from the cumulative errors then the trajectory will be diverged compared with ground truth when the moving distance is long (more than 200m, figure 6). At the same time, the positions of vehicle from GPS signals are also analyzed. The corrected positions are selected based on the measurements with odometry system. Here, one threshold is defined for these distances. If this Mahalanobis distance is smaller than this threshold then it is considered as corrected positions. After selecting the corrected position from GPS, vehicle position drifted on the odometry system will be corrected. These procedures will be interacted after 50 frames. When the car move on the tunnels or very dense building in urban the GPS may lost or jumps, the trajectory will be recovered by only odometry system. The GPS evaluation and correction process will be performed once the GPS signals appear again. The GPS position, odometry trajectory and the GPS combined odometry are shown in figure 6.

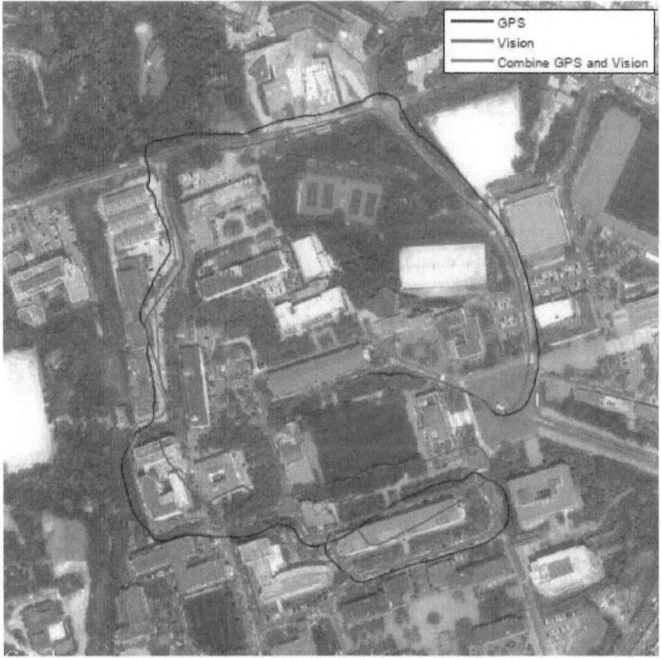

Fig. 6. Localization result

6 Conclusions

Vehicle localization and mapping by using multiple sensor fusion system in long-range motions is presented in this paper. Some advantages were pointed out through our arguments and experiments. First, this method solves cumulative errors in long-range odometry, which lead to diverged trajectory compared with ground truth. In some special cases, the trajectory or vehicle location is not completely correct in local position but the global trajectory will never lose. Second, the visual compass works accurately in complicated scene structure and difference lighting conditions of outdoor environments. The robustness and rapidity of this approach can let it be applied in other real time orientation systems. Third, the combination of local positioning system and global positioning system not only improve the local position (cumulative errors) but also can be used for GPS position correction in some cases. This system also can compensate each other when losing one of two guidance. Our future woks focus on improving this method for real time application and building high accuracy sensors fusion systems for odometry or SLAM.

Acknowledgement. This work was supported by the MKE (The Ministry of Knowledge Economy), Korea, under the Human Resources Development Program for Convergence Robot Specialists support program supervised by the NIPA(National IT Industry Promotion Agency) (NIPA-2012-H1502-12-1002).

References

1. Nistér, D., Naroditsky, O., Bergen, J.: Visual odometer. In: Proceedings of the IEEE Computer Society Conference on Computer Vision and Pattern Recognition, pp. 652–659 (2004)
2. Royer, E., Lhuillier, M., Dhome, M., Lavest, J.M.: Monocular vision for mobile robot localization and autonomous navigation. International Journal of Computer Vision 74, 237–260 (2007)
3. Garca-Garca, R., Sotelo, M.A., Parra, I., Fernndez, D., Naranjo, J.E., Gaviln, M.: 3D visual odometry for road vehicles. Journal of Intelligent and Robotic Systems 51, 113–134 (2008)
4. Konolige, K., Agrawal, M., Sol, J.: Large-scale visual odometry for rough terrain. In: Proceedings of the International Symposium on Robotics Research (2007)
5. Scaramuzza, D., Fraundorfer, F., Siegwart, R.: Real-time monocular visual odometry for on-road vehicles with 1-point RANSAC. In: IEEE International Conference on Robotics and Automation, pp. 4293–4299 (2009)
6. Scaramuzza, D., Siegwart, R.: Appearance guided monocular omnidirectional visual odometry for outdoor ground vehicles. IEEE Transactions on Robotics 24, 1015–1026 (2008)
7. Gandhi, T., Trivedi, M.: Parametric ego-motion estimation for vehicle surround analysis using an omnidirectional camera. Machine Vision and Applications 16, 85–95 (2005)
8. Tardif, J.P., Pavlidis, Y., Daniilidis, K.: Monocular visual odometry in urban environments using an omnidirectional camera. In: Proceedings of the IEEE/RSJ International Conference on Intelligent Robots and Systems, pp. 2531–2538 (2008)

9. El Najjar, M.E., Bonnifait, P.: A road-matching method for precise vehicle localization using Belief Theory and Kalman filtering. Autonomous Robots 19, 173–191 (2005)
10. Sukkarieh, S., Nebot, E.M., Durrant-Whyte, H.F.: A high integrity IMU/GPS navigation loop for autonomous land vehicle applications. IEEE Transactions on Robotics and Automation 15, 572–578 (1999)
11. Cappelle, C., El Badaoui El Najjar, M., Pomorski, D., Charpillet, F.: Intelligent geolocalization in urban areas using global positioning systems, three-dimensional geographic information systems, and vision. Journal of Intelligent Transportation Systems 14, 3–12 (2010)
12. Grimes, M., LeCun, Y.: Efficient off-road localization using visually corrected odometry. In: Proceedings of the IEEE International Conference on Robotics and Automation, pp. 2649–2654 (2009)
13. Wei, L., Cappelle, C., Ruichek, Y., Zann, F.: GPS and Stereovision-Based Visual Odometry: Application to Urban Scene Mapping and Intelligent Vehicle Localization. International Journal of Vehicular Technology (2011)
14. Stella, E., Cicirelli, G., Lovergine, F.P., Distante, A.: Position estimation for a mobile robot using data fusion. Intelligent Control, 565–570 (1995)
15. Barrow, H.G., Tenenbaum, J.M., Bolles, R.C., Wolf, H.C.: Parametric correspondence and chamfer matching: two new techniques for image matching. In: Proceedings of International Joint Conference on Artificial Intelligence, pp. 659–663 (1977)
16. Ohno, K., Tsubouchi, T., Shigematsu, B., Yuta, S.: Differential GPS and odometry-based outdoor navigation of a mobile robot. Advanced Robotics 6, 611–635 (2004)

Efficient Development of User-Defined Image Recognition Systems

Julia Moehrmann and Gunther Heidemann

Institute of Cognitive Science, University of Osnabrueck
Albrechtstr. 28, 49076 Osnabrueck, Germany
{firstname.lastname}@uni-osnabrueck.de

Abstract. Development processes for building image recognition systems are highly specialized and require expensive expert knowledge. Despite some effort in developing generic image recognition systems, use of computer vision technology is still restricted to experts. We propose a flexible image recognition system (FOREST), which requires no prior knowledge about the recognition task and allows non-expert users to build custom image recognition systems, which solve a specific recognition task defined by the user. It provides a simple-to-use graphical interface which guides users through a simple development process for building a custom recognition system. FOREST integrates a variety of feature descriptors which are combined in a classifier using a boosting approach to provide a flexible and adaptable recognition framework. The evaluation shows, that image recognition systems developed with this framework are capable of achieving high recognition rates.

1 Introduction

Current image recognition systems are developed for highly specialized purposes. Experts have to identify suitable feature descriptors and learning methods for the task at hand. Additionally, a tremendous amount of ground truth data has to be gathered, on which the classifier is trained. All in all, the development process is expensive and produces inflexible and highly specialized systems. While this is justified for commercial applications where high accuracy is crucial, it prevents the adoption of image recognition systems for a large variety of applications. With the growing importance of social networks and communities, there is a large potential for image recognition systems to exploit the huge amounts of freely available image data. Face recognition is currently being implemented in social networks and introduces image recognition to a worldwide community. Our goal is to provide a flexible tool for this community, which allows the simple development of custom, i.e., user-defined, image recognition systems. Such systems could support the further application of image recognition in everyday tasks and would typically access publicly available webcams, e.g., notifying the user if a window was left open at the office or if a parking space is available.

We present a flexible object recognition system (FOREST) which aims to overcome the limitations of current specialized image recognition systems. It is a

J.-I. Park and J. Kim (Eds.): ACCV 2012 Workshops, Part I, LNCS 7728, pp. 242–253, 2013.
© Springer-Verlag Berlin Heidelberg 2013

simple-to-use framework for building custom image recognition systems and does not require any knowledge of image recognition methods. In this work, we distinguish two different components:

1. The framework (FOREST) for developing custom image recognition systems
2. Custom recognition systems developed using FOREST

The terms *custom recognition system* and *user-defined recognition system* are used interchangeably in this work and refer to an image recognition system which was developed for a specific application by a non-expert user. The actual recognition task is defined solely through the annotation of the ground truth data for the recognition task. The FOREST framework, which provides the functionality for developing such custom recognition systems, is therefore required to be flexible. Although individual custom recognition systems are likely to solve rather simple recognition tasks, the variety of possible recognition tasks is unlimited. The required flexibility is achieved through the integration of a variety of existing keypoint detectors and feature descriptors which extract large amounts of image features independently of the recognition task at hand. Discriminative features are then identified in the process of building a boosting classifier, based on the annotations provided by the user.

Although there has been some research on generic image recognition, it focused on individual aspects, like, for example, the development of feature descriptors. In contrast, we present a complete framework, which guides the user through every step of the development process, from image acquisition to the final image recognition system, including the annotation of ground truth data. The system is not designed as a generic image recognition system in the sense that it includes the object-specific recognition functionality out of the box. Instead, it is a framework for creating application-specific recognition systems with little effort, and thereby aims to exploit publicly available image data sources, like webcams. More specifically, the contributions of FOREST compared to state-of-the-art specialized image recognition systems are the following:

– It requires no prior knowledge about the context of the recognition task
– It guides non-expert users through every step of the development process

This work is structured as follows: The next section describes the related work for this research, Sections 3 and 4 describe the development process from a user's perspective and the architecture of the framework. Afterwards, we provide an evaluation of the recognition capabilities in Section 5 and conclude with a summary and an outlook on future work.

2   Related Work

A lot of research in computer vision has been invested in the development of robust keypoint detectors [15,23] and feature descriptors, like SIFT [11], GLOH [14], SURF [3], Shape contexts [4], steerable filters [7] and color and texture descriptors [13]. Although such features are state of the art for many applications, they are, individually, not powerful enough to solve arbitrary recognition tasks.

Bag-of-keypoints approaches have been popular in recent years for object categorization tasks [5,18,12]. Such approaches extract feature descriptors from image patches, cluster them in feature space and calculate histograms. Zhang et al. [27] evaluated several feature descriptors and combinations thereof in a bag-of-keypoints approach. Although these approaches are popular in generic object categorization tasks, they are not directly applicable to recognition tasks.

Generic object recognition was investigated by Agarwal et al. [1], who employed a vocabulary of visual words and the spatial relations between these to create sparse feature vector representations. The sparseness was then exploited by the sparse network of Winnows (SnoW) used for classifier training. Recently, boosting has been successfully employed in generic recognition tasks. Viola and Jones [24] applied rectangle features together with boosting for object recognition. Opelt et al. extended the boosting algorithm to handle heterogeneous feature types, like SIFT, moment invariants, and textual moments [19,20] to achieve correct recognition even in complex and cluttered scenes. A similar classifier was applied by Hegazy et al. [8] on feature sets consisting of SIFT and color features. Zhang et al. [26] employed a two-layer boosting algorithm, which used shape context and SIFT feature sets in the first stage and the spatial relations between features in the second classifier stage.

Although generic object recognition has been a major research topic for years, it has not yet been considered within a real application environment. Especially the creation of ground truth data for such systems has been neglected. However, the task of image annotation has been investigated in other research areas, such as human computation [21], which has led to the development of so called Games with A Purpose (GWAP), with their best known example, the ESP Game [2], which gathers annotations for images. Russell et al. developed a web-interface called LabelMe for community based image annotation [22]. Objects or parts of an image can be annotated by anyone willing to contribute. Similarly, Yao et al. provide an annotation tool for the creation of a large, general purpose data set [25]. All of these approaches share the same limitation, which is that they create general purpose data sets. Such data sets are suitable for the evaluation of new image recognition or classification algorithms but they cannot be used as a training data set for custom, user-defined image recognition systems. In the area of content based image retrieval, Koskela et al. [9] developed a system for the automated semantic annotation of images based on the PicSOM system [10] using self-organizing maps (SOM), where an existing ground truth data set is used to annotate other images in the data set. Also based on self-organizing maps, the authors previously developed a graphical user interface which displays image sets as ordered by a SOM [16,17]. The system was developed for the annotation of image data sets, with a focus on creating ground truth data.

3   Development Process

The process of developing a custom image recognition system using FOREST is given in Table 1. The graphical user interface (GUI) shown in Figure 1 guides

Table 1. Development process using FOREST from a user's point of view. Brackets ([[..]]) indicate an additional step for expert users only, italicized text indicates automatic processing by FOREST.

Development process
1.   Select image data source
2.   *if* image data is not prerecorded
3.   Set time interval or end date for image acquisition
fi
4.   Select region to be observed
5.   [[Select operators and adjust parameters]]
6.   *Image acquisition, keypoint detection, and feature extraction*
7.   Annotate images
8.   *Classifier calculation*
9.   *if* results of cross validation are satisfying
10.   Calculation finished
else
11.   Add training data (goto 2) or annotate regions (see Section 4.2)
fi

the user through the complete development process. Users may enter a name and description for the custom image recognition system that is being developed at any time during the development process. This information is always displayed in the GUI (see Figure 1) to provide orientation in case users develop more than one recognition system concurrently. Figure 1a) shows the first screen that is displayed when a new process is initiated (steps 1-4 in Table 1). The user may select a local image directory or enter a webcam URL. An exemplary image is retrieved from the specified data source and displayed. The user may specify an observation region within the image, e.g. doors or meeting areas in interior scenarios. If a webcam is chosen as the data source, the user has to specify the duration for image acquisition, either by setting the time interval, the date and time when it ends, or simply the number of images that should be captured. The user may then start the image acquisition and processing (detection and feature extraction) with the default configuration or move on to the expert option panel (step 5), shown in Figure 1b, where individual operators can be enabled or disabled and the configuration of each operator can be adapted. This step is, as indicated in Table 1, meant for expert users only. However, we expect expert users to work with this framework only for evaluating new operators or rapid prototyping, since usually experts will demand more and direct access to the processing functionality. The image acquisition, keypoint detection and feature extraction run automatically without any need for user interaction.

After the automatic feature extraction is complete, the user is asked to annotate the ground truth data using the integrated annotation interface described in Section 4.2 (step 7). The annotated ground truth data is afterwards used for calculating a boosting classifier (step 8).

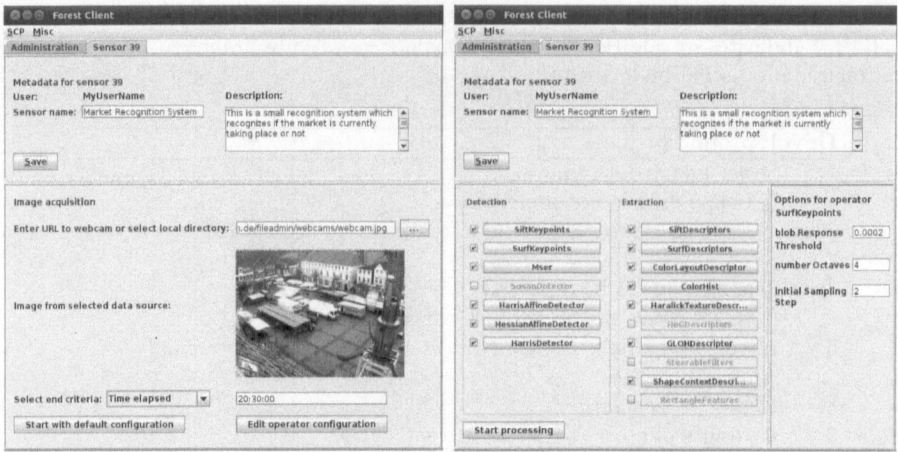

Fig. 1. Exemplary screens of the UI: a) image acquisition screen, b) expert option screen. Information about the recognition system is displayed in the upper panel.

The validation result of the calculated classifier is displayed to the user, together with information about the number of distinct feature types used (step 9). This provides basic feedback about the computational costs and might encourage users to choose a slightly inferior classifier for their task. Feature types which are not considered by the boosting classifier will not be extracted from the images in recognition mode, thereby speeding up the image processing. If the results are insufficient, the user may simply gather more training data and retrain the custom recognition system, or the user might choose to annotate image regions, as described in Section 4.3.

4   Forest

The architecture of FOREST is depicted in Figure 2. The basic design follows a three-stage architecture, which includes keypoint detection, feature extraction and classification. Image data is passed to the keypoint detection and feature extraction stages. The resulting feature vectors are passed on to the classification stage. In the training phase, the image annotation stage receives image data and feature vectors for calculating a clustering on the image data. The classifier learns discriminative features based on the annotations provided by the user. The individual stages are described in detail in the following sections.

The main functionality of FOREST, i.e., the image processing functionality, is implemented in Matlab, due to its efficiency. Various operator implementations are also already available in Matlab or can easily be integrated via its C and Java interfaces. Operators implement a custom interface and new ones can be added to extend the functionality of FOREST during runtime. A Java wrapper for the Matlab functionality of FOREST was implemented, which provides an API for the GUI.

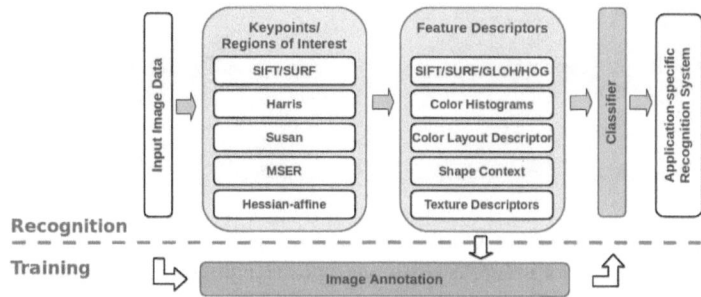

Fig. 2. Basic architecture of FOREST with exemplary operators given for the detection and extraction stage. In recognition mode, the image annotation step is obsolete.

4.1 Keypoints and Feature Descriptors

Although individual operators, like SIFT, perform well for specific tasks, they cannot achieve the same performance in substantially different scenarios. Therefore, a flexible system without prior knowledge cannot rely on single keypoint or feature descriptors. Combinations of feature types have been used previously in generic recognition systems to improve recognition rates in comparable scenarios. However, these systems were often restricted in the number of feature types they use. There is no technical limit to the integration of further operators in FOREST. Extensibility is an important aspect, since use of a variety of different operators is crucial for the flexibility of the framework.

A lot of research has been invested in the development of robust and flexible operators, therefore existing operators for keypoint detection and feature extraction were integrated into FOREST. If open source implementations existed for these operators they were integrated, e.g., SIFT[1] and affine-invariant features[2], otherwise the operators were reimplemented.

4.2 Image Annotation

Most image recognition systems rely on established ground truth data sets. By contrast, since the purpose of FOREST is to build custom recognition systems, annotation has to be performed by users themselves. We consider the prospect of having to annotate thousands of images individually as the major obstacle preventing users from using a system like FOREST. However, we overcome this problem with the image annotation tool developed by the authors [16,17]. It simplifies the image annotation task by calculating a self-organizing map (SOM) on the images and displaying this clustered structure in a zoomable user interface. The annotation interface was found to speed up the annotation process and requires no special computer skills. Images are displayed in clusters, which may be

---

[1] http://www.vlfeat.org

[2] http://www.robots.ox.ac.uk/~vgg/research/affine/index.html

Fig. 3. Positive (left, showing windsurfers) and negative (right) training images[3] highlighting salient regions of images, which were positively annotated in the first annotation stage.

annotated by users at once. Due to the fact that efficiency with respect to user interaction is important, FOREST applies weakly supervised learning, i.e. each image is annotated as a whole, not individual regions.

The annotation tool relies on features extracted from the images, therefore, the user is asked to annotate images after the feature extraction is completed. Currently, the annotation tool relies on color histograms or color layout descriptors for SOM calculation. Since there is no distance metric for comparing different feature types, SOM calculation is confined to using one individual feature type. Although color based descriptors might not lead to semantically meaningful clusters, they produce a clustering result which is intuitive and experienced as reasonable, especially by non-expert users.

4.3   Region Annotation

Custom recognition systems developed with the proposed framework by non-expert users are capable of achieving high recognition rates, as can be seen in Section 5. However, the concept of high quality training data sets may not be familiar to non-experts. Users might choose to capture training data during a short period of time and obtain data sets which do not capture enough variation, or obtain data sets with extreme skews, e.g., much more negative than positive training data. Such constellations favor the learning of unintended correlations by the classifier, especially since FOREST employs weakly supervised learning. Precise annotations of image regions help to overcome this problem, but such annotations are expensive with respect to the effort required for their creation.

We therefore extended FOREST to obtain more precise annotations of regions within the training images using the existing annotation tool. The keypoints and regions of interest detected by FOREST are used to find salient regions in positive training images. These salient regions are then used to create a new representation which is annotated by the user. The new representations are shown in Figure 3. The salient region and its immediate surrounding is drawn opaque while the rest of the image is semi-transparent. A rectangle indicates the salient region. Annotation trials showed that this visualization is well suited for directing the user's attention to the area of interest, whereas cropped versions resulted

---

[3] http://www.wsce.de/Kamera2/webcam_big.jpg

in lack of context information. Feature descriptors which were extracted from regions annotated negatively in this second annotation step will be handled appropriately in the classifier training process.

Although the annotation effort of this second annotation step is comparable to the initial annotation, it is preferable to employ only one annotation step whenever possible.

4.4   Classification

As mentioned above, FOREST initially applies weakly supervised learning. Each image produces a large, arbitrary number of feature descriptors of different types and dimensionality. Since no prior knowledge for the recognition task is available, any kind of quantization would introduce problems regarding the comparability of keypoints throughout the data set and lead to loss of important information. Opelt et al. [19,20] applied a boosting classifier for identifying discriminative feature descriptors from a heterogeneous feature data set. Since the problem statement is similar and the classifier has been found to be successful, the boosting approach was applied in FOREST.

As already stated by Opelt et al. the calculation of the distance matrix between all feature descriptors is computationally very expensive. The advantage is that the distance matrix needs to be calculated only once. The calculation of weak hypotheses from this distance matrix, as well as the evaluation of the strong hypothesis is computationally efficient. We do not consider the computational costs as a major drawback. User interaction is required at specific steps within the development process. These are compact periods and no interaction is required in between. The time needed for processing data does not keep the user occupied, therefore a long training phase is no limitation to the system.

5   Evaluation

The evaluation of FOREST focuses on two aspects, the first is the evaluation of FOREST on real world data sets, the second is the impact of a second annotation step on the final recognition rates. The impact of using multiple feature types instead of individual feature types has been evaluated in [20,26] and will not be considered here.

The first real world data set used for the evaluation is taken from a webcam, which shows a public area where a market takes place a few times a week, which the system is supposed to recognize (referred to as market data set). Images were acquired in the course of ten days. This data set is an example for an arbitrary recognition task which can be performed on live and freely available image data. The second data set is taken from an interior camera, directed towards a row of windows, acquired over the course of two weeks. This data set is an example for image recognition systems for private use, since the detection of open windows in the ground floor of a building is of interest to prevent theft. Following this motivation, the system is meant to recognize if any of the windows is open.

Table 2. Recognition rates on Caltech [6] airplanes, motorbikes, and faces, as well as the three webcam data sets. Caltech recognition rates were calculated using 10-fold cross validation. The market, window and lake recognition rates were calculated on the complete test set. Default configuration for all operators was used in all tests.

	Caltech airplanes	Caltech Motorbike	Caltech Faces	Market	Windows	Lake
Correct rec. rate (%)	87.8	93.0	95.5	97.64	92.05	95.4
# training images	60	60	60	720	300	1000
# test images	60	60	60	935	2000	2000
# weak hypotheses	35	35	15	20	25	25

Figure 4 shows some exemplary images from both data sets. The third data set is taken from a webcam showing a lake where windsurfers can be observed (see Figure 3). Due to seasonal and time constraints, this is an example for a data set with extreme skew, i.e. containing a total of only 52 positive samples. Gathering more training data might not be possible, in this case, because of a change in weather conditions. The goal of the recognition task using this lake data set was to detect windsurfers. In order to evaluate the general applicability of the framework we tested the performance of custom recognition systems on three exemplary categories of the Caltech-101 data set [6]. To enable comparability to other recognition systems trained on these data sets, 30 images of each category were used for training and testing in a 10-fold cross validation on the airplanes, motorbikes and faces categories.

Recognition rates for three data sets from the Caltech database and the three real world data sets are given in Table 2. Except for the airplane example all recognition systems achieve over 90%, with the market example achieving the best result with 97.6% correct recognition rate. It has to be noted that the classification errors in the market example resulted from uncertainties in the annotations. When the market is built up or ends, there is a certain number of cars and wagons still around (see Figure 4). It is difficult to determine at which point it exactly ends, both for a recognition system and human users. The window example had difficulties with windows that were only slightly open. Still, a number of correct features are matched near the open window (rectangles over window in middle, Figure 4), which indicates that retraining the system could increase the true positive rate.

Recognition systems for the three Caltech categories achieved high and comparable recognition rates. Discriminative features selected by the boosting classifier were of the types SIFT, SURF, color layout descriptor, shape context, and Haralick texture descriptor.

Although the recognition rate for the lake data set seems OK, the true positive rate reveals the consequence of the extreme skew. Using a second annotation stage on this data set increases the true positive rate dramatically. The additional annotation stage took approximately eight minutes to accomplish. Of course, the training set still captures limited variation (angles of sails etc.) but

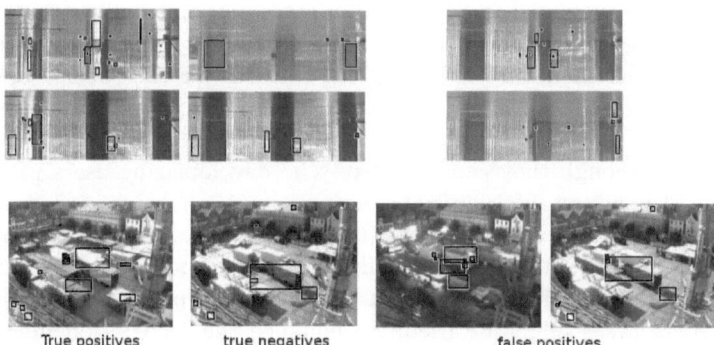

True positives true negatives false positives

Fig. 4. Exemplary images for correct classification and false positives, with matched features indicated by rectangles. False positives in the market data set were due to uncertainties in the annotations. Windows which can be opened are actually the darker areas, not the glass panes.

Table 3. Recognition rates for systems using one- and two-stage annotation

	Lake 1-stage	Lake 2-stage	Window 1-stage	Window 2-stage
Recognition rate (%)	95.4	97.7	92.05	93.55
True pos. rate (%)	17.86	62.5	88.01	92.55

it efficiently reduced correlations concerning weather conditions. The results for the lake and windows data sets using one-level and two-level annotation are given in Table 3. Applying a second annotation stage to the balanced windows data set, which already achieved high recognition rates with one annotation stage, further improved the recognition rates.

6   Conclusion

We presented FOREST, a framework for simple development of custom image recognition systems. The approach is similar to that of a generic image recognition system, since the framework is not restricted to a specific recognition application. It provides the possibility of creating custom image recognition systems, i.e., recognition systems developed for a specific application. The recognition application is defined solely through the annotation of the ground truth data for the task. To overcome the burden of the laborious annotation process, a sophisticated user interface for simple image annotation is employed and extended to provide region annotations in a second, optional, annotation step.

The system was designed to run as autonomously as possible with little user interaction, to simplify the development of image recognition systems as far as possible. To achieve satisfying recognition rates without explicit context information, a variety of keypoint detectors and feature descriptors are applied. The boosting classifier identifies discriminative features for the specific application intended by the user automatically. Evaluations on real world data sets show that image recognition systems developed using FOREST achieve good recognition rates, although they can completely be developed by users without any knowledge of image recognition whatsoever.

Future work will include the further optimization of user interfaces, with a focus on the image annotation, as well as the evaluation of the overall usability and acceptability of the framework. Integration of publicly available, multimodal data sources might be useful to enhance the recognition functionality. Complementary recognition systems could be trained on specific data depending on the situation, e.g. the weather. Additional data sources may also be used to estimate the quality of training data sets. This knowledge could be used to enhance the user's understanding of important criteria for good training sets or to automatically gather missing training image data.

References

1. Agarwal, S., Roth, D.: Learning a Sparse Representation for Object Detection. In: Heyden, A., Sparr, G., Nielsen, M., Johansen, P. (eds.) ECCV 2002, Part IV. LNCS, vol. 2353, pp. 113–127. Springer, Heidelberg (2002)
2. von Ahn, L., Dabbish, L.: Labeling Images With a Computer Game. In: Proceedings of the SIGCHI Conference on Human Factors in Computing Systems, pp. 319–326 (2004)
3. Bay, H., Tuytelaars, T., Van Gool, L.: SURF: Speeded Up Robust Features. In: Leonardis, A., Bischof, H., Pinz, A. (eds.) ECCV 2006. LNCS, vol. 3951, pp. 404–417. Springer, Heidelberg (2006)
4. Belongie, S., Malik, J., Puzicha, J.: Shape Context: A New Descriptor for Shape Matching and Object Recognition. In: NIPS, pp. 831–837 (2000)
5. Csurka, G., Dance, C.R., Fan, L., Willamowski, J., Bray, C.: Visual Categorization With Bags of Keypoints. In: Workshop on Statistical Learning in Computer Vision, ECCV, pp. 1–22 (2004)
6. Fei-Fei, L., Fergus, R., Perona, P.: Learning Generative Visual Models From Few Training Examples: An Incremental Bayesian Approach Tested on 101 Object Categories. In: Workshop on Generative-Model Based Vision (2004)
7. Freeman, W., Adelson, E.: The Design and Use of Steerable Filters. IEEE Trans. on Pattern Analysis and Machine Intelligence 13, 891–906 (1991)
8. Hegazy, D., Denzler, J.: Boosting Colored Local Features for Generic Object Recognition. Pattern Recognition and Image Analysis 18, 323–327 (2008)
9. Koskela, M., Laaksonen, J.: Semantic Annotation of Image Groups with Self-organizing Maps. In: Leow, W.-K., Lew, M., Chua, T.-S., Ma, W.-Y., Chaisorn, L., Bakker, E.M. (eds.) CIVR 2005. LNCS, vol. 3568, pp. 518–527. Springer, Heidelberg (2005)
10. Laaksonen, J., Koskela, M., Laakso, S., Oja, E.: PicSOM - Content-Based Image Retrieval With Self-Organizing Maps. Pattern Recognition Letters 21, 1199–1207 (2000)

11. Lowe, D.G.: Distinctive Image Features From Scale-Invariant Keypoints. Intl. J. of Computer Vision 60, 91–110 (2004)
12. Lu, F., Yang, X., Lin, W., Zhang, R., Yu, S.: Image Classification With Multiple Feature Channels. Optical Engineering 50, 057210 (2011)
13. Manjunath, B., Ohm, J.R., Vasudevan, V., Yamada, A.: Color and Texture Descriptors. IEEE Trans. on Circuits and Systems for Video Technology 11, 703–715 (2001)
14. Mikolajczyk, K., Schmid, C.: A Performance Evaluation of Local Descriptors. IEEE Trans. on Pattern Analysis & Machine Intelligence 27, 1615–1630 (2005)
15. Mikolajczyk, K., Tuytelaars, T., Schmid, C., Zisserman, A., Matas, J., Schaffalitzky, F., Kadir, T., Gool, L.: A Comparison of Affine Region Detectors. Intl. J. of Computer Vision 65, 43–72 (2005)
16. Moehrmann, J., Bernstein, S., Schlegel, T., Werner, G., Heidemann, G.: Improving the Usability of Hierarchical Representations for Interactively Labeling Large Image Data Sets. In: Jacko, J.A. (ed.) HCI International 2011, Part I. LNCS, vol. 6761, pp. 618–627. Springer, Heidelberg (2011)
17. Moehrmann, J., Heidemann, G.: Efficient Annotation of Image Data Sets for Computer Vision Applications. In: Proceedings of the Intl. Workshop on Visual Interfaces for Ground Truth Collection in Computer Vision Applications, pp. 2:1–2:6 (2012)
18. Nowak, E., Jurie, F., Triggs, B.: Sampling Strategies for Bag-of-Features Image Classification. In: Leonardis, A., Bischof, H., Pinz, A. (eds.) ECCV 2006. LNCS, vol. 3954, pp. 490–503. Springer, Heidelberg (2006)
19. Opelt, A., Fussenegger, M., Pinz, A., Auer, P.: Weak Hhypotheses and Boosting for Generic Object Detection and Recognition, pp. 71–84 (2004)
20. Opelt, A., Pinz, A., Fussenegger, M., Auer, P.: Generic Object Recognition with Boosting. IEEE Trans. on Pattern Analysis and Machine Intelligence 28, 416–431 (2006)
21. Quinn, A.J., Bederson, B.B.: Human Computation: A Survey and Taxonomy of a Growing Field. In: Proceedings of the Annual Conference on Human Factors in Computing Systems, pp. 1403–1412 (2011)
22. Russell, B.C., Torralba, A., Murphy, K.P., Freeman, W.T.: LabelMe: A Database and Web-Based Tool for Image Annotation. Intl. J. of Computer Vision 77, 157–173 (2008)
23. Tuytelaars, T., Mikolajczyk, K.: Local Invariant Feature Detectors: A Survey. Foundations and Trends in Computer Graphics and Vision 3, 177–280 (2008)
24. Viola, P., Jones, M.: Rapid Object Detection Using a Boosted Cascade of Simple Features. In: Proceedings of the IEEE Conference on Computer Vision and Pattern Recognition, vol. 1, p. 511 (2001)
25. Yao, B., Yang, X., Zhu, S.-C.: Introduction to a Large-Scale General Purpose Ground Truth Database: Methodology, Annotation Tool and Benchmarks. In: Yuille, A.L., Zhu, S.-C., Cremers, D., Wang, Y. (eds.) EMMCVPR 2007. LNCS, vol. 4679, pp. 169–183. Springer, Heidelberg (2007)
26. Zhang, W., Yu, B., Zelinsky, G., Samaras, D.: Object Class Recognition Using Multiple Layer Boosting With Heterogeneous Features. In: Computer Vision and Pattern Recognition, vol. 2, pp. 323–330 (2005)
27. Zhang, J., Marszalek, M., Lazebnik, S., Schmid, C.: Local Features and Kernels for Classification of Texture and Object Categories: A Comprehensive Study. In: Computer Vision and Pattern Recognition Workshop, p. 13 (2006)

Transforming Cluster-Based Segmentation for Use in OpenVL by Mainstream Developers

Daesik Jang[1], Gregor Miller[2], and Sidney Fels[2]

[1] Kunsan National University, Gunsan, South Korea
[2] Human Communication Technologies Laboratory,
University of British Columbia, Vancouver, Canada
dsjang@kunsan.ac.kr,
{gregor,ssfels}@ece.ubc.ca

Abstract. The majority of vision research focusses on advancing technical methods for image analysis, with a coupled increase in complexity and sophistication. The problem of providing access to these sophisticated techniques is largely ignored, leading to a lack of application by mainstream applications. We present a feature-based clustering segmentation algorithm with novel modifications to fit a developer-centred abstraction. This abstraction acts as an interface which accepts a description of segmentation in terms of properties (colour, intensity, texture, etc.), constraints (size, quantity) and priorities (biasing a segmentation). This paper discusses the modifications needed to fit the algorithm into the abstraction, which conditions of the abstraction it supports, and results of the various conditions demonstrating the coverage of the segmentation problem space. The algorithm modification process is discussed generally to help other researchers mould their algorithms to similar abstractions.

1 Introduction

Research into computer vision techniques has far outpaced the research of interfaces (e.g. Application Programming Interfaces) to support the accessibility of these techniques, especially to those who are not experts in the field such as mainstream developers or system designers. Advances in the robustness of vision methods have led to a surge in real-world applications, from face detection on consumer cameras to articulated human body modelling for natural user interfaces. The algorithms capable of performing these feats are in the domain of experts, even if implementations are provided, due to the understanding required to: tune the parameters, which are often poorly documented and relate directly to variables in the mathematics of the method; form the input, which may include complicated templates for detection or pre-processed images (e.g. foreground-background separated); choose this method for the problem being solved - there are usually many methods, and it is a challenge even for experts to select the right algorithm given the conditions of the problem.

We argue that a simpler, higher-level interface can be provided to developers in order for them to utilise sophisticated vision methods. Our contribution in this

J.-I. Park and J. Kim (Eds.): ACCV 2012 Workshops, Part I, LNCS 7728, pp. 254–265, 2013.

paper is an algorithm modified to fit a segmentation abstraction and a mapping of its specific algorithmic parameters to the abstraction's interface.

Developing an abstraction for computer vision is important for many reasons: 1) Developers may focus on their applications main task, rather than the algorithms; 2) Advances in the state-of-the-art can be incorporated into existing systems without re-implementation; 3) Hardware acceleration of algorithms may be used transparently; 4) The limitations of a particular platform can be taken into account automatically e.g. mobile devices may require a set of low-power consuming algorithms; 5) Computer vision expertise can be more readily adopted by researchers in other disciplines and general developers. If any abstraction is used to access vision methods, hardware and software developers of the underlying mechanisms are free to continually optimise and add new algorithms. This idea has been applied successfully in many other fields, notably OpenGL in graphics [1], but none has yet been successful within computer vision.

There has been a recent industry push to define standards for access to computer vision: the standards group Khronos have organised a working group to develop a hardware abstraction layer (tentatively titled CV HAL) to accelerate vision methods and provide simpler access mechanisms.[1] Khronos are proposing a layer beneath libraries such as OpenCV [2] in order to accelerate existing library calls (much like projects such as OpenVIDIA[2]).

We believe this abstraction layer has been targeted at too low a level to be useful for general developers. We propose an additional higher-level layer using a task-based abstraction to hide the details of algorithms, platforms and hardware acceleration from developers and allow them to focus on developing applications. The algorithm we present in this paper is tailored to an abstraction to provide developers with simpler access to segmentation results.

2   Related Work

Various surveys provide excellent overviews of the versatile approaches used for image segmentation. Shaw *et al.* surveyed important methods for segmentation based on intensity, colour and texture properties [3]. Skarbek *et al.* categorised various approaches more in depth focussing on colour segmentation [4] . Chan *et al.* showed some recent developments in variational image segmentation[5]. Zhangas surveyed unsupervised methods for image segmentatin[6]. Raut *et al.* added some modern approaches as well [7]. From these analyses we can summarise the important approaches of segmentation as follows:

Thresholding: These are generally used for greyscale images and are simple to implement [8]. Some methods use multi-dimensional histograms to extend this approach to include colour and texture properties for the segmentation [7].

Region: Region growing and region splitting-merging are the main procedures in this approach [9–11]. The region growing method groups pixels or sub-regions

---

[1] http://www.khronos.org/vision

[2] http://openvidia.sourceforge.net

into large regions based on pre-defined criteria. Regions are grown from an initial set of seed points, based on comparing neighbouring pixels' properties to that of the seed. Selection of seed points is therefore critical for colour images, and the result is highly dependent on these initial seeds.

Boundary: Edge detection is by far the most common approach for detecting meaningful discontinuities in grey level images [10]. In practice, edge-based techniques using sets of pixels seldom characterise an edge completely due to noise and non-uniform illumination which creates spurious intensity discontinuities. Hence edge detection algorithms need additional post processing by using linking procedures to assemble edge pixels into meaningful edges.

Graphing: The image is modelled as a weighted undirected graph [12]. Each pixel is a node in the graph, and an edge is formed between every pair of pixels. The weight of an edge is a measure of the similarity between the pixels. The image is partitioned into disjoint sets by removing the edges connecting the segments. The optimal partitioning of the graph is the one that minimises the weights of the edges that were removed. Shi's algorithm seeks to minimise the *normalised cut*, which is the ratio of the 'cut' to all of the edges in the set [13].

Morphology: The Watershed transformation considers the gradient magnitude as a topographic surface [14]. Pixels with the highest gradient magnitude intensities (GMIs) correspond to watershed lines (which represent the region boundaries) - 'water' placed on any pixel enclosed by a common watershed line flows downhill to a common local intensity minima. The method is initialised with markers to avoid over-segmentation due to noise and local gradient irregularities.

Clustering: Clustering for colour segmentation is especially effective with multiple features and one-dimensional methods (e.g. thresholding) cannot be applied. Colour is generally represented as multiple features, such as red, green and blue (RGB) or hue, saturation and intensity (HSI) [4]. Many techniques have been proposed in the literature of cluster analysis [10]. A classical technique for colour segmentation is k-means [15], extended to a probabilistic modelling using a fuzzy c-means algorithm [16]. There are various other approaches for segmentation via clustering, such as ISODATA (Iterative Self-Organizing Data Analysis Techniques) [10] and the mean shift algorithm [17, 18]. Connected-component labelling methods are used to compute the final segmentation based on the clusters [19, 20]. Clustering-based approaches are useful when the clusters of features are normal and easily distinguishable. If the features are cluttered among objects, this approach can not be guaranteed to give a good segmentation.

Automatic selection: Some automatic methods to select algorithms and parameters based on metrics or case-based learning have been tried recently. These approaches are meaningful in the sense that they can select an optimal algorithm and parameters adaptive to the characteristics of images to process. One methodology involves a generic framework for segmentation evaluation using a metric based on the distance between segmentation partitions [21]. Case-based reasoning was introduced to select an algorithm and parameters depending on the image characteristics [22]. The cases have image characteristics similar to those of the current input image, and the segmentation parameters associated with the most similar case is applied to the input image. Yong *et al.* [23] proposed a simulation system designed to select the optimal segmentation algorithm from four candidates for synthetic images. Martin *et al.* [24] proposed a scheme to automatically select segmentation algorithm and tune theirs key parameters using a preliminary supervised learning stage. Nickisch *et al.*[25] proposed a new evaluation and learning method with user supervision.

While cluster-based methods for segmentation have drawbacks such as over-segmentation in the presence of high detail, they are extremely effective for isolating known regions. This is the case for developers designing applications with segmentation, where we envision the majority of use-cases are known in advance looking for a particular set of objects. We present a modified algorithm designed to accommodate a segmentation abstraction, which we present first.

3   Developer-Centred Segmentation

The central part of a segmentation framework provided to developers is a higher-level abstraction which hides algorithmic detail (the algorithm used and the parameters it uses) but still provides a powerful and flexible interface to segmentation results. We use a task-centred description for the interface, through which developers may describe the segmentation problem they need to solve.

For the abstraction we use a relatively simple definition of *segmentation*: producing a set of distinct regions (*segments*) within the image. We apply the concept of *properties* to measure distinctiveness. A property is measurable over a region of the image, which leads to an extensive list of possibilities, such as colour, intensity, texture, shape, contour, etc. Conceptually, a segment is bounded by a smooth, continuous contour, and is not dependent on pixels or any other discrete representation. Developers must specify at least one property to define the segmentation of the image: segment properties allow developers to decompose the image based on what they consider to be important to their problem, and provide us with the information required to produce a corresponding segmentation.

Each property is associated with a *distinctiveness* to allow the developer to define how distinct the segments should be with respect to that property. Due to the range of possible methods of segmentation, the term 'distinct' was chosen as the best abstraction of the conceptual meaning. This was in preference to terms such as *threshold* or *distance* which may be used in other methods but would not be applicable in all cases. The description also allows multiple properties for a single segmentation. Conceptually this will lead to segments which are distinct based on all specified properties. The advantage of the task-based description is the details of how this is performed are hidden from the developer, and so they do not need to take this into account when developing an application.

When defining the available set of properties we attempt to make sure each is orthogonal to the others, to avoid repetition in the description space and encourage completeness. Our eventual goal is to create a unified space for vision

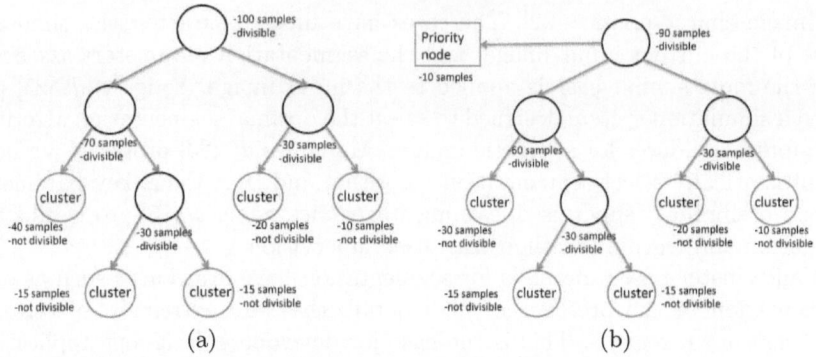

Fig. 1. Illustration of the use of a binary decision tree to create a set of clusters for segmentation. In (a), the initial feature space has 100 samples and is divided into two child nodes with 70 samples and 30 samples. The red nodes are determined to be cluster nodes since they are not divisible according to the end conditions. In (b), a priority node is added to capture use requirements for the grouping of similar pixels.

descriptions, to apply to all problems, which can be interpreted into algorithms and parameters to provide the developer with a solution. The description space should be kept as small as possible while still maintaining a wide coverage to help minimise the complexity as the description language is extended.

The last aspect of the description is the use of *priorities*: the developer can define volumes in property space towards which the segmentation should be biased, which is useful in applications such as chroma-keying or skin-colour detection.

The properties and priorities together form what we define as the *requirements* of the segmentation. The last component we need to complete our description is *constraints*. Constraints introduce some additional complexity to the operation, because they are capable of overriding the distinctiveness requirement. The three constraints we provide are *size*, *quantity* and *regularity*. Size governs the final area of the segments, quantity the number, and regularity the level of variation allowed in the gradient of the segment's contour. Size and quantity are related and must trade off against one another; Regularity constrains the overall shape of the segments: a regularity of 0 does not constrain the shape at all and a value of 1 constrains the shape of every segment to be the same.

4   Transforming Cluster-Based Segmentation

Cluster-based segmentation is one of the most well-known and useful approaches for image segmentation. It is relatively simple to understand, practical for many use cases (especially when multivariate features such as RGB colours are used) and also benefits from good performance for general purpose segmentation. The major drawback is the difficulty for non-experts to understand how it works and the configuration required to achieve their required result. The parameters have a significant effect on the result of clustering and they should be carefully determined by experts to meet the requirements of each application. However, it is often difficult even for experts to match the parameters with the requirements of diverse applications. In this section a cluster-based segmentation method for colour images is transformed to work with a developer-centred abstraction. The problem conditions of image segmentation can be described by developers instead of requiring expert knowledge of the algorithm and its parameters.

4.1   Method Overview

A cluster-based segmentation uses two main steps: feature-space clustering and region labelling. Various algorithms such as *k-means*, *fuzzy c-means* and ISO-DATA may be used to find clusters in the feature space. We use k-means with *RGB* colour, followed by connected component analysis for region labelling. A conventional k-means algorithm is as follows:

1. Place K points in RGB feature space (cluster centroids).
2. Assign each sample (RGB value) to the cluster that has the closest centroid.
3. When all samples have been assigned, recalculate the centroids based on the newly assigned samples.
4. Repeat steps 2 and 3 until the centroids are static.

This produces a separation of the samples into clusters ready for post-processing: using the distance between two clusters as the metric, we can decide whether to sub-divide clusters or not (this is discussed further below).

One of the drawbacks of k-means is the requirement for a known number of centroids and the provision of each cluster with a good initial centroid. When the number of clusters is not appropriate for the input image, the segmentation can be over- or under-sampled. Variations of clustering such as ISODATA were developed to adjust the number of clusters by merging those that are similar, but it is still sensitive to the choice of initial centroids.

This weakness of k-means also makes it challenging to transform the method into a developer-centric framework. The clustering algorithm should adjust its parameters according to the description of segmentation to produce results satisfying the developer's requirements. The k-means algorithm is very rigid: its parameters do not neatly map to a developer-level description. To begin, we propose the parameters be adjusted as follows:

– Maximum number of clusters (K_{MAX}): determined according to the desired quantity of segments.
– Minimum distance of clusters (D_{MIN}): determined according to the desired distinctiveness of segments. If any pair of clusters are too close each other, they are not distinctive enough.

To adjust the algorithm to match our segmentation abstraction, a binary decision tree is combined with k-means to make these parameters adjustable. Instead of applying the k-means algorithm to the whole feature space, it is applied to a binary tree representing the feature space. This partitions the feature space of each node into two clusters, e.g. the root node contains the original feature space and it is divided into two clusters, then the samples constituting the original feature space are divided into two subspaces based on the distance to cluster centroids. Two child nodes are generated within these two subspaces respectively and attached to the root node. K-means is applied to these two child nodes again in the same way. With this approach the initial K centroids are no longer necessary: K can be determined by the framework through tree generation. Some conditions are required to stop the subdivision and control the size of the tree. The detailed algorithm for this new clustering method is:

1. Make a root node with the samples of the original feature space.
2. For each node that is not classified as a cluster node:
   - Partition the node with k-means into two clusters.
   - Check the condition of the node with the provided parameters to determine whether it is divisible.
   -- If the node is divisible: divide the samples in the node into two subgroups and generate two child nodes.
   -- If the node is not divisible: it is classified as a cluster node.
3. Repeat step 2 until there is no node divisible.

Figure 1 shows the concept of using a binary decision tree for segmentation, and illustrates an example tree generated with this process. The conditions to determine the divisibility of a node use the following parameters:

- K_{MAX}: if the number of cluster nodes generated exceeds this parameter, all terminal nodes are marked as clusters and the process stops.
- D_{MIN}: if the distance between two clusters in a node is less than this parameter, then the node is determined not to be divisible and it becomes a cluster node.

Based on the identified clusters, a two-pass connected component labelling algorithm is used to generate segments (regions of the image) corresponding to the clusters. The two parameters of the clustering method are mapped to the description of segmentation in terms of properties and constraints. The details for this mapping are explain in the following section.

4.2   Parameter Mapping

The mapping of segmentation abstraction parameters to our method are:

- *Distinctiveness*: The distinctiveness of produced segments is linked to D_{MIN}.
- *Quantity*: The quantity of segments to produce is linked to K_{MAX}.

When D_{MIN} is large, potentially divisible clusters may not be divided and the distinctiveness of clusters is decreased. For high distinctiveness, the parameter should be small enough to produce clusters with smaller gaps. K_{MAX} affects the quantity of segments: for large values the tree will contain more branches (and more clusters), therefore more regions are segmented. Table 1 shows the mapping between the clustering parameters and the segmentation description.

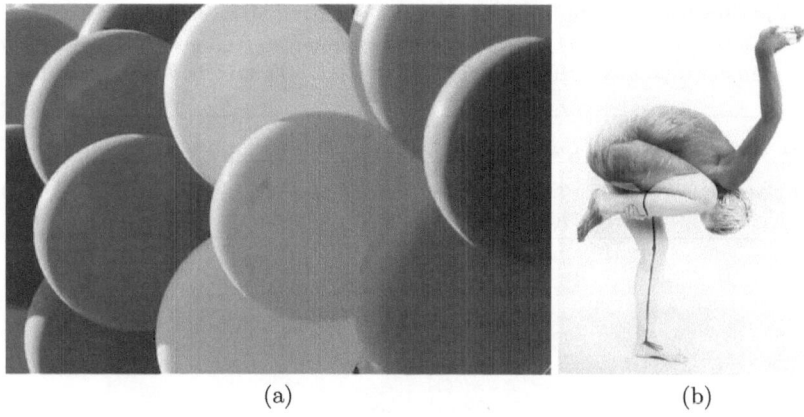

(a)   (b)

Fig. 2. Sample images used for illustration of the results. Image (a) has dimensions 400×265, (b) has dimensions 553×720.

4.3   Priorities

The last property-related aspect of the description is *priorities*. These are supplied to let the developer specify volumes in property space which should bias the segmentation. This is important in the definition of boundaries: for example if a single-colour sphere is illuminated from one angle, the colour will have a gradient - the sphere's colour can be prioritised to segment the ball into a single region. With our segmentation method, we can accommodate priorities by inserting a new subspace defined by a volume in feature space; this can form a cluster and produce segments corresponding to the developer's requirements.

To implement *priorities*, a subspace corresponding to a developer-defined priority is expressed as a range of colours. This range is represented as a pair of RGB colour values and it constitutes a cubic subspace in the feature space. This subspace is represented as a special node in the binary decision tree and is attached to the root node. The samples which fall into the subspace are excluded from the root node so that the prioritized subspace is not considered for further clustering. Multiple priorities can be defined by adding additional priority nodes to the root node. Figure 1b shows the binary decision tree when a priority is defined, and an example of the clusters in feature space compared to the same space without a priority is shown in Figure 4.

5   Results

Our method was implemented (within the abstraction) in C++ and used OpenCV for utility functions; it was tested on a MacBook Pro Retina quad-core 2.6GHz with the images presented in Figure 2. To illustrate the use of the abstraction-level *distinctiveness*, Figure 2 was segmented with *Low* (Figure 3(a), $D_{MIN} = 0.5$) and *High* (Figure 3(b), $D_{MIN} = 0.01$) distinctiveness, both with *High* set for *quantity* ($K_{MAX} = 20$). For *quantity* control, the results shown in Figure 5 have a *Low*

Table 1. Parameter mapping from the developer-centred abstraction to the clustering algorithms parameters. This is an example set of numbers given an RGB feature space and approximate measures (*High, Medium, Low*) for distinctiveness and quantity.

Description	Clustering Parameter	Mapped Values
Distinctiveness	D_{MIN}	High : 0.01; Medium: 0.3; Low: 0.5;
Quantity	K_{MAX}	High : 20; Medium: 10; Low: 5;

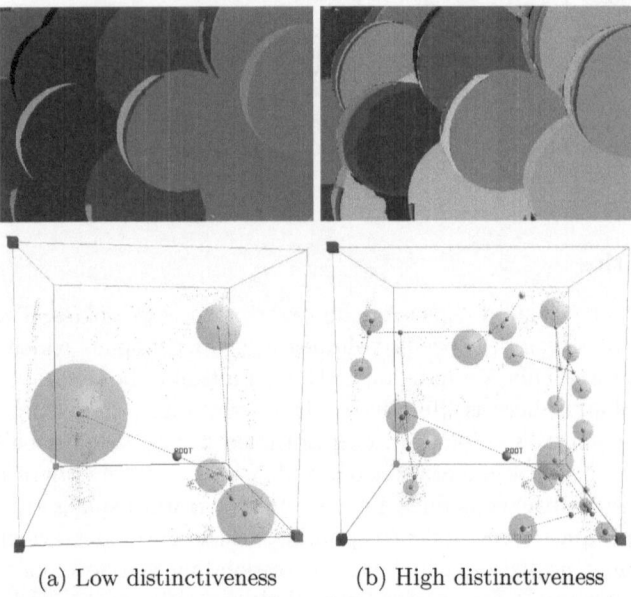

(a) Low distinctiveness   (b) High distinctiveness

Fig. 3. A feature-space visualization of a binary decision tree for clustering with the results shown above

quantity while leaving the distinctiveness constant (the images can be compared to the same distinctiveness with *High* quantity in Figure 3(b) and Figure 4(a)). A priority-based segmentation result is shown in Figure 4(b), with the associated cluster tree with the priority volume (and cluster) shown in the top right of the feature space. A priority was given using a volume defined by the RGB range $(1.0, 0.0, 0.6) - (0.8, 0.2, 0.8)$ to hint to the segmentation method which parts of the feature space are important. The result shows the reddish regions of the image have been assigned the same label, a very different result from the over-segmented image in (a) with no priorities given. In all cases the method takes approximately one second to provide a result. Please note the implementation is not optimised to use accelerated hardware processing, and is intended as a proof-of-concept to fit the abstraction defined in Section 3. The images demonstrate a close match between developer-provided parameters through the higher-level abstraction and the result produced by our segmentation method.

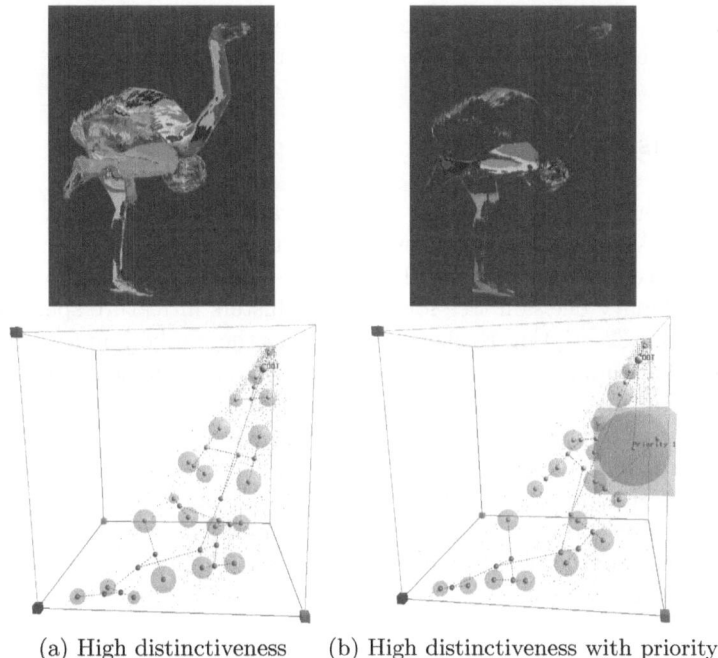

(a) High distinctiveness   (b) High distinctiveness with priority

Fig. 4. A feature-space visualization of a binary decision tree for clustering, comparing the tree with (1) and without (2) priorities, and the results shown above

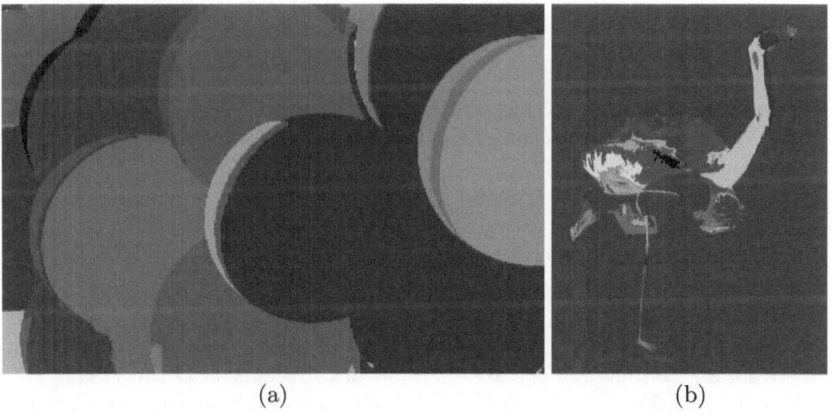

(a)   (b)

Fig. 5. The use of quantity (with *High* distinctiveness): *quantity* is set to *Low* ($K_{MAX} = 5$), and can be compared to the results in Figure 3(b) and Figure 4(a)

There is a relationship between D_{MIN} and quantity, and K_{MAX} and distinctiveness. The abstraction methodology is set up such that size, quantity and distinctiveness are related. To get very few segments, the developer could request a *Low* distinctiveness and a *Low* quantity - all parameters are used in the process, which provides the developer with greater control over the result.

6   Conclusions

We have presented a segmentation clustering algorithm which has been transformed to work with a developer-level abstraction, allowing non-experts access to sophisticated segmentation results. This has been achieved through the inclusion of a binary decision tree for creating clusters in feature space, mapping the abstraction description to the parameters of the method and modifying the clustering algorithm to allow segmentation biases to be included. Results demonstrate the clear mapping between the description a developer provides into the parameters used and the segmented images provided.

The abstraction and method will need to be modified to make it more clear to developers the impact on using distinctiveness and quantity as measures of segmentation (since they are linked); this may involve modifying the abstraction directly or making the results of using both for segmentation very clear, either with documentation or feedback from the abstraction framework.

Acknowledgements. We gratefully acknowledge support from NSERC (grant provided for *"Diving experiences: wayfinding and sharing experiences with large, semantically tagged video"*), Bell Canada, Avigilon Corporation and Vidigami Media Inc.

References

1. Shreiner, D., Woo, M., Neider, J., Davis, T.: OpenGL(R) Programming Guide: The Official Guide to Learning OpenGL(R), Version 2, 5th edn. Addison-Wesley Professional (2005)
2. Bradski, G., Kaehler, A.: Learning OpenCV: Computer Vision with the OpenCV Library, 1st edn. O'Reilly Media, Inc. (2008)
3. Shaw, K.B., Lohrenz, M.C.: A survey of digital image segmentation algorithms. Final Technical Report ADA499374, Naval Oceanographic and Atmospheric Research Lab (1995)
4. Skarbek, W., Koschan, A.: Colour image segmentation - a survey. Technical report, Institute for Technical Informatics, Technical University of Berlin (1994)
5. Chan, T., Sandberg, B., Moelich, M.: Some recent developments in variational image segmentation. In: Proceedings of the International Conference on PDE-Based Image Processing and Related Inverse Problems, pp. 175–210. Springer (2005)
6. Zhang, H., Fritts, J.E., Goldman, S.A.: Image segmentation evaluation: A survey of unsupervised methods. Computer Vision and Image Understanding 110, 260–280 (2008)
7. Raut, S., Raghuvanshi, M., Dharaskar, R., Raut, A.: Image segmentation: A state-of-art survey for prediction. In: Proceedings of International Conference on Advanced Computer Control, pp. 420–424. IEEE Computer Society, New York (2009)

8. Otsu, N.: A threshold selection method from gray-level histograms. IEEE Transactions on Systems, Man and Cybernetics 9, 62–66 (1979)

9. Lucchese, L., Mitra, S.K.: Advances in color image segmentation. In: Proceedings of Global Telecommunications Conference, pp. 2038–2044. IEEE Computer Society, Berkeley (1999)

10. Bow, S.T.: Pattern Recognition and Image Preprocessing, 2nd edn. CRC Press (2002)

11. Pavlidis, T., Liow, Y.T.: Integrating region growing and edge detection. IEEE Transactions on Pattern Analysis and Machine Intelligence 12, 225–233 (1990)

12. Felzenszwalb, P.F., Huttenlocher, D.P.: Efficient graph-based image segmentation. International Journal of Computer Vision 59, 167–181 (2004)

13. Shi, J., Malik, J.: Normalized cuts and image segmentation. IEEE Transactions on Pattern Analysis and Machine Intelligence 22, 888–905 (2000)

14. Roerdink, J.B.T.M., Meijster, A.: The watershed transform: definitions, algorithms and parallelization strategies. Fundamenta Informaticae-Special issue on mathematical morphology 41, 187–228 (2000)

15. MacQueen, J.B.: Some methods for classification and analysis of multivariate observations. In: Proceedings of the 5th Berkeley Symposium on Mathematical Statistics and Probability, pp. 281–297. University of California Press, Berkeley (1967)

16. Eumt, K.-B., Lee, J., Venetsanopoulos, A.N.: Color image segmentation using a possibilistic approach. In: IEEE International Conference on Systems, Man, and Cybernetics - SMC, vol. 2, pp. 1150–1155. IEEE Computer Society, New York (1996)

17. Comaniciu, D., Meer, P.: Robust analysis of feature spaces: Color image segmentation. In: Proceedings of the IEEE Computer Society Conference on Computer Vision and Pattern Recognition, pp. 750–755. IEEE Computer Society, New York (1997)

18. Comaniciu, D., Meer, P.: Mean shift: A robust approach toward feature space analysis. Transactions on Pattern Analysis and Machine Intelligence 24, 603–619 (2002)

19. Wang, W.: Color image segmentation and understanding through connected components. In: IEEE International Conference on Systems, Man, and Cybernetics, vol. 2, pp. 1089–1093. IEEE Computer Society, New York (1997)

20. Samet, H., Tamminen, M.: Efficient component labeling of images of arbitrary dimension represented by linear bintrees. Transactions on Pattern Analysis and Machine Intelligence 10, 579–586 (1988)

21. Cardoso, J., Corte-Real, L.: Toward a generic evaluation of image segmentation. IEEE Transactions on Image Processing 14, 1773–1782 (2005)

22. Frucci, M., Perner, P., Sanniti di Baja, G.: Case-based-reasoning for image segmentation. Pattern Recognition and Artificial Intelligence 22, 829–842 (2008)

23. Yong, X., Feng, D., Rongchun, Z., Petrou, M.: Learning-based algorithm selection for image segmentation. Pattern Recognition Letters 26, 1059–1068 (2005)

24. Martin, V., Maillot, N., Thonnat, M.: A learning approach for adaptive image segmentation. In: Proceedings of the Fourth IEEE International Conference on Computer Vision Systems (ICVS 2006). IEEE Computer Society (2006)

25. Nickisch, H., Kohli, P., Rother, C., Rhemann, C.: Learning an interactive segmentation system. In: Proceedings of the 7th Indian Conference on Computer Vision, Graphics and Image Processing, pp. 274–281. ACM, New York (2010)

Efficient GPU Implementation
of the Integral Histogram

Mahdieh Poostchi, Kannappan Palaniappan, Filiz Bunyak,
Michela Becchi, and Guna Seetharaman

Dept. of Computer Science, University of Missouri-Columbia, Columbia, Missouri
Air Force Research Laboratory, Rome, NY 13441, USA

Abstract. The integral histogram for images is an efficient preprocessing method for speeding up diverse computer vision algorithms including object detection, appearance-based tracking, recognition and segmentation. Our proposed Graphics Processing Unit (GPU) implementation uses parallel prefix sums on row and column histograms in a cross-weave scan with high GPU utilization and communication-aware data transfer between CPU and GPU memories. Two different data structures and communication models were evaluated. A 3-D array to store binned histograms for each pixel and an equivalent linearized 1-D array, each with distinctive data movement patterns. Using the 3-D array with many kernel invocations and low workload per kernel was inefficient, highlighting the necessity for careful mapping of sequential algorithms onto the GPU. The reorganized 1-D array with a single data transfer to the GPU with high GPU utilization, was 60 times faster than the CPU version for a $1K \times 1K$ image reaching 49 fr/sec and 21 times faster for 512×512 images reaching 194 fr/sec. The integral histogram module is applied as part of the likelihood of features tracking (LOFT) system for video object tracking using fusion of multiple cues.

1 Introduction

The integral histogram extends the integral image method for scalar sums to vector (*i.e.* histogram) sums and enables multiscale histogram-based search and analysis in constant time after a linear time preprocessing stage [1, 2]. The integral histogram is a popular method to speed up computer vision tasks, especially sliding window based methods for object detection, tracking, recognition and segmentation [1–12]. Histogram-based features are widely used in image analysis and computer vision due to their simplicity and robustness. Histogram is a discretized probability distribution where each bin represents the frequency or probability of observing a specific range of feature values for a given descriptor such as intensity, color, edginess, texture, shape, motion, etc. Robustness to geometric deformations makes histogram-based feature representation appealing for many applications. One major drawback of sliding window histograms is their high computational cost, limiting their use for large scale applications such as content-based image retrieval with databases consisting of millions of images or

J.-I. Park and J. Kim (Eds.): ACCV 2012 Workshops, Part I, LNCS 7728, pp. 266–278, 2013.
© Springer-Verlag Berlin Heidelberg 2013

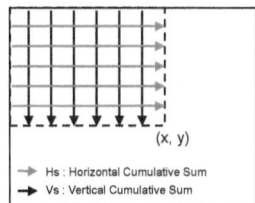
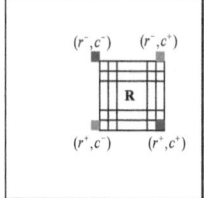

Fig. 1. (a) Computation of the histogram up to location (x, y) using a cross-weave horizontal and vertical scan on the image. (b) Computation of the histogram for an arbitrary rectangular region R (origin is the upper-left corner with y-axis horizontal.)

full motion video archives with billions of frames. Any improvement that leads to a speed-up in integral histogram calculation is imperative especially due to fast trend towards extreme-scale and high-throughput data analysis. Mapping image analysis and computer vision algorithms onto many-core and multicore architectures has benefits ranging from faster processing, deeper search, greater scalability, and better performance especially for recognition, retrieval and reconstruction tasks [13–19]

As far as we know this is the first detailed description and performance characterization of a parallel implementation of the integral histogram for GPU architectures. Previously, a parallelization of the integral histogram for the eight-core IBM Cell/B.E. processor was described in [20] and the scalar *integral image* computation was parallelized for the GPU [21]. Although both the integral image and integral histogram follow the same strategy, the integral histogram uses high memory since the histogram needs to be maintained for every pixel and leads to a 3D array data structure that is difficult to manage on small 48KB on-chip shared memory per stream multiprocessor available on GPUs. This paper presents two parallel implementations of the integral histogram computation, that we have developed, for many-core GPU architectures, using the CUDA programming model [13, 22, 23]. Both methods, GPU Integral Histogram using Multiple Scan-Transpose-Scan (GIH-Multi-STS) and GPU Integral Histogram using Single Scan-Transpose-Scan (GIH-Single-STS) use CUDA SDKs for parallel cumulative sums of rows and columns (prescan) based on a cross-weave scanning mode, 2-D or 3-D transpose kernels and communication-aware data management.

The contributions of this work are designing the best data structure and its layout in GPU memory, finding the kernel configuration that maximizes the resource utilization of the GPUs and minimizes the data movement. Section 2 presents a short review of the integral histogram algorithm. Section 3 explores our parallel integral histogram implementations on GPUs, followed by experimental results including application to tracking and conclusions.

2   Integral Histogram Description

The integral histogram is a recursive propagation preprocessing method used to compute local histograms over arbitrary rectangular regions in constant time [1].

Algorithm 1. Sequential Integral Histogram

Input : Image $\mathbf{I}$ of size $h \times w$
Output : Integral histogram tensor $\mathbf{H}$ of size $h \times w \times b$
1: **Initial H:**
   $\mathbf{H} \leftarrow 0$
2: **for** z=1:b **do**
3:   **for** x=1:w **do**
4:     **for** y=1:h **do**
5:       $H(x,y,z) \leftarrow H(x-1,y,z) + H(x,y-1,z)$
            $-H(x-1,y-1,z) + Q(I(x,y),z)$
6:     **end for**
7:   **end for**
8: **end for**

The efficiency of the integral histogram approach enables real-time histogram-based exhaustive search in vision tasks such as object recognition and tracking. The integral histogram is extensible to higher dimensions and different bin structures. The integral histogram at position (x,y) in the image holds the histogram for all the pixels contained in the rectangular region defined by the top-left corner of the image and the point (x,y) as shown in Figure 1. The integral histogram for the region defined by the spatial coordinate (x,y) and bin variable b is defined as:

$$H(x,y,b) = \sum_{r=0}^{x}\sum_{c=0}^{y} Q(I(r,c),b) \tag{1}$$

where $Q(I(r,c),b)$ is the binning function that evaluates to 1 if $I(r,c) \in$ bin b, and evaluates to zero otherwise. Sequential computation of integral histograms is described in Algorithm 1. Given the image integral histogram $\mathbf{H}$, computation of the histogram for any test region R delimited by points $\{(r^-,c^-),(r^-,c^+),(r^+,c^+),(r^+,c^-)\}$ reduces to the combination of four integral histograms:

$$h(R,b) = H(r^+,c^+,b) - H(r^-,c^+,b) - H(r^+,c^-,b) + H(r^-,c^-,b) \tag{2}$$

Figure 1 illustrates the notation and accumulation of integral histograms using vertical and horizontal cumulative sums (prescan), which is used to compute regional histograms.

3   Parallelization Using Parallel Prefix-Sum (Exclusive Scan)

One basic pattern in parallel computing is the use of independent concurrently executing tasks. The recursive sequential Algorithm 1 is a poor approach to parallelize since row $(r+1)$ cannot be executed until row r is completed, with only intra-row parallelization. The cross-weave scan mode (Fig. 1), enables cumulative sum tasks over rows (or columns) to be scheduled and executed independently allowing for inter-row and column parallelization. The GPU Integral Histogram using Multiple Scan-Transpose-Scan (GIH-Multi-STS) is shown in Algorithm 2.

Algorithm 2. GIH-Multi-STS: GPU Integral Histogram using Multiple Scan-Transpose-Scan

---

Input : Image **I** of size $h \times w$
Output : Integral histogram tensor **IH** of size $b \times h \times w$
 1: **Initialize IH**
       $\mathbf{IH} \leftarrow 0$
       $\mathbf{IH}(\mathbf{I(w, h)}, \mathbf{w}, \mathbf{h}) \leftarrow 1$
 2: **for** z=1 to b **do**
 3:    **for** x=1 to h **do**
 4:       //horizontal cumulative sums (prescan, size of rows)
          $IH(x, y, z) \leftarrow IH(x, y, z) + IH(x, y - 1, z)$
 5:    **end for**
 6: **end for**
 7: **for** z=1 to b **do**
 8:    //transpose the bin-specific integral histogram
          $IH^T(z) \leftarrow \text{2-D Transpose}(IH(z))$
 9: **end for**
10: **for** z=1 to b **do**
11:    **for** y=1 to w **do**
12:       //vertical cumulative sums (prescan, size of columns)
          $IH(x, y, z) \leftarrow IH^T(y, x, z) + IH^T(y, x - 1, z)$
13:    **end for**
14: **end for**

This approach combines cross-weave scan mode with an efficient parallel prefix sum operation and an efficient 2-D transpose kernel. The SDK implementation of *all-prefix-sums* operation using the CUDA programming model is described by Harris, *et al.* [24]. We apply prefix-sums to the rows of the histogram bins (horizontal cumulative sums or prescan), then transpose the array and reapply the prescan to the rows to obtain the integral histograms at each pixel.

3.1   Parallel Prefix Sum Operation on the GPU

The core of the parallel integral histogram algorithm for GPUs is the parallel prefix sum algorithm [24]. The *all-prefix-sums* operation (also refered as a scan) applied to an array generates a new array where each element k is the sum of all values preceding k in the scan order. Given an array $[a_0, a_1, ..., a_{n-1}]$ the prefix-sum operation returns,

$$[0, a_0, (a_0 + a_1), ..., (a_0 + a_1 + ... + a_{n-2})] \tag{3}$$

The parallel prefix sum operation on the GPU consists of two phases: an *up-sweep* (or reduce) phase and a *down-sweep* phase (see Fig. 2). *Up-sweep* phase builds a balanced binary tree on the input data and performs one addition per node. Scanning is done from the leaves to the root. In the *down-sweep* phase the tree is traversed from root to the leaves and partial sums from the up-sweep phase are aggregated to obtain the final scanned (prefix summed) array. Prescan requires only $O(n)$ operations: $2 * (n - 1)$ additions and $(n - 1)$ swaps.

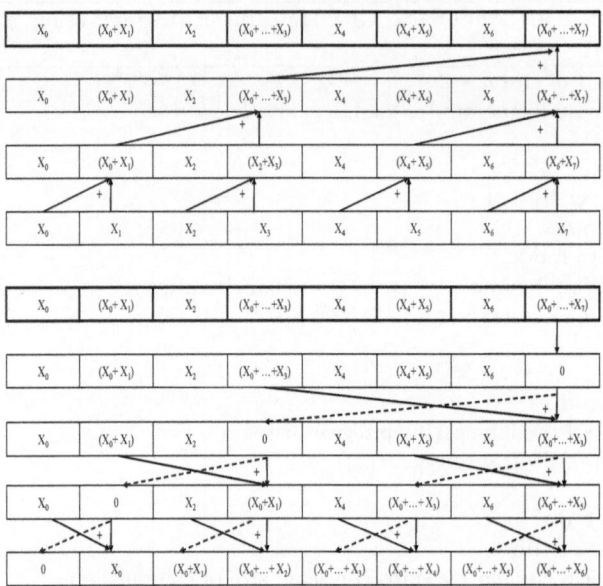

Fig. 2. Parallel prefix sum operation, commonly known as exclusive scan or prescan [24]. Top: Up-sweep or reduce phase applied to an 8-element array. Bot: Down sweep phase.

The GPU-based prefix sum (prescan) operation moves data from CPU memory to off-chip global GPU memory then exploits the on-chip shared memory for each row operation [24].

3.2   GPU-Based Transpose Kernel

The integral histogram computation requires two prescans over the data. First, a horizontal prescan that computes cumulative sums over rows of the data, followed by a second vertical prescan that computes cumulative sums over the columns of the first scan output. Taking the transpose of the horizontally prescanned image histogram, enables us to reapply the same (horizontal) prescan algorithm effectively on the columns of the data. We used the optimized transpose kernel described in [25] that uses zero bank conflict shared memory and guaranties that global reads and writes are coalesced. Figure 3 shows the data flow in the transpose kernel. A tile of size BLOCK_DIM $*$ BLOCK_DIM is written to the GPU shared memory into an array of size BLOCK_DIM $*$ (BLOCK_DIM $+ 1$). This pads each row of the 2-D block in shared memory so that bank conflicts do not occur when threads address the array column-wise. Each transposed tile is written back to the GPU global memory to construct the full histogram transpose. The SDK 2-D transpose kernel needs to be launched from the host b times in order to transpose the integral histogram tensor. In order to allow a single transpose operation, we transform the existing 2-D transpose kernel into a 3-D transpose kernel by using the bin offset in the indexing. The 3-D transpose kernel

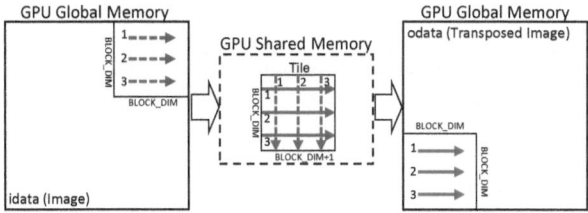

Fig. 3. Data flow between GPU global memory and shared memory while computing the coalesced transpose kernel; stage 1 in red, stage 2 blue, reads are dashed lines, writes are solid lines.

is launched using a 3-D grid of dimension $(b, w/\text{BLOCK\_DIM}, h/\text{BLOCK\_DIM})$, where BLOCK_DIM is the maximum number of banks in shared memory (32 for all graphics card used).

3.3 Data Structure and Implementation Strategy

An image with dimensions $h \times w$ produces an integral histogram tensor of dimensions $h \times w \times b$, where b is the number of bins in the histogram. This tensor can be represented as a 3-D array which in turn can be mapped to an 1-D row major ordered array for efficient access as shown in Figure 4. Both implementations, GIH-Multi-STS and the improved GPU Integral Histogram using Single Scan-Transpose-Scan (GIH-Single-STS), start by prescanning each row. Since the maximum number of threads per block is 1024 and each thread processes two elements, each row can be divided into segments up to 2048 pixels. If the size of row is smaller than 2048 then the size of the thread block will be reduced to the $w/2$. The GIH-Multi-STS implementation uses the 3-D data structure. Exclusive prefix sum (prescan) kernel (see Section 3.1) is applied to the data one row at a time. This approach suffers from many kernel invocations in the horizontal/vertical scan and 2-D transpose phases, from little work per kernel and eventually GPU under-utilization (Algorithm 2). To reduce the total number of kernels invocations from $(w + h)b + b$ to only 3 invocations, the GIH-Single-STS implementation uses a 1-D row ordered format array and launches the prescan kernel *once* using a 1-D grid of size $(b * h * w)/(2 * \text{Num\_Threads})$. Padding is applied to shared memory addresses to avoid bank conflicts by adding an offset of 32 to each shared memory index. After prescanning each row (horizontal scan),

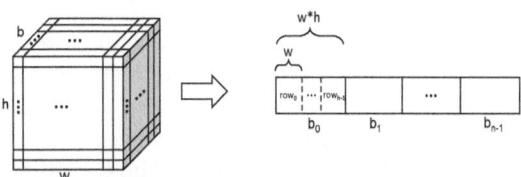

Fig. 4. Integral histogram tensor represented as 3-D array data structure (left), and equivalent 1-D array mapping (right)

Algorithm 3. GIH-Single-STS: GPU Integral Histogram using Single Scan-Transpose-Scan

---

Input : Image **I** of size $h \times w$, number of bins b
Output : Integral histogram tensor **IH** of size $b \times h \times w$
1: **Initialize IH**
      $\mathbf{IH} \leftarrow 0$
      $\mathbf{IH}(\mathbf{I(w, h)}, \mathbf{w}, \mathbf{h}) \leftarrow 1$
2: **for** all $b \times h$ blocks in parallel **do**
3:     //horizontal cumulative sums
4:     **Prescan**(IH)
5: **end for**
6: //transpose the histogram tensor
      $IH^{T} \leftarrow$ 3D_Transpose(IH)
7: **for** all $b \times w$ blocks in parallel **do**
8:     //vertical cumulative sums
9:     **Prescan**(IH^{T})
10: **end for**

---

the prescanned array is transposed to compute (column) cumulative sums in the second pass using a 3-D transpose kernel (Algorithm 3). We implemented and evaluated two parallel GPU integral histogram computation approaches: parallel GIH-Multi-STS, and parallel GIH-Single-STS and compared them to a sequential CPU-only implementation. Our experiments were conducted on a 2.0 GHz Quad Core Intel CPU (Core i7-2630QM) and two GPU cards: an NVIDIA Tesla C2070 and an NVIDIA GeForce GTX 460. The former is equipped with fourteen 32-core SMs and has about 5GB of global memory, 48KB shared memory with compute capability 2.0. The latter consists of seven 48-core SM and is equipped with 1GB global memory, 48KB shared memory with compute capability 2.1.

The parallel GIH-Multi-STS implementation exploits the work efficient prescan operation to calculate for each bin the cumulative sums of rows, one row at a time. Therefore, the scan kernel is launched $b \times h$ times for horizontal scan and $b \times w$ times for vertical scan. The efficient 2-D transpose kernel is launched b times to transpose the integral histogram tensor after horizontal scan. The GIH-Multi-STS is based on many kernel invocations, each of them performing a small amount of work and therefore greatly under-utilizing the many-cores on the GPU. In addition, the all-prefix-sum kernel works very well only on very large array consisting of millions of elements. Therefore, we proposed the GIH-Single-STS to increase the amount of work performed by each kernel invocation and reduce the number of scan kernel invocations by a factor of $(h + w)b$. This can be easily achieved by modifying the kernel configuration without rewriting the kernel code (array indices are derived from block and thread indices). Since the maximum number of threads per block is 1024 and each thread processes two elements, each row can be divided into segments up to 2048 pixels. If the size of row is smaller than 2048 then the size of the thread block will be reduced to the $w/2$ for horizontal scan and $h/2$ for vertical scan as well. Therefore, the number of blocks for horizontal scan will be $((b \times h \times w)/(2 \times threadblock))$.

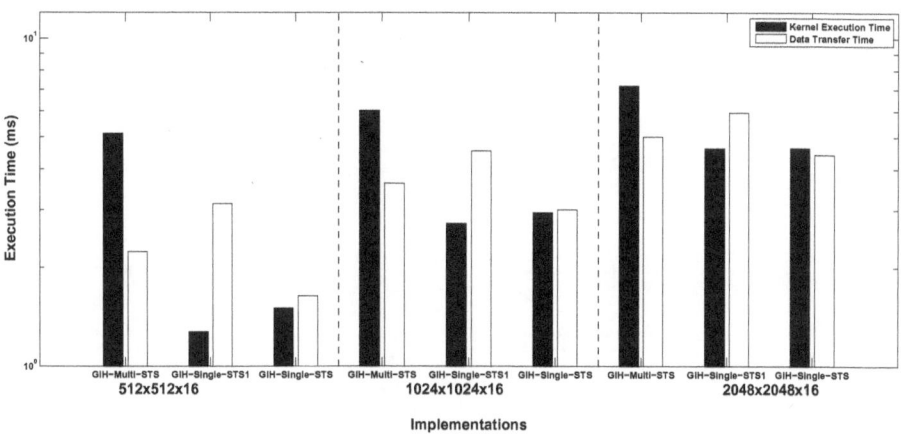

Fig. 5. Kernel execution time versus data transfer time for different image sizes

GIH-Single-STS also benefits from the modified 2-D transpose kernel which performs a single 3-D transpose operation by using the bin offset in the indexing. GIH-Single-STS is divided into three phases: a single horizontal scan, a 3-D transpose, and a vertical scan.

The initial implementation of GIH-Single-STS had several unnecessary data transfer between host and device after each phase. In the first implementation, the integral histogram tensor was being transferred to the GPU before invoking the kernel and then sent back to the CPU before launching the next kernel; these extra data transfers lead to reduced performance (referred to as GIH-Single-STS1). However, the GPU is specialized for compute-intensive, highly parallel computation and the overhead of communication between host and device cannot be hidden or double-buffered by non data-intensive kernels. In the improved GIH-Single-STS implementation, the integral histogram computations start after transferring the image to the GPU, complete the calculation of the integral histogram on the GPU then transfer the final integral histogram tensor back to the CPU, removing the extra communication overhead. In addition, the number of threads is automatically determined based on the image size to ensure maximum occupancy per kernel.

Figure 5 shows the kernel execution time versus data transfer time for GIH-Multi-STS, GIH-Single-STS1 (implementation with extra data transfers) and GIH-Single-STS for different image sizes. We see that the GIH-Multi-STS is compute bound (that is, the kernel execution time is larger than the CPU to GPU data transfer time), this method under utilizes the GPU, whereas the GIH-Single-STS1 is data-transfer-bound. The results show that the data transfer time for GIH-Single-STS1 is on average five times worse than GIH-Single-STS. The final GIH-Single-STS implementation shows a balance between data transfer and kernel execution time (Fig. 5).

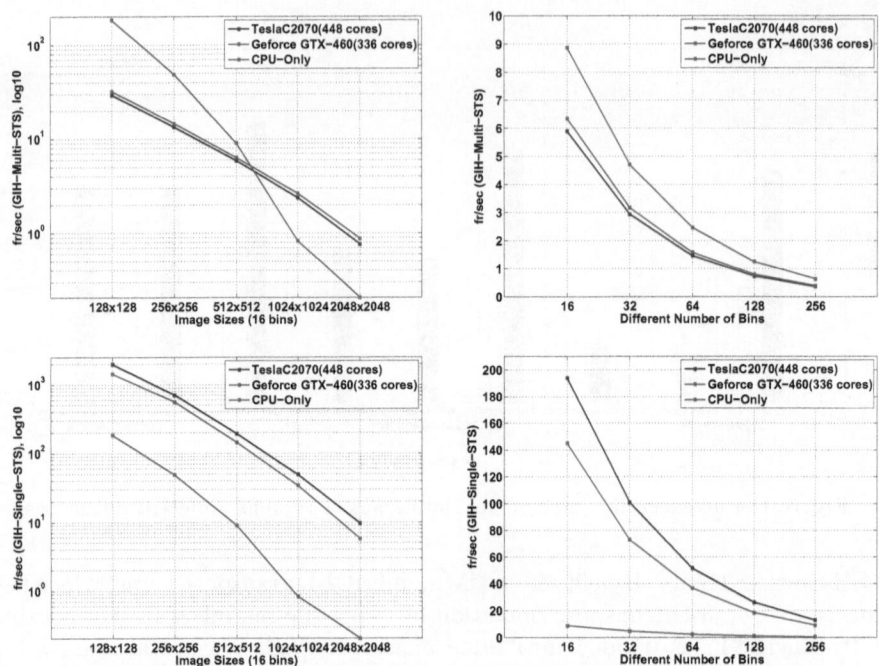

Fig. 6. Frame rate of GIH-Multi-STS, GIH-Single-STS and CPU-only integral histogram implementations: (UL) GIH-Multi-STS frame rate for different image sizes, (UR) GIH-Multi-STS frame rate for different number of bins, (LL) GIH-Single-STS frame rate for different image sizes, (LR) GIH-Single-STS frame rate for different number of bins for 512x512 image size.

Figure 6 summarizes the frame rate performance of the two GPU implementations compared to the sequential CPU-only implementation. The frame rate is defined as the maximum number of images processed per second. Since we use double buffering, the frame rate equals $1/(kernel\_execution\_time)$ for compute-bound cases, or $1/(data\_transfer\_time)$ for data-transfer-bound cases. Considering double buffering timing, our GIH-Single-STS achieves 194 fr/sec to compute 16-bin integral histograms for a 512×512 image and 94 fr/sec for $1K \times 1K$ image using the NVIDA Tesla C2070 GPU.

Figure 7 reports the speedup of our GPU implementations of the integral histogram compared to a sequential CPU implementation. The speedup takes into consideration the overlapping of computation and communication used by double buffering. The speedup of the improved GIH-Single-STS for a 16-bin integral histogram for a $1K \times 1K$ image is 60 times on an NVIDIA Tesla C2070 GPU and varies between 15 times to 25 times for a 512×512 image depending on the number of bins and the type of GPU.

Figures 8 shows feature maps for the target and search window with corresponding likelihood maps produced by the integral histogram-based likelihood

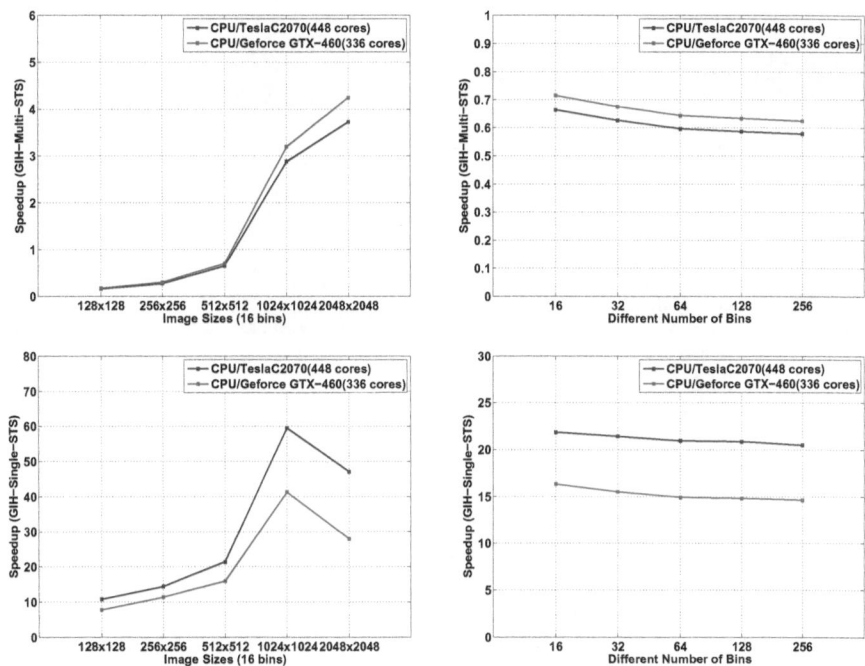

Fig. 7. Speedup of the two GPU designs over CPU on two NVIDA graphic cards: (UL) Speedup of GIH-Multi-STS (with respect to CPU-only) with different image sizes, (UR) Speedup of GIH-Multi-STS with varying number of bins, (LL) Speedup of GIH-Single-STS for different image sizes, (LR) Speedup of GIH-Single-STS with varying number of bins for 512x512 image size.

estimation approach. Figure 9 shows sample tracking results and fused likelihood maps for sample frames from an aerial wide area image sequence.

4   Conclusions

We have presented two parallel implementations of the integral histogram using the cross-weave scanning approach for GPU architectures, utilizing the CUDA programming model. The poor performance of the GIH-Multi-STS (*prescan*) implementation which was slower than the sequential version and the first implementation of GIH-Single-STS, clearly demonstrates that in parallelizing sequential image analysis algorithms on the GPU, data structures, GPU utilization and communication patterns need careful consideration. The GIH-Single-STS (*efficient communication*) implementation reduced the severe communication overhead bottleneck, by transferring an image size 1-D array instead of an integral histogram 3-D array and increased the GPU utilization. The GIH-Single-STS exploits an efficient prescan and 3-D transpose operation with maximum occupancy per kernel. It achieved frame rate of 185 for standard images 640×480 for 16 bins integral histogram computations which outperforms results presented

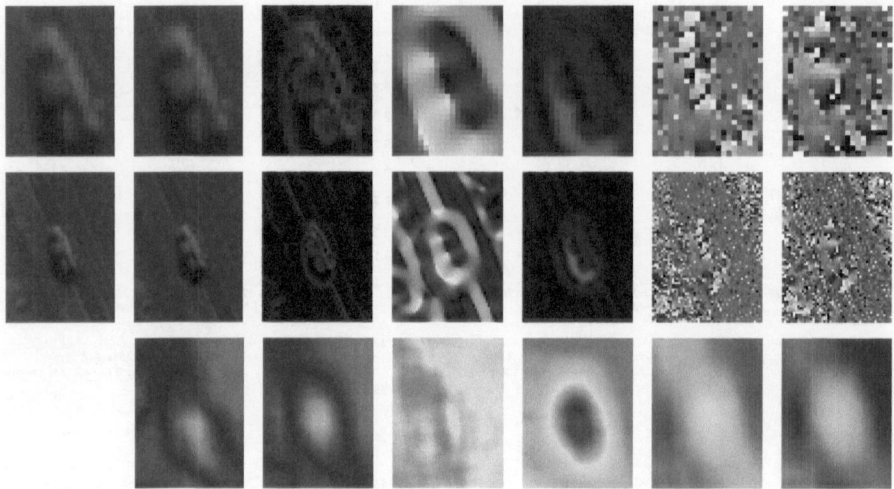

Fig. 8. Top row shows the car template and associated raw target features for intensity, gradient magnitude, Hessian shape index, normalized curvature index, Hessian eigen-vector orientations, and oriented gradient angles. Row 2 shows the predicted search window and associated raw features. Row 3 shows the corresponding likelihood maps combining target template with the associated search window features using integral histogram.

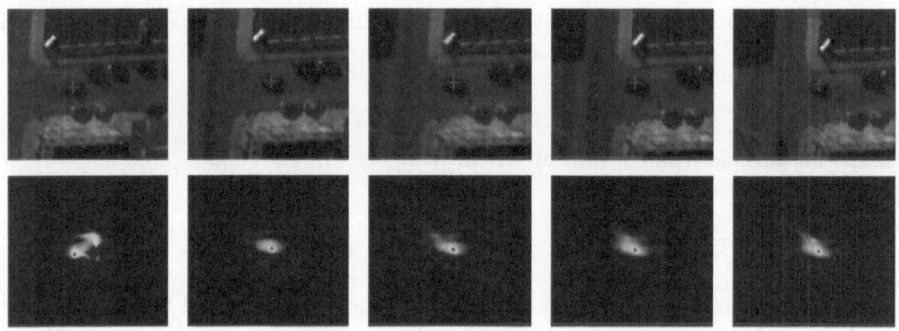

Fig. 9. LOFT tracking results are shown for the first five frames for car C4_1_0 from CLIF aerial wide-area motion imagery [26]. Top row shows the tracked car locations and the bottom row shows the fused likelihood maps used by LOFT [8] to determine the best target location in each corresponding frame.

for 8 SPEs (120 fr/sec for cross-weave scan and 172 for wavefront scan mode) in [20]. However, in most cases our performance is data-transfer-bound. One approach to further improve the time and memory efficiency of the GPU-based integral histogram method is to develop our custom parallel scan kernel for the horizontal and vertical cumulative sum computations without transpose phase for each tile of integral histogram tensor.

Acknowledgement. This research was partially supported by U.S. Air Force Research Laboratory (AFRL) under agreement AFRL FA8750-11-1-0073. Approved for public release (case 88ABW-2012-1016). The views and conclusions contained in this document are those of the authors and should not be interpreted as representing the official policies, either expressed or implied, of AFRL or the U.S. Govt. The U.S. Government is authorized to reproduce and distribute reprints for Govt. purposes notwithstanding any copyright notation thereon.

References

1. Porikli, F.: Integral histogram: A fast way to extract histograms in cartesian spaces. In: IEEE CVPR, vol. (1), pp. 829–836 (2005)
2. Sizintsev, M., Derpanis, K.G., Hogue, A.: Histogram-based search: A comparative study. In: IEEE CVPR, pp. 1–8 (2008)
3. Viola, P., Jones, M.J.: Robust real-time face detectin. Int. J. Computer Vision 57, 137–154 (2004)
4. Wei, Y., Tao, L.: Efficient histogram-based sliding window. In: IEEE CVPR, pp. 3003–3010 (2010)
5. Zhu, Q., Yeh, M.C., Cheng, K.T., Avidan, S.: Fast human detection using a cascade of histograms of oriented gradients. In: IEEE CVPR, vol. (2), pp. 1491–1498 (2006)
6. Adam, A., Rivlin, E., Shimshoni, I.: Robust fragments-based tracking using the integral histogram. In: IEEE CVPR, pp. 798–805 (2006)
7. Palaniappan, K., et al.: Efficient Feature extraction and likelihood fusion for vehicle tracking in low frame rate airborne video. In: 13th Conf. Information Fusion, pp. 1–8 (2010)
8. Pelapur, R., Candemir, S., Bunyak, F., Poostchi, M., Seetharaman, G., Palaniappan, K.: Persistent target tracking using likelihood fusion in wide-area and full motion video sequences. In: 15th Int. Conf. Information Fusion, pp. 2420–2427 (2012)
9. Erdem, E., Dubuisson, S., Bloch, I.: Fragments Based Tracking with Adaptive Cue Integration. Computer Vision and Image Understanding (7), 827–841 (2012)
10. Mosig, A., Jaeger, S., Chaofeng, W., Ersoy, I., Nath, S.K., Palaniappan, K., Chen, S.S.: Tracking cells in live cell imaging videos using topological alignments. Algorithms in Molecular Biology (4), 10 (2009)
11. Kolekar, M.H., Palaniappan, K., Sengupta, S., Seetharaman, G.: Semantic concept mining based on hierarchical event detection for soccer video indexing. Special Issue on Multimodal Information Retrieval (4), 298–312 (2009)
12. Palaniappan, K., Rao, R., Seetharaman, G.: Wide-area persistent airborne video: Architecture and challenges. In: Distributed Video Sensor Networks: Research Challenges and Future Directions, pp. 349–371 (2011)
13. Park, I.K., et al.: Design and performance evaluation of image processing algorithms on GPUs. IEEE Parallel and Distributed Systems 22(1), 91–104 (2011)
14. Grauer-Gray, S., Kambhamettu, C., Palaniappan, K.: GPU implementation of belief propagation using CUDA for cloud tracking and reconstruction. In: 5th IAPR Workshop on Pattern Recognition in Remote Sensing (ICPR), pp. 1–4 (2008)
15. Palaniappan, K., et al.: Parallel flux tensor analysis for efficient moving object detection. In: Int. Conf. Information Fusion, pp. 1–8 (2011)
16. Palaniappan, K., Bunyak, F., Nath, S.K., Goffeney, J.: Parallel Processing Strategies for Cell Motility and Shape Analysis. In: High-Throughput Image Reconstruction and Analysis, vol. (3), pp. 39–87 (2009)

17. Kumar, P., Palaniappan, K., Mittal, A., Seetharaman, G.: Parallel Blob Extraction Using the Multi-core Cell Processor. In: Blanc-Talon, J., Philips, W., Popescu, D., Scheunders, P. (eds.) ACIVS 2009. LNCS, vol. 5807, pp. 320–332. Springer, Heidelberg (2009)
18. Palaniappan, K., Vass, J., Zhuang, X.: Parallel robust relaxation algorithm for automatic stereo analysis. In: SPIE Proc. Parallel and Distributed Methods for Image Processing II, vol. (3452), pp. 958–962 (1998)
19. Palaniappan, K., Faisal, M., Kambhamettu, C., Hasler, A.F.: Implementation of an automatic semi-fluid motion analysis algorithm on a massively parallel computer. In: 10th IEEE Int. Parallel Processing Symp., pp. 864–872 (1996)
20. Bellens, P., Palaniappan, K., Badia, R.M., Seetharaman, G., Labarta, J.: Parallel Implementation of the Integral Histogram. In: Blanc-Talon, J., Kleihorst, R., Philips, W., Popescu, D., Scheunders, P. (eds.) ACIVS 2011. LNCS, vol. 6915, pp. 586–598. Springer, Heidelberg (2011)
21. Bilgic, B., Horn, B.K.P., Masaki, I.: Efficient integral image computation on the GPU. In: IEEE Intelligent Vehicles Symposium (IV), pp. 528–5338 (2010)
22. Kirk, D.: Nvidia CUDA software and GPU parallel computing architecture. In: ACM Proc. 6th Int. Symp. Memory Management (ISMM), pp. 103–104 (2007)
23. Nvidia Corp.: CUDA C Programming Guide 4.0 (2011)
24. Harris, M., Sengupta, S., Owens, J.D.: Parallel prefix sum (scan) with CUDA GPU. In: Gems, vol. 3, ch. 39, pp. 851–876 (2007)
25. Ruetsch, G., Micikevicius, P.: Optimizing matrix transpose in CUDA Nvidia CUDA SDK Application Note (2009)
26. Air Force Research Laboratory: Columbus Large Image Format (CLIF) dataset over Ohio State University (2007)

Play Estimation with Motions and Textures in Space-Time Map Description

Kyota Aoki and Takuro Fukiba

Utsunomiya University
Yoto, Utsunomiya, 321-8585 Japan

Abstract. It is easy to retrieve the small size parts from small videos. It is also easy to retrieve the middle size part from large videos. However, we have difficulties to retrieve the small size parts from large videos. We have large needs for estimating plays in sport videos. Plays in sports are described as the motions of players. This paper proposes the play retrieving method based on both of the motion compensation vector and normal color frames in MPEG sports videos. In MPEG videos, there are motion compensation vectors. Using the motion compensation vectors, we do not need to estimate the motion vectors between adjacent frames. This leads to decrease the huge computations about motion estimations. This work uses the 1-dimensional degenerated descriptions of each motion image between 2 adjacent frames. Connecting the 1-dimensional degenerated descriptions on time direction, we have the space-time map. This space-time map describes a sequence of frames as a 2-dimensional image. Using this space-time map on motion compensation vector frames and normal color frames, this work shows the method to retrieve a small number of plays in a huge number of frames. Our experiment records 0.93 as recall, 0.81 as precision and 0.86 as F-measure on 139 plays in 132503 frames.

1 Introduction

There are many videos about sports. There is a large need for content-based video retrievals. The amount of videos is huge, so we need an automatic indexing method [9]. We proposed the method that retrieves shots including a similar motion, based on the similarity of the motion with a sample part of videos [1].

We propose the method to retrieve the plays using only motion compensation vectors in MPEG videos with the 1-dimensional degeneration named Space-Time map. Many works try to index sport videos using the motions in the videos. Many of the works use the motion vectors in MPEG videos. They succeed to find camera works. They are zoom-in, zoom-out, pan and, etc. However, no work retrieves a play of a single player from only motions directly. Off cause, camera works have an important role in understanding videos. Sound also have some role in understanding videos. Many works use camera works and sound for understanding sport videos. Those feature-combining methods have some successes about retrieving home-runs and other plays. However, those works did

J.-I. Park and J. Kim (Eds.): ACCV 2012 Workshops, Part I, LNCS 7728, pp. 279–290, 2013.
© Springer-Verlag Berlin Heidelberg 2013

not success to retrieve plays from only motions. Recently, many works focused on retrieval combining many features and their relations[5,14].

There are some works proposed the method that retrieves play only from motions. The method can works with textures, sound, and camera works. However, the works proposed the method to retrieve play only from motions in motion compensation vectors in MPEG videos. The method gets the motions from motion compensation vectors in MPEG videos, and makes the 1-dimensional projections from the X direction motion and Y direction motion. The 1-dimensional projection represents the motions between a pair of adjacent frames as a 1-dimensional color strip. The method connects the strips in the temporal direction and gets an image that has 1 space dimension and 1 time dimension. The resulting image has the 1-dimensional space axis and the 1-dimensional time axis as the temporal slice [4,7,11,12]. Our method carries information about all pixels, but the temporal slice method does only about the cross-sections. We call this image as space-time map. Using the images, the method retrieves parts of videos as fast as image retrievals do.

We propose a video retrieval method based on the motions and textures for sports videos with Space-Time map. The proposed method includes making same compact description of a sequence of a frame describing MPEG motion compensation vectors and texture, similar video retrieval using the compact descriptions.

2   Motions and Textures in MPEG Videos

For retrieving a similar play in sport videos, the motions of players is important. Some works use only the mpeg motion compensation vectors and the correlations between the template and a part of videos. In mpeg motion compensation vectors of our experiments, pitchers share only 2×6 pixels. In the texture frames, pitchers share 27×107 pixels. In frames, the region representing a pitcher is 1% of a frame. It is difficult to retrieve the pitches in texture frames. In motion compensation vector frames, also the region representing a pitcher is 1% of a motion frame. However, in the motion frames, there is no move except for a pitcher. In the case, it is easy to find small motions of a pitcher. In motion frames, our proposed method works well to retrieve pitches. The smallness of the motion regions leads that the small discrimination power in the texture frames. However, it is easy to discriminate the environments.

This paper uses both of play discrimination with motion compensation vector and environment discrimination with texture frames for pitch retrievals. In motion frames, our method uses only correlations. In color frames, our method uses correlation and color differences for retrieving similar parts of videos.

3   Space-Time Map

Temporal slice is one answer to decreasing the amount of huge videos. The temporal slice is a simple selection of parts of frames. The temporal slice is sensitive about the movements of frames. This feature is not fit for our objective. We need the descriptions that do not sensitive about the movements of frames. We introduce the simple description as the temporal slice.

In mpeg video, we have the motion vector (2-dimensional) on every motion compensation block. The amount of information is $2/16 \times 16 \times 3$ of the original color video. This is very small comparing with the original color frames. The base-ball games can long about 2 hours.

Using the textures in frames, we have $16 \times 16 \times 3$ times larger descriptions than one of motion compensation vectors. The needs for the small descriptions increase in normal color frames.

The experimental video has 200K frames. If we compare frame by frame, there needs a huge computation. There is a large difficulty to retrieve similar parts of a video. We can retrieve similar parts of videos using classical representative frame-wise video retrieve method. However, it is difficult to retrieve similar part of videos based on the player's motions, because the motion leads a change of subsequent frames.

We can use many feature extraction methods to retrieve similar part of videos. However, the applicability of the method depends on the features selected to use. The generality of the method may be lost using specified features. The temporal slice is also the selections of features. In other words, the temporal slice is the selections of small number of pixels in frames. The temporal slice does not represent any information outside of the temporal slice. This paper uses the 1-dimensional degeneration for reducing the amount of information without lost generality [8].

Figure 1 shows the process to create a Space-time map from frames. We make 1-dimensional degenerations at each direction on each frame as in the top of Fig. 1. We have 2 1-dimensional descriptions. Then, we connect the 2 1-dimensional descriptions into a single 1-dimensional description. And last, we connect the 1-dimensional descriptions on time direction. The resulting description is 2-dimensional description.

This method does not select any parts in frames nor any frames in videos. The average is a major descriptive statistic of a set of numerical data. The proposed 1-dimensional description has information about all pixels. The temporal slice has no information about unselected pixels.

$$I_{1dx}(x) = \frac{\sum_{y \in [0, Y_{max}]} I_{2d}(x, y)}{Y_{max} + 1} \tag{1}$$

$$I_{1dy}(y) = \frac{\sum_{x \in [0, X_{max}]} I_{2d}(x, y)}{X_{max} + 1} \tag{2}$$

Equation (1) makes 1-dimensional degenerated description of X direction from a 2-dimensional image. (2) makes 1-dimensional degenerated description of Y direction from a 2-dimensional image. In (1) and (2), $I_{2d}(x, y)$ stands for the intensity at the pixel (x, y). (X_{max}, Y_{max}) is the coordinate of the right-upper corner.

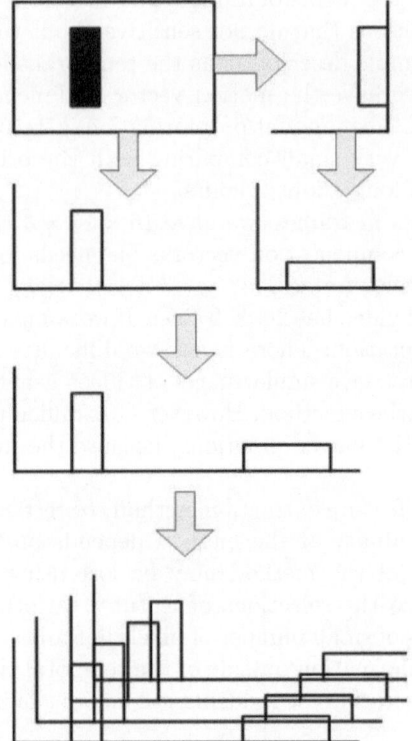

Fig. 1. Process making Space-time map

$$I(j) = \begin{cases} j \leq X_{max} \rightarrow I_{1dx}(j) \\ j > X_{max} \rightarrow I_{1dy}(j - X_{max} - 1) \end{cases} \tag{3}$$

The resulting 1-dimensional degenerated description is defined as (3).

There are 2 directions to make a 1-dimensional degeneration. We use both 2 directions that are X-axis and Y-axis using (1) and (2). In each color plane, we have a 1-dimensional degenerated description. We connect the 2 degenerated descriptions onto X-axis and the transposed projection onto Y-axis as from the second to the third of Fig. 1. We represent the X-direction motion in red, and Y-direction motion in green. There is no value in blue. Then, we have a 1-dimensional degenerated color strip from the motion compensation vector. In the color strip, red represents the X-direction motion and green does the Y-direction motion. For the convenience, we set 255 in blue when both of X and Y direction motions are 0.

We connect the 1-dimensional color strips describing motion frames on time passing direction as the bottom of Fig. 1. Connecting 1-dimensional color strips, we have a color image that has 1 space axis and 1 time axis. In this paper, the image is described as ST (Space-Time) image. In the following experiments, we use the 320×240 pixels half size frames. In the MPEG format, each 16×16 pixels block holds a motion compensation vector. This leads to reduce the amount of information into $1/256$. The resulting motion image is 20×15 pixels. The 1-dimensional degenerated description is $7/60$ of the original 20×15 pixels image. As a result, the usage of the Space-time map of motion compensation vector in MPEG video reduces the amount of information into $1/2200$. In normal color frames, the original size is 320×240. In resulting Space-time maps, a normal color frame is represented as $320 + 240$ pixels stripe. The resulting description is $1/137$ of an original color frame.

3.1   Similarity on Space-Time Map

The similar motion retrieval estimates what kind of motions exists on a place. It is same as the cost of the retrieval on images to retrieve similar part of videos on a Space-time map. There are many similar image retrieval methods. They can be applied in Space-time maps describing motions of videos. This paper uses the correlation between two images in motion compensation vectors. In normal color frames, we use both of a color difference and correlations in 3 color planes.

We normalize the resulting correlations for compensating the variance among videos. We have 2 independent correlations between two space-time maps from each color plane. They are an X-direction motion and a Y-direction motion.

3.2   Matching between Template Space-Time Map and Retrieved Space-Time Maps

All Space-time maps have same space direction size. The original frame is $X \times Y$ pixels. Then, the size of the space axis of Space-time maps is $X + Y$ pixels. For computing the correlations between the template Space-time map and any part of retrieved Space-time map, there is no freedom on space axis. There is only the freedom on time axis. If a template Space-time map is $S \times t$ and a retrieved Space-time map is $S \times T$, the computation cost of correlations is $S \times t \times T$. In a baseball game, the length t of interesting play of a video is short. So the computational cost of correlations is small enough to be able to apply large scale video retrieval. Because of the shortness of the retrieved part, there is no need to compensate the length of the part. There is no very slow pitch or no very fast one. There is no very slow running or no very fast one.

In normal color frames, it is difficult to retrieve pitches precisely. Our method uses normal color Space-time map for deciding the environment caught with a frame. In a pitching sequence of frames, there is no move of camera directions. We have no need to check the all frames in a pitching scene. We use only a single frame of a pitching scene for a template. As a result, the computation cost is $S \times T$. In normal color frames, S is 16 times larger than one in the motion compensation vector frames. A pitching scene is about 60 frames. As a result, the matching computation in normal color frames is about $1/4$ in motion compensation vector frames.

3.3 Similarity Measure in Motion Space-Time Map

We use the mutual correlation as the measure of similarity in motion Space-time maps. We have 2 dimensional correlation vectors. They are X-direction motion and Y-direction motion. If there is a similar motion between the template and the retrieved part of Space-time map, both of the 2 correlations are large. We use the similarity measure in motion Space-time maps shown in (4).

$$S(I_0, I_1) = \min\{Col(I_{0P}, I_{1P}) - T_P, P = X, Y\} \qquad (4)$$

In (4), Col is the correlation between I_{0P} and I_{1P}. I_0 and I_1 are Space-time maps. T_P is the threshold. P is one of x and y that represent the X-direction motion and Y-direction motion. This similarity measure is scalar.

3.4 Similarity Measure in Normal Color Space-Time Map

We have 2 methods about the similarity in normal color Space-time maps. They are RGB color plane-wise correlations and the color difference. The correlation in each color plane does not depend on the absolute color. It only depends on the change within the color plane. Otherwise, the color difference depends on the absolute color. We try both methods in our experiments.

Correlation in each Color Plane. We can use the mutual correlation as the measure of similarity. We have 3 dimensional correlation vectors in each color plane. They are red plane, green plane and blue plane. If there is an similar environment between the template and the retrieved part of Space-time map, all of the 3 correlations are large. We use the similarity measure shown in (5).

$$S(I_0, I_1) = \min\{Col(I_{0C}, I_{1C}) - T_C, C = red, green, blue\} \qquad (5)$$

In (5), Col is the correlation between I_{0C} and I_{1C}. I_0 and I_1 are Space-time maps in normal color frames. T_C is the threshold at each color plane. C is one of red, green and blue that represent red plane, green plane and blue plane.

This similarity measure is scalar. It is same as the similarity measure as in motion compensation vector frames.

Color Difference in Normal Color Space-Time Map. We also use the color difference as the measure of difference. In the case, we have a scalar color difference at each pixel in normal color Space-time map. We use the average square of color differences as the measure of the difference between the template and the retrieved part of a video.

From RGB color space to L*a*b color space, we use 2 steps. The first one is the conversion from RGB color space to XYZ color space as (8). The second one is the conversion from XYZ color space to L*a*b* color space as in (11). In L*a*b* color space, the color difference is defined as (12). In the equations, I_{0L} shows the $L*$ in frame 0. We define the similarity in normal color frames as (13) using the average square of color differences. In (13), E_p is the color difference at pixel p. For making the similarity measure, we invert the average square of color differences in (13).

$$X = 0.412453 \times R + 0.35758 \times G + 0.180423 \times B \tag{6}$$

$$Y = 0.212671 \times R + 0.71516 \times G + 0.072169 \times B \tag{7}$$

$$Z = 0.019334 \times R + 0.119193 \times G + 0.950227 \times B \tag{8}$$

$$L* = 116 \left(\frac{Y}{Y_n} \right)^{\frac{1}{3}} - 16 \tag{9}$$

$$a* = 500 \left(\left(\frac{X}{X_n} \right)^{\frac{1}{3}} - \left(\frac{Y}{Y_n} \right)^{\frac{1}{3}} \right) \tag{10}$$

$$b* = 200 \left(\left(\frac{Y}{Y_n} \right)^{\frac{1}{3}} - \left(\frac{Z}{Z_n} \right)^{\frac{1}{3}} \right) \tag{11}$$

$$E = \sqrt{(I_{0L} - I_{1L})^2 + (I_{0a} - I_{1a})^2 + (I_{0b} - I_{1b})^2} \tag{12}$$

$$S_{CD} = - \left(\frac{\sum_{p \in Allpixels} E_p^2}{|Allpixels|} \right) - Th_{CD} \tag{13}$$

4   Experiments on Baseball Game and Evaluation

4.1   Baseball Game

This paper treats baseball game MPEG videos. In baseball games, players uniforms change between half innings. There is large number of pitches. This paper uses a single play of a pitch as a template. Using this template, the proposed method retrieves large number of pitches using similar motion retrieval. Motion based similar video retrieval can find many types of plays based on the template. There are a few repeated plays that are not pitches. This paper distinguishes a pitch and other plays.

4.2   Experimental Objects

This paper uses a whole base-ball game for experiments. The game is 79minutes, 132485 frames in a video. In the game, there are right-hand pitchers and a left-hand pitcher. There are 168 pitches. There are 31648 frames that represent the camera work that catches the pitching scenes.

Figure 2 shows the example of pitches in our experimental video at each 5 frames distance. We use the left one as the template of a pitch. In Fig. 2, the center one and the right one are retrieved parts of a video using the template. The center one and right one differ from the template at the uniforms. The right one differs from the template one at left-hand and right-hand.

Fig. 2. Examples of pitches

4.3   Experiment Process

The experimental videos are recorded from Japanese analog TV to DVD. Then, the recorded videos are reduced into 320 × 240 pixels and monochrome and encoded MPEG1 format. Most plays of pitches are very short. So there is no reduction on time direction. There are 30 frames in 1S. There are all parts including telops, sportscasters and CG overlays. The first step of our experiment is the extraction of motion compensation vectors. The motion compensation vector is at each 16 × 16 pixel blocks. In every motion compensation block, we have a motion compensation vectors. Similar shot detection in motion frames uses 20 frames of the left pitch in Fig. 2 as a template and retrieves the shots including pitches of videos. In Fig. 3, the left one shows the part of the Space-time map based on the motions from motion compensation vectors in a MPEG video. And, the center shows the Space-time map based on the original colors in frames. The frames of this part are shown in the right. In the color-based Space-time map, we can see the place of the player. In the motion-based Space-time map, we can see the stripes representing the players motion.

Similar environment detection in normal color frames uses only 1 frame in the pitching template. There is no camera works in the template frames. As a result, there is no difference in selection of 1 frame from 20 frames in the template shot. We use the first frame of the template shot.

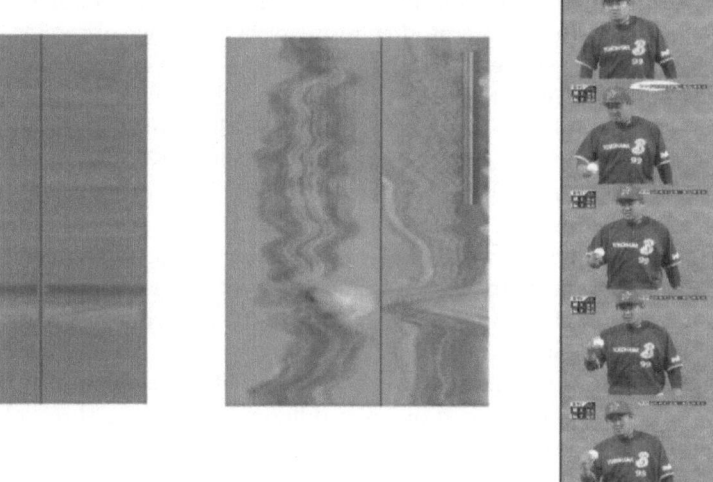

Fig. 3. Motion Space-time map, Texure Space-time map and Original frames

F-Measure

$$F - measure = \frac{2 \times precision \times recall}{precision + recall} \tag{14}$$

$$= \frac{R}{\frac{1}{2}(N + C)} \tag{15}$$

F-measure is defined as (15). In (15), N is the number of retrieved items. R is the number of retrieved proper items. C is the number of proper items.

4.4 Motion Frame Template Experiments in Pitching Retrieval

These experiments use a single template image of the length 20 frames. When we use the pitch near 83000th frame for the template play, we can use 82924th frames to 83062th frame. This sequence has 139 frames. Our pre-experiments using some length of template Space-time maps show that the 20 frames template Space-time map starting from 83001th frame is best. We control the thresholds that make the F-measure as maximum. In this experiment, precision is 0.686 and recall is 0.601. This experiment allows the multiple retrievals of a pitch. The single template Space-time map experiment shows 0.64 at F-measure. In the number, there are 147 retrievals, 36 error retrieval, and 101 recalls.

4.5 Normal Color Frame Experiments in Environment Retrieval

In normal color frames, it cannot detect precise sequences of frames that are pitches. In our experiments, we only try to retrieve the frames that catch the pitchers from right-rear direction. In the case, the caught environments are similar in every frame. There is no need to retrieve the sequence of frames. We only need to retrieve every single frame that catches the same environment.

The experiments have 2 methods that use the color difference and the correlations in each color planes. They also control the thresholds as the F-measures made as maximum. In the experiment using correlations in each color-planes, the resulting F-measure is 0.89. The recall is 0.891. The correctness is 0.891.

This performance is much better than one in motion frames. However, the retrieved objects differ between this normal color frame experiment and the motion frame experiment. In this experiment, the experiment retrieves the same environment frames. Otherwise, in the motion frame experiment, the experiment retrieves the precise pitches.

The experiment using similarity based on color difference retrieves the frames that show larger similarity based on color differences than the threshold. In the experiment, there is a single threshold. The resulting F-measure is 0.9. The recall is 0.907. The correctness is 0.906.

Table 1. Simple scene retireval with normal color frames

	Correlation	Color difference
Recall	0.891	0.907
Correctness	0.891	0.906
F-measure	0.89	0.90

In those 2 experiments, color difference experiment shows better performance.

4.6 Complex Retrieval Experiment Both of Motion and Normal Color Frames

For retrieving the precise pitches in frames, we need to use the motion frame retrieval. With combining the motion frame retrieval and normal color frame retrieval, we can improve the performance of the precise pitch retrieval.

To combining motion frame retrieval and normal color frame retrieval, we have some difficulties to find the proper set of thresholds. In motion frame retrieval, there are 2 thresholds that work in each X and Y direction motions. In color correlation retrieval, there are 3 thresholds that work in each red, green and blue color plane. In color difference retrieval, there is a threshold that work in color difference.

To combining motion frame retrieval and color correlation retrieval, we have 5 thresholds. To combining motion frame retrieval and color difference retrieval, we have 3 thresholds.

It is difficult to optimize all thresholds at once. We divided the optimization of thresholds into 2 steps. They are a normal color frame threshold optimization and motion frame threshold optimization. We have 2 methods in normal color frame retrieval. As a result, we have 4 types of optimization results. We shows the result of our experiments in table 2. In table 2, the correlation retrieval in the normal color frame and motion retrieval is best. It marks 0.865 in F-measure. The recall is 0.928. The correctness is 0.811.

There is no difference between color-correlation+motion and motion+color-correlation with same thresholds. However, it is easy to find better set of thresholds in the color-correlation+motion case.

Table 2. Total performance of pitches retrieval in a baseball game

	Motion + Correlation in Color	Motion + Color difference	Color difference + Motion	Color correlation+ Motion
Recall	0.8	0.8	0.804	0.928
Correctness	0.851	0.828	0.780	0.811
F-measure	0.82	0.81	0.79	0.865

5   Conclusion

This paper discusses about the retrieval of plays in sport MPEG videos using similar motion retrieval and similar scene retrieval in normal color frames. For recognizing sport videos, the motions represent important meanings. In the cases, there must be similar video retrieval methods based on the motions described in the videos. The scene retrieval based on color frames works well. The proposed similar play retrieval method is the combination of motion based similar play retrieval method and color based similar scene retrieval method.

The experiment shows that the similar play retrieval works well using both of motion compensation vectors and normal color frames in MPEG videos. Classical works using MPEG motion compensation vector only uses global-scale motions. However, the proposed method utilizes local motions. The proposed combination of motion compensation vector method and normal color frame method works well in our experiments. Using both methods, the proposed method gets some more performance than a single motion vector method.

References

1. Aoki, K.: Video retrieval based on motions. JIEICE technical report, IE–08 217, 45–50 (2008)
2. Aoki, K., Moro, S.: Shot change detection method for movies including large motions and camera works. JIEICE technical report, IE–08 127, 67–72 (2008)

3. Aoki, K.: Block-wise High-speed Reliable Motion Estimation Method Applicable in Large Motion. JIEICE technical report, IE–06 536, pp. 95–100 (2007)
4. Bouthemy, P., Fablet, R.: Motion Characterization from Temporal Cooccurrences of Local Motion-based Measures for Video Indexing. In: 14th Int. Conf. on Pattern Recognition, ICPR 1998, pp. 905–908 (1998)
5. Fleischman, M., Roy, D.: Situated Models of Meaning for Sports Video Retrieval. In: Proc. NAACLHLT 2009 Companion Volume, pp. 37–40 (2007)
6. Kamegaya, A., Kinoshita, H.: An image retrieving method using the object index and the motion. JIEICE technical report, IE–23 8, 43–48 (1999)
7. Keogh, E., Ratanamahatana, C.A.: Exact indexing of dynamic time warping. Knowledge and Information Systems 7, 358–386 (2005)
8. Matsuike, T., Ishii, R., Maeda, S., Okada, Y.: Spatio-Temporal Image Processing for Retrieving Similar Video Sequences JIEICE technical report, PRM–07 427, 299–304 (2007)
9. Miyamori, H., Kasutani, E., Tominaga, H.: Video Retrieval Method by Query Using Action Phrases. Trans. JIEICE(D-2) 80, 1590–1599 (1997)
10. Nagasaka, A., Miyatake, T., Ueda, H.: Realtime Video Scene Detection based on Shot Sequence Encoding. Trans. JIEICE(D-2) 79, 531–537 (1996)
11. Ngo, C., Pong, T., Zhang, H.: On Clustering and Retrieval of Video Shots Through Temporal Slices Analysis. IEEE Trans. on Multimedia 4, 446–458 (2002)
12. Ngo, C., Pong, T., Chin, R.T.: Video Partitioning by Temporal Slice Coherency. IEEE Transactions on Circuits and Systems for Video Technology 11, 941–953 (2001)
13. Nobe, M., Ino, Y., Aoki, K.: Pixel-wise Motion Estimation Based on Multiple Motion Estimations in Consideration of Smooth Regions. JIEICE technical report, IE06 423, 7–12 (2006)
14. Sebe, N., Lew, M.S., Zhou, X., Huang, T.S., Bakker, E.M.: The State of the Art in Image and Video Retrieval. In: Bakker, E.M., Lew, M., Huang, T.S., Sebe, N., Zhou, X.S. (eds.) CIVR 2003. LNCS, vol. 2728, pp. 1–8. Springer, Heidelberg (2003)
15. Yaginuma, Y., Suzuki, M., Shimizu, Y.: Retrieval of Educational Image Contents Based on Color Features and Keywords. JIEICE technical report. E–06 507, 111–116 (2007)

A Benchmark Dataset for Outdoor Foreground/Background Extraction

Antoine Vacavant[1,2], Thierry Chateau[3],
Alexis Wilhelm[3], and Laurent Lequièvre[3]

[1] Clermont Université, Université d'Auvergne, ISIT, BP10448, F-63000
Clermont-Ferrand
[2] CNRS, UMR6284, BP10448, F-63000 Clermont-Ferrand
[3] Pascal Institute, Blaise Pascal University, CNRS, UMR6602, Clermont-Ferrand

Abstract. Most of video-surveillance based applications use a foreground extraction algorithm to detect interest objects from videos provided by static cameras. This paper presents a benchmark dataset and evaluation process built from both synthetic and real videos, used in the BMC workshop (Background Models Challenge). This dataset focuses on outdoor situations with weather variations such as wind, sun or rain. Moreover, we propose some evaluation criteria and an associated free software to compute them from several challenging testing videos. The evaluation process has been applied for several state of the art algorithms like gaussian mixture models or codebooks.

1 Introduction

The ability to detect objects in videos is an important issue for a number of computer vision applications like intrusion detection, object tracking, people counting, *etc.* In the case of a static camera, a foreground extraction algorithm is a popular operation to point out objects of interest in the video sequence. Although modeling background seems simple, challenging situations occur in classic outdoor environments such as variation of illumination conditions or local appearance modifications resulting to wind or rain. In order to handle such situations, many background/foreground adaptive models have been proposed in the last fifteen years. An important issue is to provide a way to evaluate and compare most popular models according to standard criteria.

Although the evaluation of background subtraction algorithms (BSA) is an important issue, the impact of relevant papers that handle with both benchmarks and annotated dataset is limited [1,10]. Moreover, many authors that propose a novel approach use [11] as a gold-standard, but rarely compare their method with recent related work. This paper proposes a set of both synthetic and real video and several performance evaluation criteria in order to evaluate and rank background/foreground algorithms. Popular methods are then evaluated and ranked according to these criteria.

The next section (Section 2) presents the annotated datasets we have proposed for the BMC (Background Models Challenge), composed of 20 synthetic

J.-I. Park and J. Kim (Eds.): ACCV 2012 Workshops, Part I, LNCS 7728, pp. 291–300, 2013.

videos and 9 real videos. We also define the quality metrics available in the benchmark, and computable with a free software (BMCW). In Section 3, we conduct a complete evaluation of six classic background subtraction algorithms of the literature, thanks to the benchmark of BMC.

2   Datasets and Evaluation Criteria

2.1   Learning and Evaluation Videos

In the contest BMC (Background Models Challenge) [1], we have proposed a complete benchmark composed of both synthetic and real videos. They are divided into two distinct sets of sequences: learning and evaluation.

The benchmark is first composed of 20 urban video sequences rendered with the SiVIC simulator [4]. With this tool, we are also able to render the associate ground truth, frame by frame, for each video (at 25 fps). Two scenes are used for the benchmark:

1. a street;
2. a rotary.

For each scene, we propose 5 event types:

1. cloudy, without acquisition noise;
2. cloudy, with noise;
3. sunny, with noise;
4. foggy, with noise;
5. wind, with noise.

For each configuration, we have two possible use-cases:

1. 10 seconds without objects, then moving objects during 50 seconds;
2. 20 seconds without event, then event (*e.g.* sun uprising or fog) during 20 seconds, finally 20 seconds without event.

The *learning* set is composed of the 10 synthetic videos representing the use-case 1. Each video is numbered according to presented event type (from 1 to 5), the scene number (1 or 2), and the use-case (1 or 2). For example, the video 311 of our benchmark describes a sunny street, under the use-case 1 (see Figure 1). In the learning phase of the BMC contest, authors use these sequences in order to set the parameters of their BSA, thanks to the ground truth of each image that is available, and to a software of computation of quality criteria (see next section).

The *Evaluation* set first contains the 10 synthetic videos with use-case 2. In Figure 1, the video 422, presenting a foggy rotary under use-case 2, is depicted. This set is also composed of real videos acquired from static cameras in video-surveillance contexts (see Figure 2). This dataset has been built in order test

---

[1] http://bmc.univ-bpclermont.fr

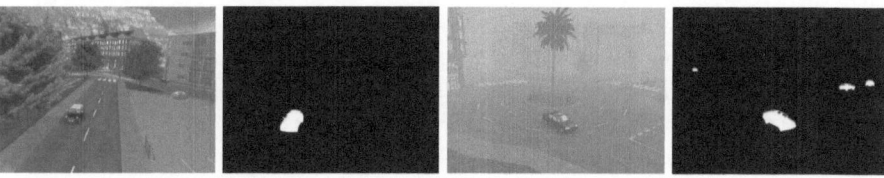

Fig. 1. Examples of synthetic videos and their associated ground truth in our dataset. Left: scene 1, configuration 3, use-case 1 (learning phase). Right: scene 2, configuration 4, use-case 2 (evaluation phase)

the algorithms reliability during time and in difficult situations such as outdoor scenes. So, real long videos (about one hour and up to four hours) are available, and they may present long time change in luminosity with small density of objects in time compared to previous synthetic ones. This dataset allows to test the influence of some difficulties encountered during the object extraction phase. Those difficulties have been sorted according to:

1. the ground type (bitumen, ballast or ground);
2. the presence of vegetation (trees for instance);
3. casted shadows;
4. the presence of a continuous car flow near to the surveillance zone;
5. the general climatic conditions (sunny, rainy and snowy conditions);
6. fast light changes in the scene;
7. the presence of big objects.

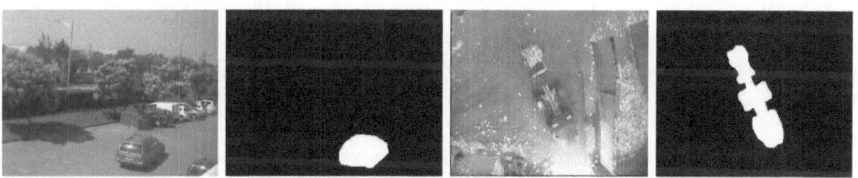

Fig. 2. Examples of real videos and their associated ground truth in our dataset (evaluation phase)

For each of these videos have been manually segmented some representative frames that can be used to evaluate a BSA. In the evaluation phase of the BMC contest, no ground truth image is available, and authors should test their BSA with the parameters they have set in the learning phase.

2.2 Quality Assessment of a Background Subtraction Algorithm

In our benchmark, several criteria have been considered, and represents different kinds of quality of a BSA.

Static Quality Metrics. Let S be the set of n images computed thanks to a given BSA, and G be the ground truth video sequence. For a given frame i, we denote by TP_i and FP_i the true and false positive detections, and by TN_i and FN_i the true and false negative ones. We first propose to compute the F-measure, defined by:

$$F = \frac{1}{n} \sum_{i=1}^{n} 2 \frac{Prec_i \times Rec_i}{Prec_i + Rec_i}, \qquad (1)$$

with

$$Rec_i(P) = TP_i/(TP_i + FN_i) \; ; \qquad Prec_i(P) = TP_i/(TP_i + FP_i) \qquad (2)$$
$$Rec_i(N) = TN_i/(TN_i + FP_i) \; ; \qquad Prec_i(N) = TN_i/(TN_i + FN_i) \qquad (3)$$
$$Rec_i = (1/2)(Rec_i(P) + Rec_i(P)) \; ; \; Prec_i = (1/2)(Prec_i(P) + Prec_i(P)). \quad (4)$$

We also compute the PSNR (Peak Signal-Noise Ratio), defined by:

$$PSNR = \frac{1}{n} \sum_{i=1}^{n} 10 \log_{10} \frac{m}{\sum_{j=1}^{m} ||S_i(j) - G_i(j)||^2} \qquad (5)$$

where $S_i(j)$ is the jth pixel of image i (of size m) in the sequence S (with length n). These two criteria should permit to compare the raw behavior of each algorithm for moving object segmentation.

Application Quality Metrics. We also consider the problem of background subtraction in a visual and perceptual way. To do so, we use the gray-scale images of the input and ground truth sequences (see Figure 3) to compute the perceptual measure SSIM (Structural SIMilarity), given by [14]:

$$SSIM(S, G) = \frac{1}{n} \sum_{i=1}^{n} \frac{(2\mu_{S_i}\mu_{G_i} + c_1)(2cov_{S_iG_i} + c_2)}{(\mu_{S_i}^2 + \mu_{G_i}^2 + c_1)(\sigma_{S_i}^2 + \sigma_{G_i}^2 + c_2)}, \qquad (6)$$

where μ_{S_i}, μ_{G_i} are the means, $\sigma_{S_i}, \sigma_{G_i}$ the standard deviations, and $cov_{S_iG_i}$ the covariance of S_i and G_i. In our benchmark, we set $c_1 = (k_1 \times L)^2$ and $c_2 = (k_2 \times L)^2$, where L is the size of the dimension of the signal processed (that is, $L = 255$ for gray-scale images), $k_1 = 0.01$ and $k_2 = 0.03$ (which are the most used values in the literature).

We finally use the D-Score [8], which consists in considering localization of errors according to real object position. As Baddeleys distance, it is a similarity measure for binary images based on distance transform. To compute this measure we only consider mistakes in BSA results. Each error cost depends on the distance with the nearest corresponding pixel in the ground-truth. As a matter of fact, for object recognition, short or long range errors in segmentation step are less important than medium range error, because pixels on medium range impact greatly on object's shape. Hence, the penalty applied to medium range errors is heavier than the one applied to those in a short or large range, as shown on Figure 4.

Fig. 3. To compute the SSIM, we need the intensities of pixels, in the ground truth sequence G (Left), and in the sequence computed by a BSA (Right)

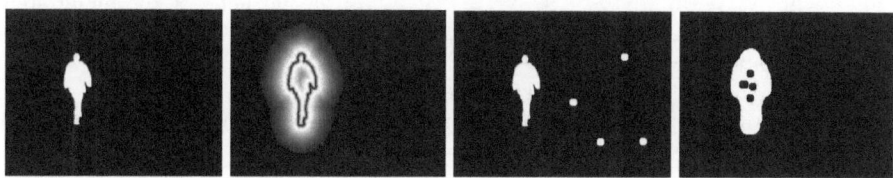

Fig. 4. Examples of computation of the D-Score. From Left to Right: a ground-truth image; cost map based on a DT; example of long ranges errors, leading to a D-Score of 0.003; omissions with medium range errors, with D-Score: 0.058

More precisely, the D-Score is computed by using:

$$D-score(S_i(j)) = \exp\left((-\log_2\left(2.DT(S_i(j)) - 5/2\right)^2\right) \tag{7}$$

where $DT(S_i(j))$ is given by the minimal distance between the pixel $S_i(j)$ and the nearest reference point (by any distance transformation algorithm). With such a function, we punish errors with a tolerance of 3 pixels from the ground-truth, because these local errors do not really affect the recognition process. For the same reason, we allow the errors that occur at more than a 10 pixels distance. Details about such metric can be found in [8]. Few local/far errors will produce a near zero D-Score. On the contrary, medium range errors will produce high D-Score. A good D-Score has to tend to 0.

3   Results and Analysis

3.1   Material and Methods

In this article, we will present the quality measures presented in the previous section for the methods depicted in Table 1. Most of those approaches are available thanks to the OpenCV library [2]. The parameters were tuned with a stochastic gradient descent to maximize the F-measure for the sequences of the learning phase.

We present the values of all the quality criteria exposed in the previous section, for the evaluation set of videos. Criteria are calculated thanks to the BMC

---

[2] http://opencv.org/

Table 1. The methods tested in this article, with their associated references

Name	Description
NA	Naive approach, where pixels differing from the first image of the sequence (under a given threshold) are considered as foreground ($threshold = 22$).
GMM 1	Gaussian mixture models from [5,11], improved by [6] for a faster learning phase.
GMM 2	Gaussian mixture models improved with [12,13] to select the correct number of components of the GMM ($history\ size = 355$, $background\ ratio = 16$).
BC	Bayesian classification processed on feature statistics [9] ($L = 256$, $N_1 = 9$, $N_2 = 15$, $L^c = 128$, $N_1^c = 25$, $N_2^c = 25$, no holes, 1 morphing step, $\alpha_1 = 0.0422409$, $\alpha_2 = 0.0111677$, $\alpha_3 = 0.109716$, $\delta = 1.0068$, $T = 0.437219$, $min\ area = 5.61266$).
CB	Codewords and Codebooks framework [7].
VM	VuMeter [3], which uses histograms of occurences to model the background ($\alpha = 0.00795629$ and $threshold = 0.027915$).

Wizard (BMCW, see a screenshot in Figure 5), which can be downloaded from the BMC website [3].

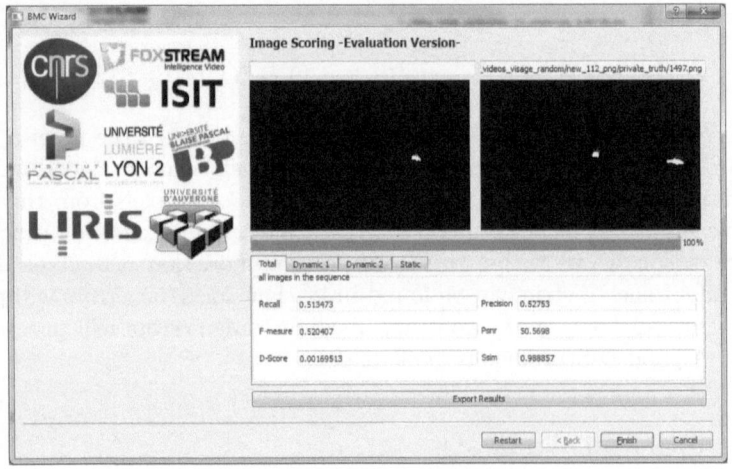

Fig. 5. The BMC Wizard, a free software to compute criteria of our benchmark

3.2   Results

Figures 6 to 10 show the global performance of each method for each evaluated score. Figure 10 can be consulted in color in the online version.

Tables 1 to 29, from the supplementary material of this article, show the performance of each method for each sequence:

---

[3] http://bmc.univ-bpclermont.fr/?q=node/7

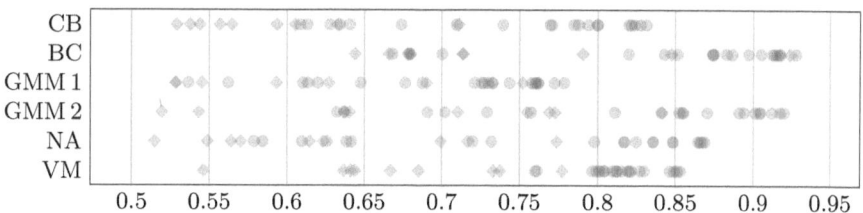

Fig. 6. F-measure for each method

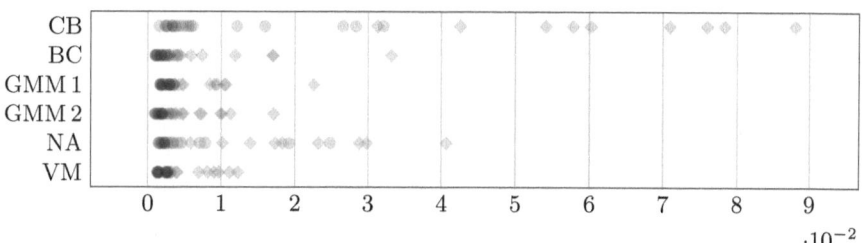

Fig. 7. D-score for each method

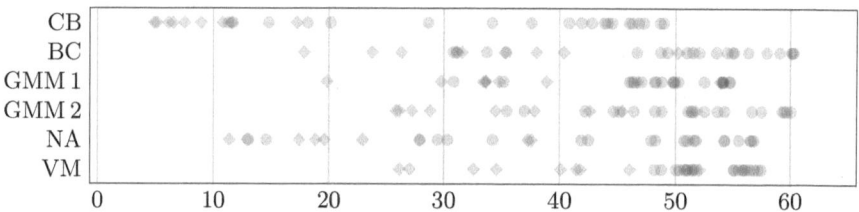

Fig. 8. PSNR for each method

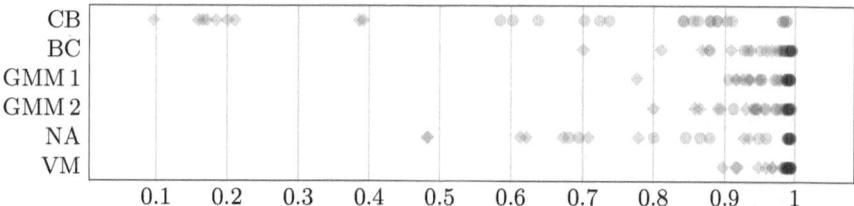

Fig. 9. SSIM for each method

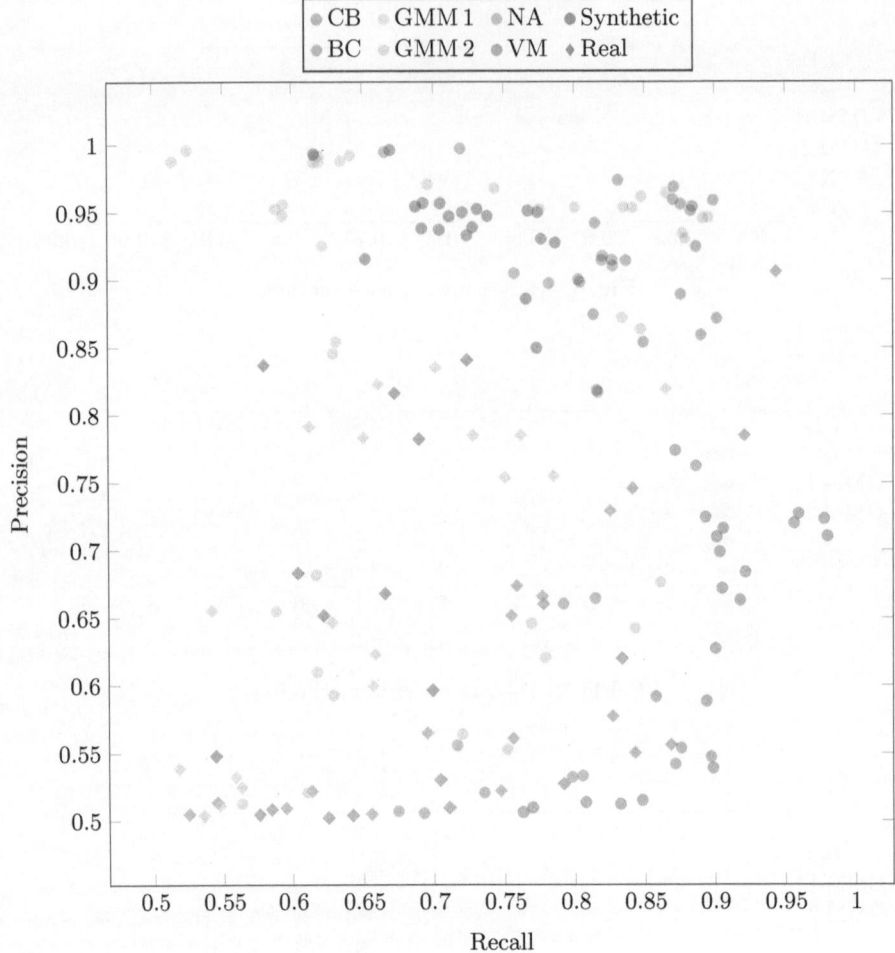

Fig. 10. Precision and Recall for each method

- Learning phase:
  - Street: tables 1 to 5;
  - Rotary: tables 6 to 10.
- Evaluation phase:
  - Street: tables 11 to 15;
  - Rotary: tables 16 to 20;
  - Real applications: tables 21 to 29.

3.3 Analysis

From a statistical point of view (Figure 6), we can notice that the best method of our tests is BC, since its F-measure has the shortest range of values, with highest values (from 0.65 to 0.93 approximately). The case of the VM method is interesting because its F-measure is focused around the interval [0.8; 0.85]. These observation can be confirmed by Figure 10, where BC and VM have the greatest numbers of points coming close the $(1, 1)$ point. GMM1 has also a similar behaviour, around the 0.75 value, and a very good precision. GMM2 has a point of focus around the 0.9 value, but has also a wide interval of F-measures. The CB approach returns a very wide range of values, which could be induced by the high variability of the parameters of the method. Figure 10 informs us that the real videos of our benchamrk are not correctly processed by CB, impacting a global bad results. This phenomenon can also be observed for the NA, in a more negative way.

As illustrated in Figure 8, the PSNR gives us equivalent general informations about the tested BSA. We can also notice an increasing feeling of non-control of the results of CB and NA. Points of focus are also observable for VM ([50; 60]) and GMM1 ([45; 55]).

From a structural point of view, the values of SSIM and D-score lead to similar conclusions: CB and NA are not constant, and not efficient on the whole benchmark. Its seems even better to choose NA (SSIM greater than 0.4) instead of CB (SSIM can be around 0.1 or 0.2).

4   Conclusion

In this article, we have proposed to test the benchmark proposed in the BMC contest, with six classic background subtraction algorithms of the literature. Thanks to the measures we have computed, we can determine several qualities of the tested methods.

We would like to propose an other contest in 2013, with maybe more real videos, containing complex contexts. The BMC website is an interesting way to keep our benchmark available to researchers who want to test their algorithm.

References

1. Benezeth, Y., Jodoin, P.-M., Emile, B., Laurent, H., Rosenberger, C.: Review and evaluation of commonly-implemented background subtraction algorithms. In: Proc. of IEEE Int. Conf. on Pat. Rec. (2008)
2. Dhome, Y., Tronson, N., Vacavant, A., Chateau, T., Gabard, C., Goyat, Y., Gruyer, D.: A benchmark for background subtraction algorithms in monocular vision: a comparative study. In: Proc. of IEEE Int. Conf. on Image Proc. Theory, Tools and App. (2010)
3. Goyat, Y., Chateau, T., Malaterre, L., Trassoudaine, L.: Vehicle trajectories evaluation by static video sensors. In: Proc. of IEEE Int. Conf. on Intel. Transp. Sys. (2006)
4. Gruyer, D., Royere, C., du Lac, N., Michel, G., Blosseville, J.-M.: SiVIC and RTMaps, interconnected platforms for the conception and the evaluation of driving assistance systems. In: Proc. of World Cong. and Exh. on Intel. Trans. Sys. and Serv. (2006)

5. Hayman, E., Eklundh, J.-O.: Statistical background subtraction for amobile observer. In: Proc. of Int. Conf. on Comp. Vis. (2003)
6. Kaewtrakulpong, P., Bowden, R.: An improved adaptive background mixture model for realtime tracking with shadow detection. In: Proc. of Eur. Work. on Adv. Video Based Surv. Sys. (2001)
7. Kim, K., Chalidabhongse, T.H., Harwood, D., Davis, L.: Real-time foreground-background segmentation using codebook model. Real-time Imag. 11(3), 167–256 (2005)
8. Lallier, C., Renaud, E., Robinault, L., Tougne, L. In: Proc. of IEEE Int. Conf. on Adv. Video and Signal-based Surv. (2011)
9. Li, L., Huang, W., Gu, I.Y.H., Tian, Q.: Foreground Object Detection from Videos Containing Complex Back-ground. In: Proc. of ACM Multimedia (2003)
10. Prati, A., Mikic, I., Trivedi, M., Cucchiara, R.: Detecting moving shadows: Algorithms and evaluation. IEEE Trans. on PAMI 25(7), 918–923 (2003)
11. Stauffer, C., Grimson, W.E.L.: Adaptative background mixture models for a real-time tracking. In: Proc. of IEEE Int. Conf. on Comp. Vision and Pat. Rec. (1999)
12. Zivkovic, Z.: Improved adaptive Gaussian mixture model for background subtraction. In: Proc. of IEEE Int. Conf. on Pat. Rec. (2004)
13. Zivkovic, Z., van der Heijden, F.: Efficient adaptive density estimapion per image pixel for the task of background subtraction. Pat. Rec. Let. 27(7), 773–780 (2006)
14. Wang, Z., Bovik, A.C., Sheikh, H.R., Simoncelli, E.P.: Image quality assessment: From error visibility to structural similarity. IEEE Trans. on IP 13(4), 600–612 (2004)

One-Class Background Model

Assaf Glazer, Michael Lindenbaum, and Shaul Markovitch

Technion – Israel Institute of Technology, Haifa, Israel
{assafgr,mic,shaulm}@cs.technion.ac.il

Abstract. Background models are often used in video surveillance systems to find moving objects in an image sequence from a static camera. These models are often built under the assumption that the foreground objects are not known in advance. This assumption has led us to model background using one-class SVM classifiers. Our model belongs to a family of block-based nonparametric models that can be used effectively for highly complex scenes of various background distributions with almost the same configuration parameters for all examined videos. Experimental results are reported on a variety of test videos from the Background Models Challenge (BMC) competition.

1 Introduction

Moving foreground objects in an image sequence from a static camera can be detected by comparing new images with a representation of the background scene. This process is called background subtraction and the representation of the background is called the background model. Background models must be able to cope with changes in the background scene that may occur over time. These include illumination changes, fluctuations of local image patterns (e.g., swaying trees and fluttering flags), flickering CRTs, and so on.

A common assumption in background modeling is that ground-truth images are not available for training. Thus, background models should be built without knowledge about what foreground objects are expected to appear [1]. With respect to the background, however, it is common to assume that *a priori* knowledge is available for training. Because the available information pertains to only one side of the problem, we propose in the following to use one-class classification tools to model background scenes.

In general, traditional supervised classifiers are trained using positive and negative examples. However, in our settings, labeled data exist for only the background class. A straightforward learning approach would be to estimate the distribution of the background. However, density estimation in high-dimensional data is hard, requiring a large number of examples, and is sensitive to outliers. One-class classifiers are an efficient alternative. Unlike the binary decisions output by traditional classifiers, the decisions output by one-class classifiers tell us whether examples were drawn from the distribution of the learned class. In this work we use *one-class SVM (OCSVM)* classifiers [2] to model the distribution of the background. Our decision to use OCSVM is motivated by its nonparametric

J.-I. Park and J. Kim (Eds.): ACCV 2012 Workshops, Part I, LNCS 7728, pp. 301–307, 2013.
© Springer-Verlag Berlin Heidelberg 2013

assumption about the expected distribution of the background, which allows us to effectively model highly complex scenes of various background distributions. In addition, its robustness to outliers during training allows us to efficiently construct models where some foreground objects may appear in the images.

In short, our model works as follows. Images are processed at three levels of resolution: *block-level*, *region-level*, and *frame-level*. We follow the approach used by Stauffer *et al.* [3] and assume that no foreground objects exist in the first n images. The first images are thus used to initialize the model. At the block-level, images are divided into b equal-sized blocks of pixels. Then, b OCSVM classifiers are independently trained on each block to model the distribution of its background. At the region-level, inter-block relationships are used to refine the OCSVM classification results. At the frame-level, an adaptive background method is used to re-initialize the model with regions that are considered with high confidence to be part of the background.

Our method performed very well on a variety of test videos in the Background Models Challenge (BMC) competition [1]. Two guidelines influenced our design and implementation decisions. First, in order that our model be effective in practice, we wanted it to rely on as few parametric assumptions as possible. Second, when faced with a trade-off between performance stability (under various background distributions) and precision, we were more likely to prefer the stable alternative. As a result, we were able to run all synthetic videos using the same configuration parameters. Almost the same parameters were used for all real videos as well.

2   Related Work

Our proposed background model belongs to a family of nonparametric models. Unlike parametric models, which make distributional assumptions about the background scene and will not perform well when the model does not fit the data, nonparametric models make no such assumptions and hence are more likely to perform well on a wider range of background distributions.

Background models can further be divided to pixel-based and block-based models. Pixel-based models, which are perhaps the most common, are used to model each pixel separately. For example, a nonparametric pixel-based model was introduced by Elgammal *et al.* [4], where kernel density estimators were used to model distributions of each pixel. This is in contrast to block-based models, which take a broader view of the problem: knowledge about groups of adjacent pixels (blocks) are used to model distributions. For example, in [5], the median and variance statistics over the background learning period are calculated for each block. A block is thus considered as a part of the background if it correlates positively with the background statistics.

Although block-based models are expected to have more stable performance than pixel-based models, they are less commonly used in practice, mainly because their resolution is limited to a block size. Indeed, this limitation may

---

[1] http://bmc.univ-bpclermont.fr

lead in some image sequences to a degradation in precision and an increased false-positive rate. However, recent improvement in efficiency of parallel computational methods and increased resolution of surveillance cameras might change this trend. Our introduction, in this work, of a nonparametric block-based model is a step in this promising direction. In addition, as far as we know, this work is the first to use OCSVM for this purpose.

3 One-Class Background Model

We now introduce the proposed one-class model to perform robust background subtraction.

3.1 Block Level

A one-class SVM classifier is built for each equal-sized block of $N \times N$ pixels [2]. We assume that a set of n block instances taken from the background scene is available to train the classifier. For color images, each block instance is represented by its intensity values as a feature vector of size $d = 3 \times N \times N$. A rough invariance to illumination changes is obtained by normalizing the feature vectors so that their sum equals one. The set of n d-dimensional feature vectors $x_1, \ldots, x_n$ is used as the training set.

The OCSVM classifier is a nonparametric approach for estimating the support vectors (SVs) of a high-dimensional distribution [2]. Suppose we use a mapping function $\Phi : \mathbb{R}^d \to \mathcal{F}$ to map the feature vectors to some other space $\mathcal{F}$ such that each mapped vector lies on a hypersphere. The basic concept is to treat the origin of this hypersphere as the only member of the second class, and to find the separating hyperplane between the classes with maximal margins. This optimization problem is solved under the condition that no more than a predefined quantile of training examples lies outside the hyperplane. At the end of the training phase, the signed distance between a mapped feature vector $\Phi(x)$ and the calculated separating hyperplane is

$$t(x) = -\sum_{i=1}^{n} \alpha_i K(x_i, x) + b \qquad (1)$$

where $\alpha_1, \ldots, \alpha_{q_c}$ are called the SVs' coefficients and $K(\cdot, \cdot)$ is a kernel function that serves as a dot product in $\mathcal{F}$, i.e., $K(x_i, x_j) = (\Phi(x_i) \cdot \Phi(x_j))_{\mathcal{F}}$ [3].

While not strictly true in theory, it is common to assume in practice that a larger distance $t(x)$ corresponds to a lower likelihood that x was drawn from the distribution of the learned class. The proposed one-class model uses this property to estimate the probability of a new block to be a part of the background. For this purpose, a cross-validation procedure over the training set is used to

---

[2] A block size of 5×5 pixels was used for all experiments.
[3] We use a Gaussian kernel $K(x_i, x_j) = e^{-\gamma(x_i - x_j)^2}$.

estimate the expected distances of background blocks. Then, given a distance $t(x)$ associated with a new block, the following nonparametric, one-tailed version of the Chebyshev inequality is used to bound the probability that this is a background block:

$$P\left(t(x) - \mu \geq k\sigma\right) \leq \frac{1}{1 + k^2} \tag{2}$$

where μ and σ are the expected mean and standard deviation of the distance of background blocks, and k is a real number greater than zero. New blocks with probability lower than a predefined threshold th are classified as foreground blocks. Since the distribution of background blocks may vary over time, a pooled variance procedure is used every fixed number of m blocks to update μ and σ with distances of new blocks that were classified as a part of the background.

3.2   Region Level

Given a binary image consisting of foreground and background pixels that were set at the block-level, we apply the following region-level process. Foreground pixels of each connected component in the image are replaced with their convex hull to get a hole-free component. Then, erosion and dilation operators are applied with a disk structuring size of 8 for erosion and 4 for dilation. We found that this process increases the recall with a reasonable decrease of precision.

3.3   Frame Level

To support model adaptation for abrupt changes in the background scene, detection results of the last n images are saved (recall that n is also the number of block instances used for training). When all n recent detections of a specific block have been classified as foreground, the model associated with this block is re-initialized by training its OCSVM classifier using these n blocks.

3.4   Technical Considerations

The number of blocks b is an external input parameter specified by the available computational resources [4]. Given b, each image is resized so that no more than b blocks of size $N \times N$ will fit into the image.

Experiments were conducted on 20 synthetic and 9 real videos taken from the BMC competition. On some of the synthetic videos [5], all blocks in the training set that correspond to the same location might be identical. This ill-conditioned input provides a meaningless solution and a constant distance is returned for all inputs. A simple ad-hoc solution that overcomes this problem is to classify a new block as foreground if it does not share the same values as in the training set. Although this trivial solution may lead to increased false-positive rates,

---

[4] Note that all blocks can be trained and run independently, suggesting straightforward, effective, parallel implementation.

[5] Synthetic videos $111, 112, 121, 122$; see Section 4 for details.

we decided to direct our time and efforts in other directions because it is very unlikely that this problem will actually occur.

For the synthetic videos, we treat the beginning of the evaluation sequences as extensions of the learning sequences. Due to abrupt changes at the beginning of most evaluation sequences, th is multiplied by a factor of 0.01 for the first n images, until the model is stabilized.

One last modification was made to remove temporal patterns from image sequences. For each pixel in a new image, its median RGB intensities over the last three images are used as inputs for the block-level instead of their original intensities. This modification slightly improved the precision for the snowy weather video (real video 5).

4   Experimental Results

Experiments were conducted on 10×2 synthetic and 9 real videos taken from the BMC competition [6]. The default parameters were as follows: The first $n = 150$ frames were used to initialize the model, thresholds were set to $th = 0.001$, and distance statistics were updated every $m = 50$ images. Exceptions are $n = 50$ for real videos $2, 3, 6, 8$ (due to the presence of foreground objects during training), and $th = 0.0005$ for real videos $1, 4, 6, 9$. For synthetic videos, images were divided into $b = 5000$ blocks of pixels. Due to runtime considerations, real videos were divided to $b = 500$ blocks [7]. Results, displayed in the format of the BMC competition, are provided in Table 1 and Table 2.

Table 1. Results for the learning phase (synthetic videos only)

-	Total				Dynamic 1 phase				Dynamic 2 phase				Static phase			
#	Rec.	Prec.	F	PSNR	Rec.	Prec.	F	PSNR	Rec.	Prec.	F	PSNR	Rec.	Prec.	F	PSNR
111	0.99	0.63	0.77	43.29	0.99	0.63	0.77	38.99	0.99	0.62	0.76	43.12	0.99	0.63	0.77	38.32
121	0.99	0.68	0.81	45.30	0.99	0.68	0.80	47.71	0.98	0.66	0.79	47.70	0.98	0.68	0.81	40.94
211	1.00	0.78	0.87	56.19	1.00	0.79	0.88	51.86	0.99	0.79	0.88	57.78	1.00	0.78	0.87	51.07
221	0.97	0.76	0.85	51.44	0.99	0.72	0.83	50.69	0.87	0.73	0.79	53.85	0.98	0.77	0.86	48.09
311	1.00	0.76	0.86	55.24	1.00	0.78	0.88	51.69	1.00	0.79	0.88	57.44	0.99	0.77	0.87	50.35
321	0.96	0.76	0.85	51.32	0.99	0.71	0.83	50.72	0.86	0.74	0.74	54.30	0.97	0.77	0.86	47.53
411	0.98	0.76	0.85	54.62	0.99	0.79	0.88	52.00	0.98	0.81	0.89	59.36	0.98	0.80	0.89	53.08
421	0.93	0.70	0.80	47.07	0.95	0.70	0.80	49.51	0.81	0.71	0.76	52.88	0.96	0.79	0.86	48.96
511	0.99	0.78	0.88	56.77	1.00	0.79	0.88	52.36	0.99	0.80	0.88	58.29	0.99	0.79	0.88	51.75
521	0.97	0.76	0.85	51.49	0.99	0.72	0.83	51.14	0.86	0.73	0.79	54.00	0.98	0.77	0.86	47.96

5   Discussion

Our decision to prefer stability over precision is illustrated in Figure 1 (blocks are bordered with white lines). It can be seen that while all foreground pixels are

---

[6] 10 synthetic videos are divided to learning and evaluation phases, 20 in total. See http://bmc.univ-bpclermont.fr/?q=node/6 for details.

[7] On a standard PC with an Intel Duo-core $2.4GHz$ 4G RAM processor, average training time for a block is 0.136 seconds (classification time is less than 1 millisecond).

Table 2. Results for the evaluation phase (synthetic and real videos)

	Synthetic video							Real video					
#	Rec.	Prec.	F	PSNR	D-Score	SSIM	#	Rec.	Prec.	F	PSNR	D-Score	SSIM
112	0.85	0.66	0.75	45.41	0.0035	0.99	Vid1	0.87	0.64	0.74	35.72	0.0126	0.97
122	0.88	0.70	0.78	42.28	0.0051	0.99	Vid2	0.67	0.63	0.65	21.58	0.0255	0.87
212	0.98	0.82	0.89	50.49	0.0019	1.00	Vid3	0.67	0.64	0.66	34.57	0.0215	0.95
222	0.89	0.82	0.85	46.65	0.0031	0.99	Vid4	0.97	0.68	0.80	39.16	0.0107	0.98
312	0.95	0.82	0.88	50.53	0.0018	1.00	Vid5	0.87	0.60	0.71	38.61	0.0085	0.97
322	0.85	0.82	0.83	46.74	0.0030	0.99	Vid6	0.84	0.64	0.73	25.65	0.0267	0.91
412	0.88	0.76	0.82	49.25	0.0020	0.99	Vid7	0.92	0.77	0.83	28.41	0.0215	0.93
422	0.81	0.75	0.78	45.18	0.0031	0.99	Vid8	0.59	0.53	0.56	24.04	0.0271	0.89
512	0.96	0.84	0.89	50.98	0.0018	1.00	Vid9	0.94	0.64	0.76	45.74	0.0069	0.99
522	0.85	0.85	0.85	47.42	0.0027	0.99							

Fig. 1. Recall: 1.0. Precision: 0.17 (Vid1)

detected, the precision drops dramatically due to misclassifications in the region surrounding the foreground object. Indeed, this drop in precision is not desired. However, we believe that regions of misclassified pixels adjacent to foreground objects are an acceptable price to pay to achieve stable results.

In general, although better results can be achieved by choosing different parameters for different videos, we preferred, for the sake of efficiency, to evaluate our model with almost the same configuration for all videos. The stability of our results can be found across almost all videos, especially the long ones where background scenes change over time (real videos 1, 5, and 9). Note that videos with degraded performance (real videos 2, 3, 6, and 8) are usually the shorter ones where foreground objects are present from the very first frame. When no semantic knowledge about foreground objects is available for modeling, it is hard to tell that these objects are not part of the background scene. In terms of its computational efforts, our model is easy to implement in real-time on standard computers.

References

1. Toyama, K., Krumm, J., Brumitt, B., Meyers, B.: Wallflower: Principles and practice of background maintenance. In: ICCV, vol. 1, pp. 255–261 (1999)
2. Schölkopf, B., Platt, J.C., Shawe-Taylor, J.C., Smola, A.J., Williamson, R.C.: Estimating the support of a high-dimensional distribution. Neural Computation 13, 1443–1471 (2001)
3. Stauffer, C., Grimson, W.: Adaptive background mixture models for real-time tracking. In: CVPR, vol. 2, pp. 637–663 (1999)
4. Elgammal, A., Duraiswami, R., Harwood, D., Davis, L.: Background and foreground modeling using nonparametric kernel density estimation for visual surveillance. Proceedings of the IEEE 90, 1151–1163 (2002)
5. Matsuyama, T., Ohya, T., Habe, H.: Background subtraction for non-stationary scenes. In: ACCV 2000, pp. 662–667 (2000)

Illumination Invariant Background Model Using Mixture of Gaussians and SURF Features

Munir Shah, Jeremiah Deng, and Brendon Woodford

Department of Information Science, University of Otago, Dunedin, New Zealand

Abstract. The Mixture of Gaussians (MoG) is a frequently used method for foreground-background separation. In this paper, we propose an on-line learning framework that allows the MoG algorithm to quickly adapt its localized parameters. Our main contributions are: local parameter adaptations, a feedback based updating method for stopped objects, and hierarchical SURF features matching based ghosts and local illumination suppression method. The proposed model is rigorously tested and compared with several previous models on BMC data set and has shown significant performance improvements.

1 Introduction

Precise localization of foreground objects is the most important building block of the higher-level computer vision applications including smart video surveillance, automatic sports video analysis, health care and interactive gaming [1]. However, an accurate foreground detection for complex visual scenes in real-time is a difficult task due to the intrinsic complexities of the real-world scenarios. The key challenges are: dynamic background, shadows, sudden illumination changes, bootstrap, camouflage and foreground aperture [2].

Recently, important research efforts have been made in developing methods and systems for detecting foreground objects for complex video streams [1, 3–5]. The Mixtures of Gaussians (MoG) is the most popular background model among the community due to its robustness for multi-model backgrounds, and gradual illumination changes [1, 3]. Thus, it is widely adopted as a basic framework in many subsequent models [6, 7]. However, it is a parametric model and in-order to get satisfactory results one has to manually tune the parameters for typical scene, which is a tedious task and makes this less attractive for real-time applications [1, 8].

In the last decade, several improvements have been proposed for MoG parameter learning. The Dirichlet prior [6], stochastic approximation procedure [9], mixture weight and particle swarm [8] based approaches are adopted for optimal setting of the number of mixture components. In [7], an adaptive learning rate for each Gaussian component is adopted. Furthermore, the MoG [3] is enhanced in [4] to address its slow learning issue. They used different updating equations for initial training and on-line updating to make it robust for dynamic scenes. Moreover, they also presented a shadow detection technique using brightness

J.-I. Park and J. Kim (Eds.): ACCV 2012 Workshops, Part I, LNCS 7728, pp. 308–314, 2013.

and chromatic distortion cue's. Nevertheless, reliable moving object detection in complex visual scenes is still an open problem [1].

In this paper, we propose an on-line adaptive background learning model for better foreground detection in complex video scenes. The main contributions are as follows: Firstly, we introduce a new localized learning algorithm using some of the recent random samples for the periodic re-learning and a local frequency of change for optimal parameters selection. Secondly, a novel background update algorithm for handling paused objects in the scene is proposed. We use temporal foreground history as feedback to adjust foreground updating to prevent incorporating sleeping objects into the background. Fourth, we introduce a new matching function by separately modeling intensity and color cue's. Fifth, a novel local illumination (shadows and lighting) and ghosts suppression method is proposed using SURF [10] features matching. We have rigorously tested and compared of our proposed model with several previous techniques on background model challenge (BMC'12) data set [1]. The propose model achieves significantly better results as compared to the previous models.

2   The Proposed Method

The main components of the propose models are described as follows.

2.1   Background Model

We have used MoG framework [3] as a base for our proposed model and built on it. In our proposed model each pixel is characterized by YUV color feature and probability of observing the current pixel value is given as follows:

$$P(X_t) = \sum_{i=1}^{K} \omega_{i,t}.\eta(X_t, \mu_{k,t}, \Sigma_{k,t}) \tag{1}$$

Here X_t is YUV color feature of the current pixel, K is the number of components, $\omega_{i,t}$ is a weight associated to the ith component, μ_k is mean, Σ_k standard deviation of pixel values and η is Gaussian probability density function. The first few video frames are used for initial training by employing the EM algorithm and very first frame is used to initialize the model by setting mean of the first component to the pixel value and variances to some higher value and weight to 1 for the first component and to 0 for all others. It should be noted that mixture weights are non-negative and add up to one. To avoid the costly matrix inversion, it is assumed that dimensions of X are statistically independent and identical (i.i.d) and represented covariance matrix as ($\Sigma_k = \sigma_{i,t}^2 I$). However, we computed separate variance for color (UV) and intensity (Y) channels.

It has been observed from several real video data that distribution of pixels' has higher variance in intensity channel and relatively lower variations in color

---

[1] http://bmc.univ-bpclermont.fr/

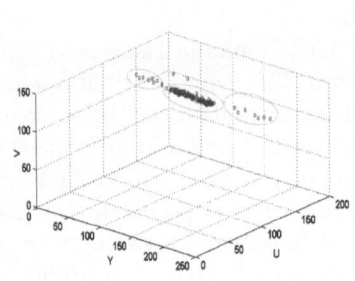

Fig. 1. Distribution of pixel data in YUV color space

Fig. 2. Example demonstrating the effectiveness of SURF features matching for ghosts suppression

channels as shown in Fig.1. This is a common trend specially if data is represented in YUV color space [2]. We propose a new match function by computing separate variances for color and intensity channels. A new match function is formulated as follows:

$$M_{i,t}^k = \begin{cases} 1, \text{if } (|\mu_{i,t}^{k,I} - X_{i,t}^I| < \lambda_{i,t}\sigma_{i,t}^{2,k,I}) \wedge (|\mu_{i,t}^{k,c} - X_{i,t}^c| < \lambda_{i,t}\sigma_{i,t}^{2,k,c}) \\ 0, \text{Otherwise} \end{cases} \quad (2)$$

where k, i and t are the indexes of Gaussian component, pixel number in a frame and frame number in a video respectively whereas $X_{i,t}^I=Y$ is an intensity of the pixel and $X_{i,t}^c=\{U,V\}$ is a color vector of the pixel. Here, the match function is 1 for the closest component to the pixel data and 0 for all others.

Background pixels appear more frequently than the foreground ones, thus the components are arranged in a descending order by the rank $R_k = \omega_k/\sigma_k$, and the first B components having cumulative posterior probability greater than the threshold T are considered background as follows:

$$B = \operatorname{argmin}_b(\sum_{k=1}^{b} \omega_k > T). \quad (3)$$

For each input video frame, pixel value is matched against already learned models. If the matching component is among the first B components then it is classified as background, otherwise as foreground.

2.2 SURF Features Based Ghost Suppression

Local illumination changes (shadows and lighting) are common phenomena in real world videos and have similar motion properties as foreground, which causes

---

[2] The YUV color space represent intensity and color information separately thus a better choice for modeling the underlying data.

large number of false positives (ghosts). In this section we introduce SURF [10] features based approach to remove the ghosts from the foreground map.

The foreground regions (bounding box of the blobs) of the current image and background image generated from the highest rank components of the background model are used for SURF feature detection and matching. For every foreground blob SURF features are matched and irrelevant features are removed using RANSAC sampling. If there is an enough evidence of similarity between background and current image we classify that blob as a background and stop further processing. Otherwise, we divide the foreground blob into 4 equal size blocks and repeat the same matching process recursively until the block is classified as background or its size reaches to some minimum area. The effectiveness of ghost removing techniques can be seen in Fig.2.

2.3   Background Model Update Algorithm

If there is no match with any of the existing component, then the least probable Gaussian is replaced by setting mean to a the pixel data, σ to some higher value (Σ_0), and weight to a small value (ω_{init}). The parameters are then updated as follows:

$$\mu_{i,t+1}^{k,c} = (1-\rho)\mu_{i,t}^{k,c} + \rho X_{i,t}^{c}, \quad (4) \qquad \mu_{i,t+1}^{k,I} = (1-\rho)\mu_{i,t}^{k,I} + \rho X_{i,t}^{I}, \qquad (5)$$

$$\sigma_{k,t+1}^{2,k,I} = (1-\rho)\sigma_{k,t}^{2,k,I} + \rho((\mu_{i,t}^{k,I} - X_{i,t}^{I})o(\mu_{i,t}^{k,I} - X_{i,t}^{I})), \qquad (6)$$

$$\sigma_{k,t+1}^{2,k,c} = (1-\rho)\sigma_{k,t}^{2,k,c} + \rho((\mu_{i,t}^{k,c} - X_{i,t}^{c})o(\mu_{i,t}^{k,c} - X_{i,t}^{c})), \qquad (7)$$

$$\rho = \frac{\alpha_{i,t}P(k|X_t,\Theta)}{\omega_{k,t+1}}. \qquad (8)$$

where $P(k/X_t,\Theta)$ is the likelihood for component k given the pixel value (X_t) and Gaussian parameters (Θ). Here $P(k/X_t,\Theta)$ is 1 for winning component and 0 for all others.

The presence of sleeping objects is an other issue that confounds traditional models. The adaptive models quickly adapt to the changing condition and therefore incorporate stopped objects into the background model due to the blind update mechanism. A new background updating technique is proposed here to protect paused objects from being incorporated into the background. Foreground blobs having a traceable history of spatial and temporal movement are detected as paused objects and the models corresponding to these objects are adapted as follows:

$$\omega_{t+1}^{k} = \begin{cases} \omega_t^k + \epsilon, & \text{if } M_{i,t}^k == 1 \\ \omega_t^k, & \text{otherwise} \end{cases} \qquad (9)$$

where ϵ is fixed global learning rate set to very small value. Our proposed technique increases the weight of winning Gaussian a little and keep the weights for all others the same. In this way proposed model is more resistive in adding foreground objects into the background model. Non-sleeping pixels are updated as follows:

$$\omega_{t+1}^k = (1 - \alpha_{i,t})\omega_t^k + \alpha_{i,t}M_{i,t}^k, \tag{10}$$

2.4   Automatic Parameter Selection

To enable automatic parameter adaptation we use a fixed length sliding window to keep the most recent N frames in order to capture ongoing statistical changes in a video. Empirically we find a setting of $10 \leq N \leq 20$ is sufficient.

Learning Rate (α): Learning rate α controls the convergence speed of the model. The optimal value for the learning rate depends on the given background scene. A static background needs a small learning rate, whereas a dynamic background scene require higher α [7]. Complex dynamic scenes are hard to model by using single global α therefore we employ an adaptive α defined locally for each pixel as a follows:

$$f_{i,t} = \begin{cases} f_{i,t-1} + 1, & \text{if } \delta_{i,t} > \varphi \\ f_{i,t-1}, & \text{otherwise} \end{cases} \tag{11} \qquad \alpha_{i,t} = \frac{\sum_{k=1}^N f_{i,k}}{K_{i,t}N}, \tag{12}$$

where δ_n is an absolute difference in pixel values of consecutive frames and φ is a threshold used to avoid changes due to noise, typically set to some small value. The local learning rate is then modeled as shown in Eq. (12) where $K_{i,t}$ is the current number of Gaussian components for pixel i. Hence the adaptive α does not depend on the initial settings but will adapt to the scene dynamics rapidly.

Deviation Threshold (λ): The deviation threshold λ is used to avoid various video acquisition noises such as sensor noise, weather condition and auto focus. We again propose to use local adaptive deviation thresholds and relate them to the local intensity difference δ_i as follows.

$$\gamma_i = \begin{cases} \delta_{i,t}, & \text{if } \delta_{i,t} \leq 2.5\lambda_{i,t-1} \\ \lambda_{i,t-1}, & \text{otherwise} \end{cases} \quad , \tag{13} \qquad \lambda_{i,t} = \frac{1}{N}\sum_{k=1}^N \gamma_{i,k} \tag{14}$$

It should be noted that only the change due to noise is used whereas large changes are considered as outliers and thus ignored. The deviation threshold is then smoothened as shown in Eq.(14).

3   Experimental Results

The proposed model is rigorously evaluated on the data-set provided for background models challenge (BMC) data-set. This data-set contains number of synthetic and real world video sequences both for indoor and out door environment. Furthermore, for performance evaluation human annotated ground truths are

Table 1. Results on synthetic video dataset

Video	Recall	Precision	F-measure	PSNR	D-score	SSIM
112	0.94773	0.86402	0.90394	51.85110	0.00128	0.99590
122	0.93307	0.89738	0.91487	48.20490	0.00176	0.99359
212	0.92327	0.89341	0.90810	52.70680	0.00108	0.99612
222	0.92826	0.91176	0.91994	48.68980	0.00157	0.99374
312	0.86582	0.90613	0.88552	53.13000	0.00113	0.99574
322	0.78298	0.93670	0.85297	49.41880	0.00193	0.99214
412	0.84390	0.74639	0.79216	49.19960	0.00105	0.99091
422	0.79753	0.74788	0.77191	45.32370	0.00174	0.98623
512	0.92118	0.76930	0.83842	49.17320	0.00123	0.99236
522	0.91955	0.89088	0.90499	48.22730	0.00169	0.99275

Table 2. Results on real video dataset

Video	Recall	Precision	F-measure	PSNR	D-score	SSIM
001	0.70782	0.70391	0.70175	35.61480	0.00859	0.96129
002	0.83300	0.78968	0.81076	30.00140	0.01102	0.94430
003	0.93272	0.86177	0.89584	46.59210	0.00445	0.98359
004	0.83701	0.89368	0.86442	51.90270	0.00304	0.99063
005	0.73716	0.72640	0.71420	37.45680	0.00284	0.95699
006	0.80093	0.81573	0.80826	35.93770	0.01111	0.95763
007	0.75143	0.75175	0.75159	27.73580	0.01453	0.95243
008	0.68647	0.64794	0.66655	39.11460	0.00723	0.95004
009	0.69182	0.70130	0.69652	53.09370	0.00284	0.99185

provided for all the sequences. We compared our proposed model "illumination invariant background model (IIBM)" with three previous techniques: MoG [3], AMoG [6] and MoG-SH [4]. These models are evaluated using the parameters values mentioned in the original papers whereas our proposed model automatically learn optimal setting of parameters from the data, therefore in that sense it can be seen as a non-parametric model. Furthermore, we used standard performance metrics Precision, Recall and F-measure for quantitative and PSNR, SSIM and D-Score for qualitative analysis. [3]

Table 1 and 2 shows the results for synthetic and real video sequences. For synthetic dataset IIBM gives on average more than 85% accuracy both for qualitative and quantitative performance measures. Whereas for real video data-set IIBM achieves more than 80% accuracy which is slightly less than synthetic data-set. Our propose model (IIBM) gives about 6% improvement above the best reported previous model (MoG-SH) as shown in Table 3. In summary, our proposed model achieve significantly better results on most of the performance evaluation metrics. The processing time for the proposed model running on an

---

[3] Peak signal-to-noise ratio (PSNR), structural similarity index measure (SSIM) and dis-similarity criteria (D-Score).

Intel (R) Core(TM) 2 (2.26 GHz Quad core CPU) machine was about 10 FPS for
BMC data-set (average). Thus, it can easily be applied in real-time application.

Table 3. Overall comparative results on BMC dataset

Video	Recall	Precision	F-measure	PSNR	D-score	SSIM
MoG [3]	0.82230	0.69273	0.74876	37.03239	0.00980	0.93598
AMoG [6]	0.82950	0.72803	0.77268	40.28893	0.00752	0.95627
MoG-SH [4]	0.70812	**0.81325**	0.75556	44.43956	0.00468	0.97206
IIBM	**0.83085**	0.81109	**0.81852**	**46.32124**	**0.00337**	**0.98363**

4   Conclusion

In this paper we presented an enhanced MoG background model by introducing
an online and self-adaptive mechanism for automatic selection of the parameters
and a novel match function. is presented. Furthermore, traceable temporal and
spatial history of foreground blobs is used as a feedback to detect and handle
paused objects by adjusting update speed. In the last step, the SURF features
matching is used to remove ghosts due to illumination changes.

References

1. Bouwmans, T., El Baf, F., Vachon, B.: In: Statistical Background Modeling for Foreground Detection: A Survey. World Scientific Publishing (2010)
2. Kentaro, T., John, K., Barry, B., Brian, M.: Wallflower: Principles and practice of background maintenance. In: Proc. IEEE International Conference on Computer Vision, CCV 1999, vol. 1, p. 255 (1999)
3. Stauffer, C., Grimson, W.: Adaptive background mixture models for real-time tracking. In: Proc. IEEE Computer Society Conference on Computer Vision and Pattern Recognition, CVPR 1999, vol. 2, pp. 246–252 (1999)
4. KaewTraKulPong, P., Rowden, R.: An improved adaptive background mixture model for real-time tracking with shadow detection. In: Proceedings of the Second European Workshop on Advanced Video Based Surveillance Systems, pp. 149–158 (2001)
5. Shah, M., Deng, J., Woodford, B.: Localized adaptive learning of mixture of gaussians models for background extraction. In: 2010 25th International Conference of Image and Vision Computing, IVCNZ, New Zealand, pp. 1–8 (2010)
6. Zivkovic, Z.: Improved adaptive gaussian mixture model for background subtraction. In: Proc. 17th International Conference on Pattern Recognition, ICPR 2004, vol. 2, pp. 28–31 (2004)
7. Lee, D.S.: Effective gaussian mixture learning for video background subtraction. IEEE Transaction on Pattern Analysis and Machine Intelligence 27, 827–832 (2005)
8. White, B., Shah, M.: Automatically tunings background subtraction parameters using particle swarm optimization. In: Proc. IEEE International Conference on Multimedia and Expo, pp. 1826–1829 (2005)
9. Cheng, J., Yang, J., Zhou, Y., Cui, Y.: Flexible background mixture models for foreground segmentation. Image and Vision Computing 24, 473–482 (2006)
10. Bay, H., Ess, A., Tuytelaars, T., Van Gool, L.: Speeded-up robust features (SURF). Comput. Vis. Image Underst. 110, 346–359 (2008)

Foreground Detection via Robust Low Rank Matrix Decomposition Including Spatio-Temporal Constraint

Charles Guyon, Thierry Bouwmans, and El-Hadi Zahzah

Laboratoire MIA (Mathematiques, Image et Applications) - University of La Rochelle
thierry.bouwmans@univ-lr.fr

Abstract. Foreground detection is the first step in video surveillance system to detect moving objects. Robust Principal Components Analysis (RPCA) shows a nice framework to separate moving objects from the background. The background sequence is then modeled by a low rank subspace that can gradually change over time, while the moving foreground objects constitute the correlated sparse outliers. In this paper, we propose to use a low-rank matrix factorization with IRLS scheme (Iteratively reweighted least squares) and to address in the minimization process the spatial connexity and the temporal sparseness of moving objects (e.g. outliers). Experimental results on the BMC 2012 datasets show the pertinence of the proposed approach.

1 Introduction

The detection of moving objects is the basic low-level operations in video analysis. This detection is usually done using foreground detection. This basic operation consists of separating the moving objects called "foreground" from the static information called "background". Recent research on robust PCA shows qualitative visual results with the background variations approximatively lying in a low dimension subspace, and the sparse part being the moving objects. First, Candes et al. [1] proposed a convex optimization problem to address the robust PCA problem. The observation matrix is assumed represented as: $A = L + S$ where L is a low-rank matrix and S must be sparse matrix with a small fraction of nonzero entries. This research seeks to solve for L with the following optimization problem:

$$\min_{L,S} \; ||L||_* + \lambda ||S||_1 \quad \text{subj} \quad A = L + S \tag{1}$$

where $||.||_*$ and $||.||_1$ are the nuclear norm (which is the L_1 norm of singular values) and l_1 norm, respectively, and $\lambda > 0$ is an arbitrary balanced parameter. Under these minimal assumptions, this approach called Principal Component Pursuit (PCP) solution perfectly recovers the low-rank and the sparse matrices.

In this paper, we propose a robust low-matrix factorization with IRLS scheme to adress the second limitation. For a data matrix A containing the sequence, we assume that a part is approximatively low-rank and product of two matrices, and

J.-I. Park and J. Kim (Eds.): ACCV 2012 Workshops, Part I, LNCS 7728, pp. 315–320, 2013.
© Springer-Verlag Berlin Heidelberg 2013

a small part of this matrix is corrupted by the outliers. Furthermore, we directly introduced a spatial term in the l_1 minimization to address the spatial connexity of the pixels. So, our contributions can be summarized as follows: 1) Addition of spatial constraint to minimization process, 2) IRLS alternating scheme for weighted the 2-parameters $||.||_{\alpha,\beta}$ for matrix low-rank decomposition. The rest of this paper is organized as follows. The Section 2 focus on IRLS method applied on vector regression problems. In Section 3, we present a robust low-rank matrix factorization which allows us to detect foreground objects in dynamic backgrounds. In Section 4, we present results on the BMC 2012 datasets[1] and the Section 5 provides the conclusion.

2 L_p Minimization with Spatial Constraint

In most applications, video surveillance data is assumed to be compose of background, foreground and noise. Regression task is a crucial part of the proposed decomposition algorithm. We consider the following minimization problem (2), where A is a dictionary matrix (row order) and b is a row vector, the second term forces the error E to be a connexe shape, through the TV (Total Variation) of the residual must be small, where the matrix ∇_s is a spatial gradient.

$$\underset{x}{\operatorname{argmin}} \ ||Ax - b||_\alpha + \lambda ||\nabla_s(Ax - b)||_1 \tag{2}$$

The left part of the problem (3) is convex for $\alpha > 1$ and the usual IRLS (Iteratively reweighted least squares) scheme for solve $\underset{x}{\operatorname{argmin}} \ ||Ax - b||_\alpha$ is given by

$$\left| \begin{array}{l} D^{(i)} = \operatorname{diag}((\varepsilon + |b - Ax^{(i)}|)^{\alpha-2}) \\ x^{(i+1)} = (A^t D^{(i)} A)^{-1} A^t D^{(i)} b \end{array} \right. \tag{3}$$

It was proven that a suitable IRLS method is convergent for $1 \leq \alpha < 3$ [2]. Since if the process is expressed with a residual formulation, we gain more numericaly stability and let us to choose freely $\alpha \in [1, \infty[$ with an adapted step size λ_{opt} on every iteration.

$$\left| \begin{array}{l} r^{(i)} = b - Ax^{(i)} \\ D = \operatorname{diag}((\varepsilon + |r^{(i)}|)^{\alpha-2}) \\ y^{(i)} = (A'DA)^{-1} A'Dr^{(i)} \\ x^{(i+1)} = x^{(i)} + (1 + \lambda_{opt})y^{(i)} \end{array} \right. \tag{4}$$

With a fixed λ_{opt}, we should choose λ_{opt} as developed [3].

Otherwise, the algorithm is twice iterative, where we try to get an optimal x and an optimal λ at each step.

$$\left| \begin{array}{l} c^{(i)} = Ay^{(i)} \\ d^{(i)} = b - A(x^{(i)} + y^{(i)}) \\ \underset{\lambda_{opt}}{\operatorname{argmin}} \ ||c^{(i)}\lambda - d^{(i)}||_\alpha \end{array} \right. \left| \begin{array}{l} \lambda^{(0)} = \Lambda(\alpha) \\ s^{(k)} = d - \lambda^{(k)}c \\ E = \operatorname{diag}((\varepsilon + |s^{(k)}|)^{\alpha-2}) \\ z^{(k)} = \frac{c^t E s^{(k)}}{c^t E c} \\ \lambda^{(k+1)} = \lambda^{(k)} + (1 + \Lambda(\alpha))z^{(k)} \end{array} \right. \tag{5}$$

---

[1] http://bmc.univ-bpclermont.fr

Only few iterations (≈ 10) is enough for acceptable approximation of λ_{opt} of the $\lambda^{(k)}$ sequence. Moreover, the convergence is usually improved by a Aitken process or an other acceleration technique. Note for case $\alpha > 2$, convergence is achieved when $0 < 1 + \lambda < \frac{2}{\alpha-1}$. Additionally, TV is particular case of the following problem:

$$\underset{x}{\text{argmin}} \ ||Ax - b||_\alpha + \lambda ||Cx - d||_\beta \tag{6}$$

By derivation, the associated IRLS scheme is,

$$\left|\begin{array}{l} r_1 = b - Ax^{(i)}, r_2 = d - Cx^{(i)}, e_1 = \varepsilon + |r_1|, e_2 = \varepsilon + |r_2| \\ D_1 = (\sum e_1^\alpha)^{\frac{1}{\alpha}-1}\text{diag}(e_1^{\alpha-2}), D_2 = \lambda(\sum e_2^\beta)^{\frac{1}{\beta}-1}\text{diag}(e_2^{\beta-2}) \\ y^{(i)} = (A'D_1A + C'D_2C)^{-1}(A'D_1r_1 + C'D_2r_2) \\ x^{(i+1)} = x^{(i)} + (1 + \lambda_{opt})y^{(i)} \end{array}\right. \tag{7}$$

More generally, we consider the following matrix regression problem with two parameters norm (α, β) and a weighted matrix (W),

$$\underset{X}{\min} \ ||AX - B||_{\alpha,\beta \atop W} \quad \text{with} \quad ||M_{ij}||_{\alpha,\beta \atop W} = (\sum_{i=1}^{n}(\sum_{j=1}^{m} W_{ij} |M_{ij}|^\beta)^{\frac{\alpha}{\beta}})^{\frac{1}{\alpha}} \tag{8}$$

The problem is solved in the same manner on matrices with a reweighted regression strategy,

$$\begin{array}{l} \text{Until X is stable, repeat on each } k\text{-columns} \\ \left|\begin{array}{l} R \ \leftarrow B - AX \\ S \ \leftarrow \varepsilon + |R| \\ D_k \ \leftarrow \text{diag}(S_{ik}^{\beta-2} \circ (\sum_j(S_{ij}^\beta \circ W_{ij}))^{\frac{\alpha}{\beta}-1} \circ W_{ik})_k \\ X_{ik} \ \leftarrow X_{ik} + (1 + \Lambda(\max(\alpha, \beta)))(A^tD_kA)^{-1}A^tD_kR_{ik} \end{array}\right. \end{array} \tag{9}$$

3   Foreground Detection via Robust Low-Rank Matrix Factorization and Temporal Constraint

The training video sequence $A \in \mathbb{R}^{n \times m}$ is stored as a matrix with a particular structure. Columns are spatial frames and rows are values of a fixed pixel over time. For $A = \{I_1, \ldots, I_m\}$, I_j denotes a vectorized frame of n pixels at j-time with m is the number of frames. $A_{y+hx,t}$ implies the pixel intensity at coordinate x, y, t. The background modeling process finds an ideal subspace of the video sequence, which describes the best as possible the (dynamic) background as shown in Fig. 2. Then, the decomposition involves the following model:

$$A = L + S = BC + S \tag{10}$$

where B is a *low-rank* matrix corresponding to the background model plus noise and C allows to approximate L by linear combination. S is a *sparse* matrix which corresponds to the foreground component obtained by subtraction.

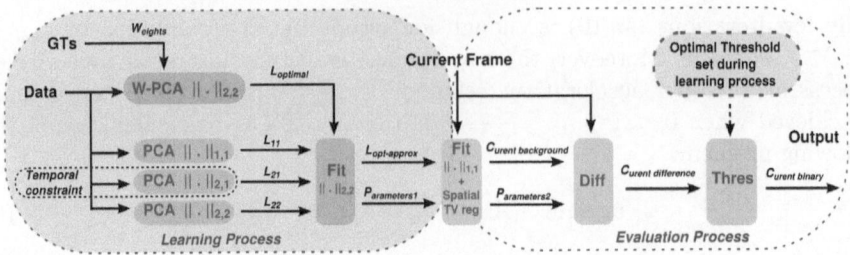

Fig. 1. Overview of the learning and evaluation process. Learning process needs GT (Groundtruth) for better fits the eigenbackground components.

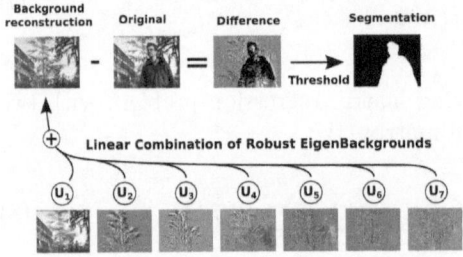

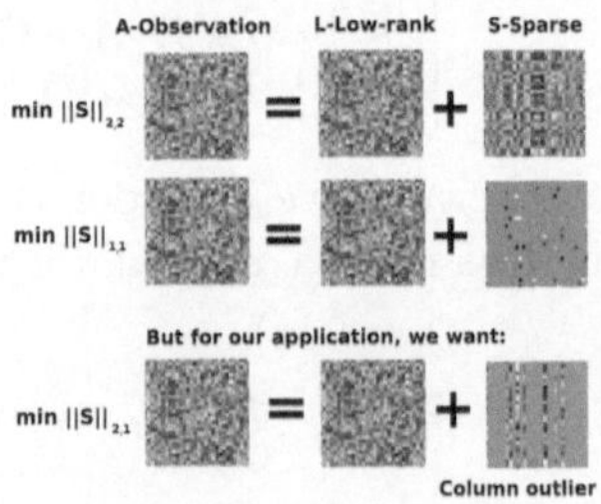

Fig. 2. At left: The common process of background subtraction via PCA (Principal Component Analysis). At final step, an adaptive threshold is used to get a binary result.

Fig. 3. At right: Using the previous decomposition on a low-rank random matrix plus noise, different kind of pattern on residual matrix emerge with the choice of the norm

The model involves the error reconstruction determined by the following constraints:

$$\min_{B \in \mathbb{R}^{n \times p}, C \in \mathbb{R}^{p \times m}} ||(A - BC) \circ W||_{\alpha,\beta} + \mu||BC||_* \qquad (11)$$

where $||.||_*$ denote the nuclear norm. The decomposition is split into two parts. Firstly, we track 1-Rank decomposition since the first eigen-vector is strongly dominant in video surveillance.

$$
\begin{array}{c|c}
R_1 = A - B_1 C_1 & \min_{B_1, C_1} ||R_1||_{1,1} \\
R = A - B_1 C_1 - B_r C_r & \min_{B_r, C_r} ||R \circ \phi(R_1)||_{2,1 \to 0}
\end{array} \qquad (12)
$$

We use $||.||_{2,1 \to 0}$ instead of usual $||.||_{1,1}$ because it forces spatial homogeneous fitting. Besides $\beta = (1 \to 0)$ means the β parameter decreases during iteration. First, we search a solution of the convex problem $||.||_{2,1}$, then use the solution as an initial guess for non-convex problem $||.||_{2,(1-\varepsilon)}$. Finally, we find a local minimum of $||.||_{2,0}$ and hope that is near of the global minimum of this problem. Furthermore, this norm enforce temporarily sparseness of outliers as shown

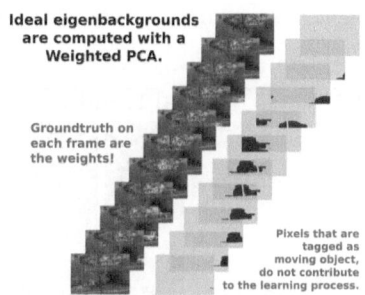

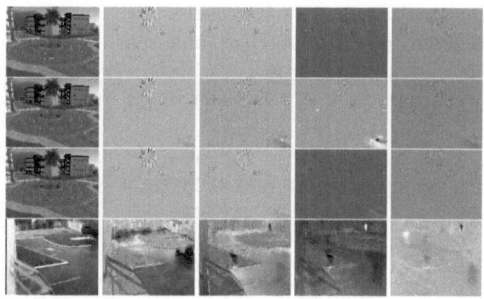

Fig. 4. At Left: Schema ofideal PCA processing. The eigenbackground are computed using a Weighted-PCA with GroundTruth.

Fig. 5. At right: First eigenBackground of the fifths sequence of Rotary (BMC) with the norms $||.||_{opt}$, $||.||_{1,1}$ and $||.||_{2,1}$. Last row shows eigenBackground on real dataset with $||.||_{2,1}$.

in Fig.4. In the case where $\alpha = \beta = 2$, the decomposition is usually solved by a SVD (Singular Value Decomposition). Thus, our SVD algorithm can be seen as an iterative regression. The proposed scheme determines alternatively the optimal coefficients, it means searching C for B fixed and searching B for C fixed.

$$C^{(k+1)} = (A^t A)^{-1} A^t B^{(k)}$$
$$\bar{C}^{(k+1)} = C^{(k+1)} \sqrt{C^{t(k+1)} C^{(k+1)}}^{-1} \qquad (13)$$
$$B^{(k+1)} = (A^t A)^{-1} A^t \bar{C}^{(k+1)}$$

Additionnaly, this alternating regression framework allows to associate a weigthed matrix W which is entrywise multiplied to the error term,

$$\min_{B,C} ||(A - BC) \circ W||_{\alpha,\beta} \qquad (14)$$

The W mask is iteratively computed and aims to enforce the fit exclusively on guessed background region.

We define a function ϕ that have two goals, smooth the error (like spatial median filtering) and transform the error for obtain a suitable weighted mask for regression.

$$W = \phi(|A - BC|) \ , \ \phi(E) = e^{-\gamma TV(E)} \qquad (15)$$

By including local penalty as a constraint in RPCA, this explicitly increases local coherence of the sparse component as filled/plain shapes (therefore moving object).

4   Experimental Results

Here, we show experimental results on the real dataset of BMC,

Table 1. Quantitative results with common criterions. Last column show the original, GT and result of the first four real video sequences.

Video	Recall	Precision	F-measure	PSNR	Visual Results
1	0.9139	0.7170	0.8036	38.2425	
2	0.8785	0.8656	0.8720	26.7721	
3	0.9658	0.8120	0.8822	37.7053	
4	0.9550	0.7187	0.8202	39.3699	
5	0.9102	0.5589	0.6925	30.5876	
6	0.9002	0.7727	0.8316	29.9994	
7	0.9116	0.8401	0.8744	26.8350	
8	0.8651	0.6710	0.7558	30.5040	
9	0.9309	0.8239	0.8741	55.1163	

5   Conclusion

In this paper, we have presented a robust matrix factorization for foreground detection. This method is conceptually simple, easy to implement and efficient. Furthermore, experiments on video surveillance datasets show that this approach is more robust than recent RPCA approaches in the presence of dynamic backgrounds and illumination changes. Further research consists in developping an incremental version to update the model at every frame and to achieve real-time requirements.

References

1. Candes, E., Li, X., Ma, Y., Wright, J.: Robust principal component analysis? International Journal of ACM 58 (2011)
2. Osborne, M.R.: Finite algorithms in optimization and data analysis. John Wiley & Sons (1985)
3. Guyon, C., Bouwmans., T., Zahzah, E.: Foreground detection via robust low rank matrix factorization including spatial constraint with iterative reweighted regression. In: International Conference on Pattern Recognition, ICPR 2012 (2012)

Temporal Saliency for Fast Motion Detection

Hamed Rezazadegan Tavakoli, Esa Rahtu, and Janne Heikkilä

Center for Machine Vision Research, University of Oulu, Oulu, Finland

Abstract. This paper presents a novel saliency detection method and apply it to motion detection. Detection of salient regions in videos or images can reduce the computation power which is needed for complicated tasks such as object recognition. It can also help us to preserve important information in tasks like video compression. Recent advances have given birth to biologically motivated approaches for saliency detection. We perform salience estimation by measuring the change in pixel's intensity value within a temporal interval while performing a filtering step via principal component analysis that is intended to suppress noise. We applied the method to Background Models Challenge (BMC) video data set. Experiments show that the proposed method is apt and accurate. Additionally, the method is fast to compute.

1 Introduction

Salient regions of a scene are regions that are important to tasks such as object recognition, surveillance, event detection, video compression and video retargeting. In video processing, common approach to detection of salient regions is background subtraction (*e.g.* [1–4]).

In videos, detection of salient region (*i.e.* saliency detection) is highly dependant on recognition of salient motion (*i.e.* the motion that attracts attention). Salient motion depends on environment's dynamics which makes saliency detection in videos a challenging problem.

Visual attention theories have had central role in psychology and neuroscience for ages. Recent advances in biologically inspired system engineering have led to development of remarkably effective methods for relating visual attention to relevant salient regions. These techniques can be applied to the problem of salient region detection in images and videos.

Method of [5] is one of the first publications in this area. Their approach is based on extracting early visual features (*e.g.* colors, orientations, edges, ...) and fusing them into a saliency map using center-surround technique. Later, many publications adapt center-surround technique because of it biological plausibility and effectiveness [6].

Itti and Baldi [7] define video saliency in terms of surprising stimulus, measured as the Kullback-Leibler (KL) divergence between posterior and prior beliefs of an observer. In [8], a stochastic method is introduced where saliency of each video sequence is treated as a Markovian process. Rahtu *et al.* [9] exploit Bayesian inference to derive a saliency detector for both images and videos.

J.-I. Park and J. Kim (Eds.): ACCV 2012 Workshops, Part I, LNCS 7728, pp. 321–326, 2013.
© Springer-Verlag Berlin Heidelberg 2013

Mahadevan *et al.* [10] introduced a spatio-temporal method which utilizes dynamic texture model of [11] and KL divergence criteria to measure saliency in a patch. Based on center-surround theory, they measure saliency in terms of KL divergence value which shows the disparity of center and surround in an image patch. Evaluating saliency in a video frame requires several patch evaluations. This makes the method a computationally intensive one.

Later, Gopalakrishnan, *et al.* [12] adapt dynamic texture model used in [10] to propose a pure temporal method. They apply observability measure [13] to determine saliency of video frames exploring pixel state-space model by assuming a video sequence follows a Multi Input Multi Output (MIMO) state-space model.

In this paper, we introduce a method of estimating motion saliency in the context of Background Models Challenge (BMC). The approach is based on temporal cues obtained using frame decorrelation. The method was evaluated thanks to the BMC data set. Our analysis shows that the proposed method is fast and accurate.

2   Method

In human vision system, it is shown that neurons adapt to small changes in visual perception in small temporal windows in order to reduce dependencies between neural responses. To replicate this phenomena, principal component analysis (PCA) with whitening can be applied to reduce amount of redundant information in image sequences to ease the realization of salient object. Moreover, assuming that the noise is Gaussian, PCA suppresses noise.

In this paper, we focus on computing temporal salience maps which presents amount of motion in a frame. Assuming that salient motion is steady, we concentrate on detection of firm movements. Hence, we apply principal component analysis (PCA) procedure to reconstruct the video buffer extracted from video sequence while suppressing background clutter and noise. Figure 1 summarizes the process as a work flow.

Let us assume that we have an image sequence $F = \{f_t, f_{t+1}, \ldots, f_{t+n}\}$, where f_i is the luminance of video frame at time i in column representation obtained in LMS color space. Initially, we subtract the mean value μ_f of each row of image sequence; it represents the static information of sequence (*i.e.* background). Consequently, $\tilde{F} = F - \mu_f$ will provide an approximation of movements (*i.e.* change over time) in the scene.

Later, we apply eigendecomposition to covariance of $\tilde{F}$, $\Sigma = EDE^T$. Approximation of image sequence is obtained by applying PCA and computing the back projected sequence $\tilde{F}_p$ as follows:

$$\tilde{F}_p = E_p E_p^T \tilde{F}. \tag{1}$$

where E_p is the first p eigenvectors of covariance matrix. Finally, we normalize the $\tilde{F}_p$ to have uncorrelated and unit variance variables (*i.e.* whiten them) denoted by $\tilde{F}_p^w$; and temporal salience measure is computed as follows:

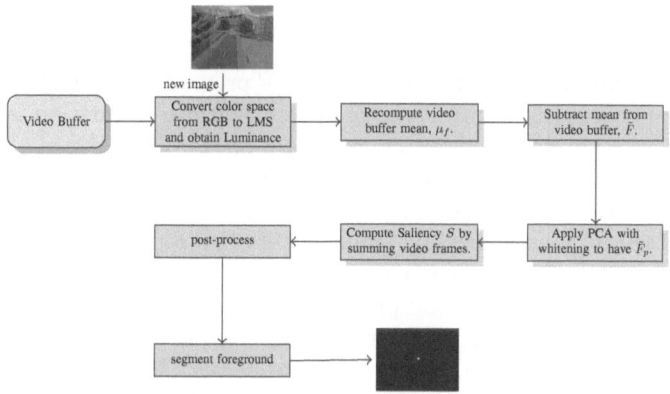

Fig. 1. Flowchart of the proposed method

$$S = \sum_i |\tilde{f}_{pi}^w|. \tag{2}$$

where $\tilde{f}_{pi}^w$ is the i_{th} image vector of $\tilde{F}_p^w$.

Eventually, the final saliency map is obtained by initially applying morphological gray-scale dilation operator to (2) with a disk structure element of size 3. Later, it is blurred with a Gaussian filter of standard deviation 5 and normalizing to the range of $[0, 1]$. In order to segment the foreground, we apply a simple threshold where pixels with saliency value exceeding 0.5 are labelled as foreground. Figure 2 depicts an example video frame, ground truth and segmented foreground object using the proposed method.

Fig. 2. Some arbitrary video frames from different sequences. From left to right, original image, corresponding ground truth and result of the proposed method.

At the first glance it may look that the proposed method has similar ideas with well-known subspace learning techniques [14], but it is different in several aspects. The major difference is that in subspace learning methods PCA is applied to model the background where m largest eigenvalues of an image sequence (usually background sequence) is used to build a background model for a given image. On the other hand, the proposed method applies PCA to provide a representation of foreground object in a temporal queue.

3   Results

We ran the proposed method on sequences provided by BMC data set [1]. Sequences are categorized into learning and evaluation. Evaluation sequences contain real world videos as well as synthetic ones. We analysed the segmented results using the software provided by BMC. It provides precision, recall, F-measure, PSNR, for both learning and evaluation sequences, and two extra measures of D-Score, and structural similarity (SSIM) for evaluation videos.

Table 1 summarizes the average performance for the learning phase sequences. The proposed method has good precision values, which means that it accurately detects the objects based on their motions. High PSNR value shows that the proposed method does not produce false positives. The same behaviour can be observed in images depicted in Figure 2.

Table 1. Performance analysis for learning phase synthetic videos

	Sequence	Recall	Precision	F-measure	PSNR
	111	0.5810	0.7967	0.6720	54.7278
	121	0.5662	0.8197	0.6697	51.2139
	211	0.5847	0.6610	0.6205	52.5032
	221	0.5689	0.7085	0.6311	50.1010
Total	311	0.5982	0.5986	0.5984	49.5130
	321	0.5903	0.7417	0.6574	50.5152
	411	0.6207	0.7957	0.6974	54.9855
	421	0.5850	0.7931	0.6734	51.1417
	511	0.5696	0.7332	0.6411	54.0039
	521	0.5686	0.7692	0.6538	50.8672

More in-depth analysis is available for evaluation sequences through the evaluation software. Table 2 depicts the results for synthetic evaluation sequences. F-measure, D-score and SSIM can be used to compare different methods while summarizing precision and recall information. In some cases, F-measure value indicates that possibly there could exist detection error. Although the F-measure is not very high in those cases, small value of D-Score suggests that the errors do not perturb object recognition. This is in compliance with the low recall value and high precision amount for those cases. Table 3 summarizes the results for real videos. Although the sequences are more difficult, we can draw the same conclusion.

We also measured the running time of the proposed method. The pure Matlab implementation of the proposed method requires 0.07 second to process a frame of size 320 × 240 on a machine with 2.4GHz CPU running 32-bit Windows 7 and Matlab 2012a. This makes the proposed method a suitable algorithm for methods that require real time motion detection.

[1] Available at: http://bmc.univ-bpclermont.fr/

Table 2. Performance analysis for evaluation synthetic videos

Sequence	Recall	Precision	F-measure	PSNR	D-score	SSIM
112	0.6051	0.8704	0.7139	53.7327	0.0013	0.9937
122	0.5705	0.8719	0.6897	49.4959	0.0023	0.9894
212	0.6123	0.8156	0.6995	53.2329	0.0013	0.9934
222	0.5760	0.8762	0.6950	49.4481	0.0023	0.9893
312	0.6041	0.8112	0.6925	53.2750	0.0013	0.9933
322	0.5811	0.8669	0.6958	49.4123	0.0023	0.9893
412	0.6166	0.5595	0.5867	44.3411	0.0024	0.9838
422	0.5796	0.8925	0.7028	49.5036	0.0023	0.9891
512	0.6061	0.7745	0.6800	52.8273	0.0013	0.9929
522	0.5770	0.8761	0.6957	49.4432	0.0023	0.9893

Table 3. Performance analysis for evaluation real videos

Sequence	Recall	Precision	F-measure	PSNR	D-score	SSIM
001	0.6423	0.6403	0.6413	39.7397	0.0064	0.9744
002	0.5409	0.8633	0.6651	26.5448	0.0084	0.9231
003	0.5510	0.7359	0.6302	38.5132	0.0134	0.9647
004	0.5816	0.6891	0.6308	44.1028	0.0042	0.9825
005	0.5230	0.5443	0.5358	45.2341	0.0048	0.9825
006	0.5708	0.8080	0.6690	33.3523	0.0091	0.9551
007	0.5318	0.6961	0.6029	27.2633	0.0089	0.9277
008	0.5507	0.6688	0.6040	37.1402	0.0092	0.9653
009	0.6027	0.6662	0.6329	52.9102	0.0027	0.9918

4   Conclusion

In this paper we introduced a saliency detection mechanism for motion detection. We measured salience by estimating the change in pixel's intensity value within a temporal interval while performing a filtering step via principal component analysis that is intended to replicate sensory adaptation in human neurons. Moreover, it suppress noise under assumption of Gaussian noise. The method is fully unsupervised and requires no training. We applied the method to Background Models Challenge (BMC) video data set. Experiments showed that the proposed method has good performance, and is fast enough to compute in real-time.

References

1. Elgammal, A., Harwood, D., Davis, L.: Non-parametric Model for Background Subtraction. In: Vernon, D. (ed.) ECCV 2000. LNCS, vol. 1843, pp. 751–767. Springer, Heidelberg (2000)
2. Monnet, A., Mittal, A., Paragios, N., Ramesh, V.: Background modeling and subtraction of dynamic scenes. In: 2003 Proceedings. Ninth IEEE International Conference on Computer Vision, vol. 2, pp. 1305–1312 (2003)

3. Cucchiara, R., Grana, C., Piccardi, M., Prati, A.: Detecting moving objects, ghosts, and shadows in video streams. IEEE Transactions on Pattern Analysis and Machine Intelligence 25, 1337–1342 (2003)
4. Zivkovic, Z., van der Heijden, F.: Efficient adaptive density estimation per image pixel for the task of background subtraction. Pattern Recogn. Lett. 27, 773–780 (2006)
5. Itti, L., Koch, C., Niebur, E.: A model of saliency-based visual attention for rapid scene analysis. IEEE Transactions on Pattern Analysis and Machine Intelligence 20, 1254–1259 (1998)
6. Gao, D., Mahadevan, V., Vasconcelos, N.: On the plausibility of the discriminant center-surround hypothesis for visual saliency. Journal of Vision 8 (2008)
7. Itti, L., Baldi, P.: Bayesian surprise attracts human attention. Vision Research 49, 1295–1306 (2009); Visual Attention: Psychophysics, electrophysiology and neuroimaging
8. Miyazato, K., Kimura, A., Takagi, S., Yamato, J.: Real-time estimation of human visual attention with dynamic bayesian network and mcmc-based particle filter. In: IEEE International Conference on Multimedia and Expo, ICME 2009, pp. 250–257 (2009)
9. Rahtu, E., Kannala, J., Salo, M., Heikkilä, J.: Segmenting Salient Objects from Images and Videos. In: Daniilidis, K., Maragos, P., Paragios, N. (eds.) ECCV 2010, Part V. LNCS, vol. 6315, pp. 366–379. Springer, Heidelberg (2010)
10. Mahadevan, V., Vasconcelos, N.: Spatiotemporal saliency in dynamic scenes. IEEE Transactions on Pattern Analysis and Machine Intelligence 32, 171–177 (2010)
11. Doretto, G., Chiuso, A., Wu, Y., Soatto, S.: Dynamic textures. International Journal of Computer Vision 51, 91–109 (2003)
12. Gopalakrishnan, V., Hu, Y., Rajan, D.: Sustained Observability for Salient Motion Detection. In: Kimmel, R., Klette, R., Sugimoto, A. (eds.) ACCV 2010, Part III. LNCS, vol. 6494, pp. 732–743. Springer, Heidelberg (2011)
13. Tarokh, M.: Measures for controllability, observability and fixed modes. IEEE Transactions on Automatic Control 37, 1268–1273 (1992)
14. Bouwmans, T.: Subspace learning for background modeling: A survey. Recent Patents on Computer Science 2, 223–234 (2009)

Background Model
Based on Statistical Local Difference Pattern

Satoshi Yoshinaga, Atsushi Shimada,
Hajime Nagahara, and Rin-ichiro Taniguchi

Kyushu University, Fukuoka, Japan

Abstract. We present a robust background model for object detection
and report its evaluation results using the database of Background Models Challenge (BMC). Our background model is based on a statistical local feature. In particular, we use an illumination invariant local feature
and describe its distribution by using a statistical framework. Thanks
to the effectiveness of the local feature and the statistical framework,
our method can adapt to both illumination and dynamic background
changes. Experimental results, which are done thanks to the database
of BMC, show that our method can detect foreground objects robustly
against background changes.

1 Introduction

Many researchers proposed a lot of object detection methods based on background modeling [1–7]. To accurately detect foreground objects, it is necessary
to adapt to background changes, which are divided into two types: "illumination
changes" and "dynamic background changes", such as waving trees.

To handle illumination changes in the background, some local feature-based
background models [1, 2] have been proposed. However, it is difficult for them
to handle dynamic background changes, which affect the local features in the
background significantly. Statistical methods [3, 4] have been used to cope with
dynamic background changes, and they model multimodal distribution of the
previously observed intensity values of each pixel. However, it is difficult for
them to handle illumination changes, which vary intensity values rapidly and
significantly. To handle both illumination and dynamic background changes,
Tanaka *et al.* [5] used multiple different background models, and the results of
them were combined using "logical AND" operation. However, their method
tends to detect many false-negative pixels, since only positive regions from both
algorithms are accepted and all other regions are rejected. On the other hand,
Zhaoa *et al.* [6] used a local feature defined by multiple point pairs that exhibit
a stable statistical intensity relationship as a background model. However, their
method is not suitable for on-line surveillance since it needs to scan the entire
input sequence to analyze the stability between point pairs.

In this paper, we present a background model [7], in which the concepts of a
local feature-based and a statistical approaches are integrated into a single framework. Our target scenes are mainly "long shot" scenes in the outdoors, and our

J.-I. Park and J. Kim (Eds.): ACCV 2012 Workshops, Part I, LNCS 7728, pp. 327–332, 2013.

method is not intended for "close-up shot" scenes such that a foreground object is very large. To verify the effectiveness of our method, we report its evaluation result using the database of Background Models Challenge (BMC[1]). The experimental results show that our method can detect the foreground objects robustly against both illumination and dynamic background changes.

2   Background Model Based on Statistical Local Feature

We apply a Gaussian mixture model (GMM) to a local feature called the *Local Difference* (LD) to get a statistical local feature called the *Statistical Local Difference* (SLD). Finally, we define *Statistical Local Difference Pattern* (SLDP) [7] for the background model by using several SLDs.

2.1   Construction of Local Difference

A target pixel and its neighboring pixel in an observed image are described by the vectors $\boldsymbol{p}_c = (x_c, y_c)^T$ and $\boldsymbol{p}_j = (x_j, y_j)^T$, respectively. $f(\boldsymbol{p})$ represents the image intensity at pixel $\boldsymbol{p}$. We can then define a local feature called the *Local Difference* (LD) as $\boldsymbol{X}_j = f(\boldsymbol{p}_c) - f(\boldsymbol{p}_j)$. In cases where illumination changes occur, the changes in the LD are small, since the pixels in the localized region show a similar change. Therefore, the value of LD is stable under the illumination changes as shown in Fig.1(a).

2.2   Construction of Statistical Local Difference

We apply a Gaussian mixture model (GMM) to LD to represent probability density functions (PDF) for LD. This gives a statistical local feature called *Statistical Local Difference* (SLD). We define the SLD $P(\boldsymbol{X}_j^t)$ (PDF for LD) at time t by:

$$P(\boldsymbol{X}_j^t) = \sum_{k=1}^{K} w_{j,k}^t \eta(\boldsymbol{X}_j^t | \boldsymbol{\mu}_{j,k}^t, \boldsymbol{\Sigma}_{j,k}^t), \tag{1}$$

where $w_{j,k}^t$, $\boldsymbol{\mu}_{j,k}^t$ and $\boldsymbol{\Sigma}_{j,k}^t$ are the weight, the mean and the covariance matrix of the k-th Gaussian in the mixture at time t respectively, and η is the Gaussian probability density. We construct the background model by updating the GMM (SLD). The updating method for the GMM is based on the method proposed by Shimada *et al* [4]. The SLD can handle dynamic background changes, since its GMM can learn the variety of background hypotheses as shown in Fig.1(b).

2.3   Object Detection Using Statistical Local Difference Pattern

In our method, each pixel has a pattern of SLD in the background model, and we call it *Statistical Local Difference Pattern* (SLDP) [7]. The SLDP at time t

---

[1] 1st ACCV Workshop on Background Models Challenge:
http://bmc.univ-bpclermont.fr/

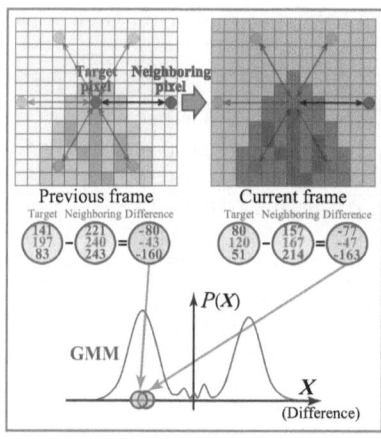 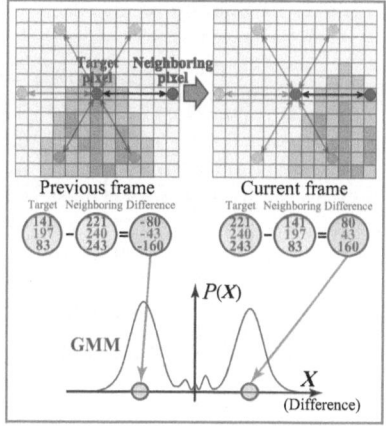

(a) Adaptivity to illumination change   (b) Adaptivity to dynamic background

Fig. 1. Adaptivities of our method to background fluctuation

is defined as $\boldsymbol{S}^t = \{P(\boldsymbol{X}_1^t), \ldots, P(\boldsymbol{X}_j^t), \ldots, P(\boldsymbol{X}_N^t)\}$ by using a target pixel $\boldsymbol{p}_c$ and N neighboring pixels $\boldsymbol{p}_j$ which radiate out from $\boldsymbol{p}_c$. Here, N represents the number of neighboring pixels (Fig.1 shows an example for $N = 6$). Note that all of the neighboring pixels lie on a circle with radius r centered at a target pixel $\boldsymbol{p}_c$.

Foreground detection using SLDP uses a voting method to judge whether a target pixel $\boldsymbol{p}_c$ belongs to the background or the foreground. When the pattern of N LDs is given as $\boldsymbol{D}^t = \{\boldsymbol{X}_1^t, \ldots, \boldsymbol{X}_j^t, \ldots, \boldsymbol{X}_N^t\}$, foreground detection based on SLDP is decided according to:

$$\varPhi(\boldsymbol{p}_c) = \begin{cases} \text{background} & \text{if } \Sigma_j \ \phi(\boldsymbol{D}_j^t, \boldsymbol{S}_j^t) \geq T_B, \\ \text{foreground} & \text{otherwise,} \end{cases} \tag{2}$$

where T_B is a threshold for determining whether a target pixel $\boldsymbol{p}_c$ belongs to the background or the foreground. In Eq.2, $\phi(\boldsymbol{D}_j^t, \boldsymbol{S}_j^t)$ is a function which returns 0 or 1, depending on whether or not the LD $\boldsymbol{X}_j^t$ matches the SLD $P(\boldsymbol{X}_j^t)$ at time t. For further details, we refer the reader to the literature[4].

3   Evaluation

We evaluated the SLDP on the database provided for the Background Models Challenge (BMC). Human annotated ground truth is also available for all videos and is used for performance evaluation. Thus, exhaustive competitive comparison of methods is possible on this database.

3.1   Parameters

All the videos were processed with a unique set of parameters which are tuned based on 10 synthetic videos for learning phase. The list of parameters and their

Table 1. Evaluation results using 10 synthetic videos for evaluation phase

Method	Measure	Street					Rotary				
		112	212	312	412	512	122	222	322	422	522
GMM [4]	Recall	0.927	0.927	0.897	0.861	0.921	0.923	0.931	0.897	0.843	0.934
	Precision	0.866	0.868	0.580	0.526	0.619	0.886	0.890	0.626	0.535	0.840
	F-measure	**0.896**	**0.896**	0.705	0.653	**0.740**	0.904	0.910	0.738	0.655	0.884
Adaptive RRF	Recall	0.843	0.889	0.866	0.848	0.878	0.856	0.897	0.853	0.836	0.894
	Precision	0.840	0.760	0.745	0.726	0.560	0.867	0.836	0.830	0.756	0.726
	F-measure	0.841	0.820	0.801	0.782	0.684	0.861	0.865	0.841	0.794	0.801
Ours (SLDP)	Recall	0.857	0.857	0.827	0.822	0.852	0.915	0.920	0.885	0.854	0.924
	Precision	0.883	0.894	0.876	0.773	0.643	0.894	0.906	0.888	0.794	0.870
	F-measure	0.870	0.875	**0.851**	**0.797**	0.733	**0.904**	**0.913**	**0.886**	**0.823**	**0.896**

value is as follows: the radial distance is $r = 20$, the number of neighboring pixels is $N = 6$ and the detection threshold for SLDP is $T_B = 5$. Although the details of GMM are not explained in Section 2.2, we also indicate the parameter settings in GMM for reproducibility: the learning rate is $\alpha = 0.01$, the initial weight is $W = 0.05$ and the threshold of choosing the background model $T = 0.7$.

3.2   Analysis of Experimental Results

To evaluate the effectiveness of the statistical and local feature-based approaches respectively, we compared the performance of foreground detection with two different approaches: the GMM method [4] and the Adaptive Radial Reach Filter (RRF). The GMM method [4] removes the local feature-based framework from our method, and is consistent with a statistical approach using Gaussian mixture model. The Adaptive RRF introduces an updating scheme into a local feature-based approach using *Radial Reach Correlation* (RRC) [2]. Table 1 and 2 show evaluation results on BMC database including 10 synthetic and 9 real videos for evaluation phase. As shown in Table 1 and 2, except real video 002, the SLDP achieves similar or higher F-measure compared to other methods. To demonstrate the effectiveness of SLDP, we also show some examples of foreground detection results in Fig.2: the first row is a scene where the illumination changes in "Street(312)", the second row is a scene where tree leaves flutter in the wind in "Street(512)", the third row is a scene where the fog is coming in "Rotary(421)" and the fourth row is a scene where both illumination and dynamic background changes are observed in "Real Applications(008)."

Fig.2 (the first and the third rows) shows that GMM method detects a lot of false-positive pixels which are affected by illumination change and the fog, and Table 1 shows Precision of GMM is low on Street(312) and Rotary(421). On the other hand, Adaptive RRF and SLDP methods which use a local feature-based framework detect few false-positive pixels. These results are typical evidence of the effectiveness of a local feature-based framework regarding illumination changes which affect a target pixel value in proportion with others. We also see that Adaptive RRF falsely detects the movement of tree leaves from Fig.2 (the second row), and its corresponding Precision is low from Table 1. Meanwhile,

Table 2. Evaluation results using 9 real videos for evaluation phase

Method	Measure	Real Applications								
		001	002	003	004	005	006	007	008	009
GMM [4]	Recall	0.949	0.680	0.959	0.929	0.854	0.880	0.791	0.823	0.928
	Precision	0.782	0.646	0.880	0.680	0.535	0.736	0.703	0.595	0.890
	F-measure	0.857	0.662	0.918	0.785	0.658	0.802	0.744	0.691	0.909
Adaptive RRF	Recall	0.849	0.819	0.870	0.894	0.835	0.832	0.722	0.764	0.756
	Precision	0.824	0.889	0.820	0.812	0.657	0.794	0.823	0.609	0.914
	F-measure	0.837	**0.853**	0.844	0.851	**0.735**	0.813	0.769	0.678	0.828
Ours (SLDP)	Recall	0.926	0.671	0.954	0.916	0.823	0.856	0.790	0.824	0.909
	Precision	0.818	0.862	0.913	0.891	0.597	0.825	0.780	0.829	0.920
	F-measure	**0.869**	0.754	**0.933**	**0.904**	0.692	**0.841**	**0.785**	**0.827**	**0.914**

GMM and SLDP methods which use a statistical framework can detect the foreground objects correctly. These results are typical evidence of the effectiveness of a statistical framework regarding dynamic background changes.

In cases of real videos, illumination changes and dynamic background changes are often observed at the same time as shown in Fig.2 (the fourth row). Fig.2 (the fourth row) shows that GMM method falsely detects the region of the sky because of illumination change, and that Adaptive RRF method falsely detects the region of grass because of dynamic background change. On the other hand, we see our method can detect foreground objects robustly against both illumination and dynamic background changes from Fig.2 (the fourth row), and both Recall

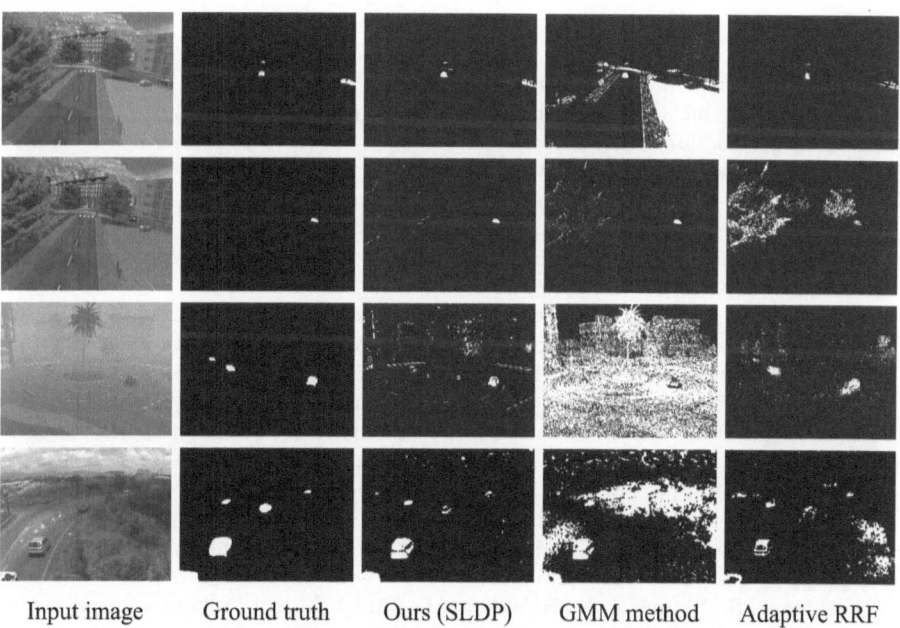

Input image   Ground truth   Ours (SLDP)   GMM method   Adaptive RRF

Fig. 2. Examples of foreground detection results

and Precision of SLDP is higher than those of other methods from Table 2 (008). This is because SLDP has the ability to tolerate the effects of illumination changes thanks to a local feature (LD), and also can learn the variety of dynamic background thanks to a statistical framework (GMM).

In the case of real video 002, Table 2 shows that Recall of SLDP is much low. This is because the SLDP does not model the background color but rather the difference between a target pixel and its neighboring pixels. In most cases of close-up shot scenes including real video 002, the background has an uniform texture, and then the change in the SLDP is hardly-detectable when an object with an uniform texture appears. Therefore, the SLDP confuses foreground objects with background in the close-up shot scenes.

4   Conclusion

In this paper, we have presented a background model based on the *Statistical Local Difference Pattern* (SLDP) by combining the concepts of a local feature-based approach and a statistical approach into a single framework. Our method can handle both illumination and dynamic changes in the background. This is because the SLDP uses illumination-invariant local features which have the ability to tolerate the effects of illumination changes, and describes their distribution by GMMs which can learn the variety of dynamic background. As a result of evaluation on BMC database, we have confirmed that SLDP can detect the foreground objects robustly against illumination changes and dynamic background changes. ·

References

1. Marko, H., Matti, P.: A Texture-Based Method for Modeling the Background and Detecting Moving Objects. IEEE Transactions on Pattern Analysis and Machine Intelligence 28, 657–662 (2006)
2. Satoh, Y., Kaneko, S., Niwa, Y., Yamamoto, K.: Robust object detection using a Radial Reach Filter (RRF). Systems and Computers in Japan 35, 63–73 (2004)
3. Stauffer, C., Grimson, W.E.L.: Adaptive background mixture models for real-time tracking. In: IEEE International Conference on Computer Vision and Pattern Recognition (CVPR), vol. 2, pp. 246–252 (1999)
4. Shimada, A., Arita, D., Taniguchi, R.: Dynamic Control of Adaptive Mixture-of-Gaussians Background Model. In: CD-ROM Proceedings of IEEE International Conference on Advanced Video and Signal Based Surveillance (2006)
5. Tanaka, T., Shimada, A., Taniguchi, R.-I., Yamashita, T., Arita, D.: Towards Robust Object Detection: Integrated Background Modeling Based on Spatio-temporal Features. In: Zha, H., Taniguchi, R.-I., Maybank, S. (eds.) ACCV 2009, Part I. LNCS, vol. 5994, pp. 201–212. Springer, Heidelberg (2010)
6. Zhaoa, X., Satohb, Y., Takaujia, H., Kanekoa, S., Iwatab, K., Ozakic, R.: Object detection based on a robust and accurate statistical multi-point-pair model. Pattern Recognition 44, 1296–1311 (2011)
7. Yoshinaga, S., Shimada, A., Nagahara, H., Taniguchi, R.: Statistical Local Difference Pattern for Background Modeling. IPSJ Transactions on Computer Vision and Applications 3, 198–210 (2011)

Author Index